AF412498

HIGH - LEVEL SYNTHESIS
Introduction to
Chip and System Design

HIGH - LEVEL SYNTHESIS
Introduction to
Chip and System Design

by

Daniel D. Gajski
Nikil D. Dutt
Allen C-H Wu
University of California/Irvine

Steve Y-L Lin
Tsing Hua University

Kluwer Academic Publishers
Boston/Dordrecht/London

Distributors for North America:
Kluwer Academic Publishers
101 Philip Drive
Assinippi Park
Norwell, Massachusetts 02061 USA

Distributors for all other countries:
Kluwer Academic Publishers Group
Distribution Centre
Post Office Box 322
3300 AH Dordrecht, THE NETHERLANDS

Library of Congress Cataloging-in-Publication Data

High-level synthesis : introduction to chip and system design / by
    Daniel D. Gajski ... [et al.].
       p.   cm
    Includes bibliographical references and index.
    ISBN 0-7923-9194-2
      1. Integrated circuits -- Very large scale integration -- Design and
construction -- Data processing. 2. Computer-aided design.
3. Silicon compilers. I. Gajski, Daniel D.
TK7874.H5242   1992
  621.39'5 -- dc20   91-41308
                                                    CIP

Printed on acid-free paper.

Printed in the United States of America

Contents

Preface

Rationale

Computer-aided design (CAD) research, and the CAD industry in particular, has been very successful and has enjoyed exceptional growth, paralleled only by the advances in IC fabrication. Since problems at lower levels of design became humanly intractable and time consuming earlier than on higher abstraction levels, CAD researchers and the industry first turned to problems such as circuit simulation, placement, routing and floorplanning. CAD tools for logic simulation and synthesis came later. As design complexities grew and time-to-market requirements shrank drastically, industry and academia started focusing on even higher levels of design than logic and layout. A higher level of abstraction reduces the number of objects that a designer needs to consider by an order of magnitude, which in turn allows the design and manufacture of larger systems in shorter periods of time. High-level synthesis is thus the natural next step in the design methodology of VLSI systems.

Another reason for the emphasis on high-level design methodologies is that high-level abstractions are closer to a designer's way of thinking. It is difficult to imagine a designer specifying, documenting and communicating a chip design in terms of a circuit schematic with 100,000 gates, or a logic description with 100,000 Boolean expressions. With increasing design complexity, it becomes impossible for a designer to comprehend the functionality of a chip or a system specified completely with circuit or logic schematics. A system described in terms of higher level components (e.g., memories, registers, ALUs and buses) and specified using higher level operations on data values over time exposes the design's functionality and allows a designer to consider alternative implementations with ease.

Research on high-level synthesis started over twenty years ago, but did not come into focus since lower level tools were not available to seriously support the insertion of high-level synthesis into the mainstream design methodology. Since then, substantial progress has been made in formulating and understanding the basic concepts in high-level synthesis. Although many open problems remain, the two most important problems are the lack of a universally accepted theoretical framework and a CAD environment supporting both automatic and manual high-level synthesis. In spite of these deficiencies, high-level synthesis has matured to the point that a book is necessary to summarize the basic concepts and results developed so far and to define the remaining open problems. Such a reference text will allow the high-level synthesis community to grow and prosper in the future.

Audience

This book in intended for three different groups in the CAD community.

First, it is intended for CAD managers and system designers who may be interested in the methodology of chip and system design and in the capabilities and limitations of high-level synthesis tools.

Second, this book can be used by CAD tool developers who may want to implement or modify algorithms for high-level synthesis. Complementary books by Camposano and Wolf [CaWo91] and Walker and Camposano [WaCa91] discuss specific research approaches to high-level synthesis.

Finally, since this book surveys basic concepts in high-level design and algorithms for automatic synthesis, it is also intended for graduate students and seniors specializing in design automation and system design.

Textbook Organization

The book is organized into nine chapters that can be divided into five parts. Chapters 1 and 2 present the basic concepts and the system design process. Chapters 2 and 3 deal with design models and quality metrics for those models. Chapters 4 and 5 deal with design description languages and design representation. Chapters 6, 7 and 8 provide a survey of algorithms for partitioning, scheduling and allocation, while Chapter 9 covers the issues of design methodology and frameworks for high-level synthesis.

Given an understanding of the concepts defined in Chapters 1 and 2, each chapter is self-contained and can be read independently. We used the same writing style and organization in each chapter of the book. A typical chapter starts with an introductory example, defines the basic concepts and describes the main problems to be solved. It follows with a description of several well known algorithms or solutions to the posed problems and explains the advantages and disadvantages of each approach. Each chapter ends with a short survey of other work in the field and some open problems.

The book is designed for use in two different courses. One course would be on system design and methodology, omitting the synthesis algorithms in Chapters 6, 7 and 8; a second course would emphasize high-level synthesis techniques and omit the material on languages and frameworks in Chapters 4 and 9.

We have included several exercises at the end of each chapter. These exercises are divided into three categories: homework problems, project problems and thesis problems. Homework problems test the understanding of the basic material in the chapter. Project problems, indicated by an asterisk, require some literature research and a good understanding of the topic; they may require several weeks of student work. Thesis problems, indicated by a double asterisk are open problems that may result in an M.S. or even a Ph.D. thesis if researched thoroughly.

We hope that this book fills the need for a unifying body of material on high-level synthesis; we welcome your comments and corrections.

Daniel Gajski, Nikil Dutt, Allen Wu, Steve Lin
Irvine, CA
September 1991

Acknowledgements

We would like to thank all of our colleagues and students with whom we discussed the basic issues in high-level synthesis over the last ten years. Without them, many issues would never be clarified and many ideas would go unchallenged.

We would also like to thank those individuals who helped us formulate the issues and focus on the material to be presented in this book. In particular, we thank Bob Grafton of NSF for encouraging research in this area, and Bob Larsen, who over the years helped us in developing quality measures and objective methods for comparative analysis.

We extend our gratitude to the following members of the U.C. Irvine CADLAB without whom this book would not have been possible: Sanjiv Narayan and Loganath Ramachandran for rewording and editing parts of and reformulating algorithms in Chapters 7 and 8; Elke Rundensteiner and Frank Vahid for help in writing three sections and editing parts of Chapters 5, 6 and 9; Roger Ang and Jim Kipps for editing and proofreading of Chapters 2 and 4; Viraphol Chaiyakul and Tedd Hadley for typing and proofreading of and figure drawing in Chapters 1, 5 and 9; and Indraneel Ghosh, Jie Gong and Pradip Jha for help with figure drawing. In addition, all of these people carefully read drafts of chapters, suggested organizational and conceptual changes and helped with corrections to make the book more understandable. We thank Tedd Hadley for his assistance in formatting and overall production of the camera-ready manuscript.

This work was partially supported by the National Science Foundation (grants MIP-8922851 and MIP-9009239) and by the Semiconductor Research Corporation (grant 90-DJ-146). The authors are grateful for their support.

Chapter 1

Introduction

1.1   The Need for Design Automation on Higher Abstraction Levels

VLSI technology reached densities of over one million transistors of random logic per chip in the early 1990s. Systems of such complexity are very difficult to design by handcrafting each transistor or by defining each signal in terms of logic gates. As the complexities of systems increase, so will the need for design automation on more abstract levels where functionality and tradeoffs are easier to understand.

VLSI technology has also reached a maturity level: it is well understood and no longer provides a competitive edge by itself. The industry has started looking at the product development cycle comprehensively to increase productivity and to gain a competitive edge. Automation of the entire design process from conceptualization to silicon has become a necessary goal in the 1990s.

The concepts of first silicon and first specification both help to shorten the time-to-market cycle. Since each iteration through the fabrication line is expensive, the first-silicon approach works to reduce the number of iterations through the fabrication line to only one. This approach requires CAD tools for verification of both functionality and design rules for the entire chip. Since changes in specification may affect the entire design team, the first-specification approach has a goal of reducing the number of iterations over the product specification to only one. A single-iteration specification methodology requires accurate modeling of the design process and accurate estimation of the product quality-measures, including performance and cost. Furthermore, these estimates are required early in the product specification phase.

There are several advantages to automating part or all of the design process and moving automation to higher levels. First, automation assures a much shorter design cycle. Second, it allows for more exploration of different design styles since different designs can be generated and evaluated quickly. Finally, if synthesis algorithms are well understood, design automation tools may out-perform average human designers in generating high quality designs. However, correctness verification of these algorithms, and of CAD tools in general, is not easy. CAD tools still cannot match human quality in automation of the entire design process, although human quality can be achieved on a single task. The two main obstacles are a large problem size that requires very efficient search of the design space and a detailed design model that requires sophisticated algorithms capable of satisfying multiple goals and multiple constraints.

Two schools of thought emerged from the controversy over solutions to these two problems. The capture-and-simulation school believes that human designers have very good design knowledge accumulated through experience that cannot be automated. This school believes that a designer builds a design hierarchy in a bottom-up fashion from elementary components such as transistors and gates. Thus, design automation should provide CAD tools that capture various aspects of design and verify them predominantly through simulation. This approach focuses on a common framework that integrates capture-and-simulation tools for different levels of abstraction and allows designers to move easily between different levels.

The describe-and-synthesize school believes that synthesis algorithms can out-perform human designers. Subscribers to this approach assume that human designers optimize a small number of objects well, but are not capable of finding an optimal design when thousands of objects are in question. CAD algorithms search the design space more thoroughly and are capable of finding nearly optimal designs. This school believes that a top-down methodology, in which designers describe the intent of the design and CAD tools add detailed physical and electrical structure, would be better suited for the future design of complex systems. This approach focuses on definition of description languages, design models, synthesis algorithms and environments for interactive synthesis in which designers' intuition can substantially reduce the search through the design space.

Both these schools of thought may be correct at some point during the evolution of the technology. For example, it is still profitable to handcraft a memory cell that is replicated millions of times. On the other hand, optimizing a 20,000-gate design while using a gate-array with 100,000 gates is not cost effective if the design already satisfies all other constraints. Thus, at this point in technological evolution, even suboptimal synthesis becomes more cost effective than design handcrafting. We believe that VLSI technology has reached the point where high-level synthesis of VLSI chips and electronic systems is becoming cost effective. This book is devoted to problems and solutions in high-level synthesis.

1.2 Levels of Abstraction

We define design synthesis, broadly speaking, as a translation process from a behavioral description into a structural description.

To define and differentiate types of synthesis, we use the Y-chart, a tripartite representation of design [GaKu83, TLWN90] (Figure 1.1). The axes in the Y-chart represent three different domains of description: behavioral, structural and physical. Along each axis are different levels of the domain of description. As we move farther away from the center of the Y, the level of description becomes more abstract. We can represent design tools as arcs along a domain's axis or between the axes to illustrate what information is used by each tool and what information is generated by use of the design tool.

We can extend this chart by drawing concentric circles on the Y. Each circle intersects the Y axis at a particular level of representation within a domain. The circle represents all the information known about the design at some point of time. The outer circle is the system level, the next is the microarchitectural or register-transfer (RT) level, followed by the logic and circuit levels. Table 1.1 lists these levels.

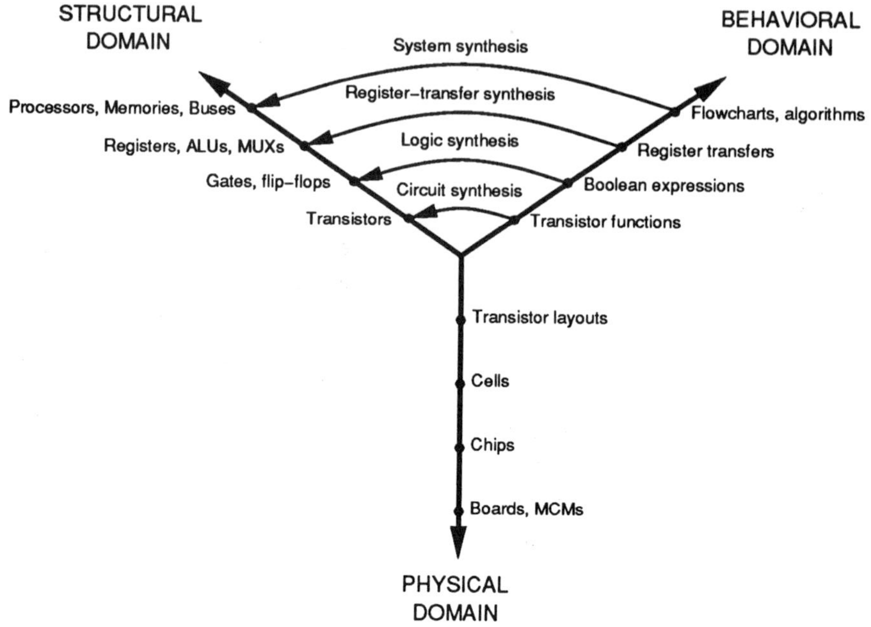

Figure 1.1: The Y-chart.

In the behavioral domain we are interested in what a design does, not in how it is built. We treat the design as one or more black boxes with a specified set of inputs and outputs and a set of functions describing the behavior of each output in terms of the inputs over time. In addition to stating functionality, a behavioral description includes an interface description and a description of constraints imposed on the design. The interface description specifies the I/O ports and timing relationships or protocols among signals at those ports. Constraints specify technological relationships that must hold for the design to be verifiable, testable, manufacturable and maintainable.

To describe behavior, we use transfer functions and timing diagrams on the circuit level and Boolean expressions and state diagrams on the logic level. On the RT level, time is divided into intervals called control states or steps. We use a register-transfer description, which specifies for each control state the condition to be tested, all register transfers to be executed, and the next control state to be entered. On the system

Level Name	Behavioral Representation	Structural Representation	Physical Representation
System level	Spec. charts Flowcharts Algorithms	Processors Controllers Memories Buses	Cabinets Boards MCMs Chips
Microarchitectural level	Register transfers	ALUs Multipliers MUXs Registers Memories	Chips Floorplans Module floorplans
Logic level	Boolean equations Sequencers	Gates Flip–flops	Modules Cells
Circuit level	Transfer functions	Transistors Connections	Transistor layouts Wire segments Contacts

Table 1.1: Design objects on different abstraction levels.

level, we use variables and language operators to express functionality of system components. Variables and data structures are not bound to registers and memories, and operations are not bound to any functional units or control states. In a system level description, timing is further abstracted to the order in which variable assignments are executed.

A structural representation bridges the behavioral and physical representation. It is a one-to-many mapping of a behavioral representation onto a set of components and connections under constraints such as cost, area, and delay. At times, a structural representation, such as a logic or circuit schematic, may serve as a functional description. On the other hand, behavioral descriptions such as Boolean expressions suggest a trivial implementation, such as a sum-of-product structure consisting of NOT, AND and OR gates. The most commonly used levels of structural representation are identified in terms of the basic structural elements used. On the circuit level the basic elements are transistors, resistors, and capacitors, while gates and flip-flops are the basic elements on the logic level. ALUs, multipliers, registers, RAMs and ROMs are used to identify register-transfers. Processors, memories and buses are used on the system level.

The physical representation ignores, as much as possible, what the design is supposed to do and binds its structure in space or to silicon. The most commonly used levels in the physical representation are polygons, cells, modules, floorplans, multi-chip modules (MCMs) and printed circuit (PC) boards.

Figure 1.2 illustrates behavioral, structural, and physical domain descriptions. The behavioral description shown is a conditional-subroutine-call instruction of a hypothetical processor. In this description, depending on the value of the third bit in the instruction register ($IR(3)$), either the "then" or "else" branch is executed. In the former case, the program counter (PC) is incremented, and in the latter case, a subroutine address ($MEM(PC)$) is moved into the PC, after the return address ($PC+1$) is pushed onto the stack ($MEM(SP)$). A more complete description of a processor would include all of the machine's instructions and addressing modes. Note that behavioral operators such as $+$ and $-$ in Figure 1.2(a) do not necessarily correspond one-to-one to functional blocks in Figure 1.2(b). Certainly, the behavior specified in the statements must be implemented, but it may be implemented using shared functional units that implement several other behavioral operators also. Thus we can see that a behavioral description contains very little structural information. At this level of behavioral description, there may be different RT representations that can implement a specific behavior. Similarly, a structural description contains very little information about physical representation. For each structural description there may be many floorplans in the physical domain. One floorplan is shown in Figure 1.2(c). Memory (MEM) and its corresponding data buffer ($DBUF$) occupy the right half of the floorplan, and the rest of the components are grouped together into a datapath laid out as a bit-slice stack with the *Address* and *Data* buses running over each bit slice. Note that extra silicon area is needed whenever a wire or a bus changes direction or makes a "T" connection.

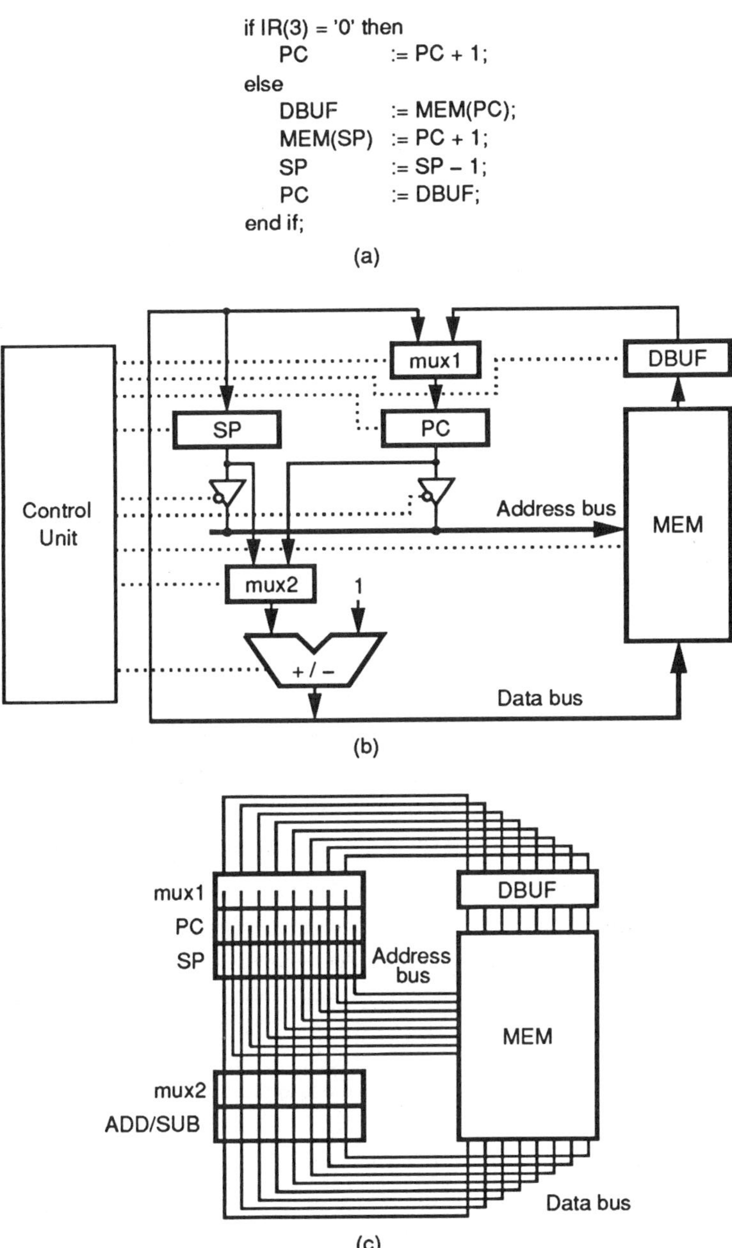

Figure 1.2: An example of three domain descriptions: (a) behavioral, (b) structural, (c) physical.

1.3   Definition of Synthesis

We define synthesis as a translation from a behavioral description into a structural description (Figure 1.1), similar to the compilation of programming languages such as C or Pascal into an assembly language. Each component in the structural description is in turn defined by its own behavioral description. The component structure can be obtained through synthesis on a lower abstraction level. Synthesis, sometimes called design refinement, adds an additional level of detail that provides information needed for the next level of synthesis or for manufacturing of the design. This more detailed design must satisfy design constraints supplied with the original description or generated by a previous synthesis step.

Each synthesis step may differ in sophistication, since the work involved is proportional to the difference in the amount of information between the behavioral description and the synthesized structural descriptions. For example, a sum-of-products Boolean expression can be trivially converted into a 2-level AND-OR implementation using AND and OR gates with an unlimited number of inputs. Substantially more work is needed to implement the same expression with only 2-input NAND gates. Although each behavioral description may suggest some trivial implementation, it is generally impossible to find an optimal implementation under arbitrary constraints for an arbitrary library of components.

We will now briefly describe the synthesis tasks at each level of the design process. Circuit synthesis generates a transistor schematic from a set of input-output current, voltage and frequency characteristics or equations. The synthesized transistor schematic contains transistor types, parameters and sizes.

Logic synthesis translates Boolean expressions into a netlist of components from a given library of logic gates such as NAND, NOR, EXOR, AND-OR-INVERT, and OR-AND-INVERT. In many cases, a structural description using one library must be converted into one using another library. To do so we convert the first structural description into Boolean expressions, and then resynthesize it using the second library.

Register-transfer synthesis starts with a set of states and a set of register-transfers in each state. One state corresponds roughly to a clock cycle. Register-transfer synthesis generates the corresponding structure in two parts: (a) a datapath, which is a structure of storage elements and functional units that performs the given register transfers, and (b) a control unit that controls the sequencing of the states in the register-transfer description.

System synthesis starts with a set of processes communicating through either shared variables or message passing. It generates a structure of processors, memories, controllers and interface adapters from a set of system components. Each component can be described by a register-transfer description.

The synthesis tasks previously described generate structures that are not bound in physical space. At each level of the design, we need another phase of synthesis to add physical information. For example, physical design or layout synthesis translates a structural description into layout information ready for mask generation. Cell synthesis generates a cell layout from a given transistor schematic with specified transistor sizes. Gate netlists are converted into modules by placing cells into several rows and connecting I/O pins through routing in the channels between the cells. A microarchitecture is converted into chip layout through floorplanning with modules that represent register-transfer components. Systems are usually obtained by placing chips on multi-chip carriers or printed circuit boards and connecting them through several layers of interconnect.

1.4   Languages, Designs and Technologies

There is a strong correlation between (a) description languages used to specify a design, (b) the design itself and (c) the technology used for implementation of that design (Figure 1.3).

Hardware description languages (HDL) are used to describe the behavior or structure of systems either on a chip level or board level. The target technology introduces a set of constraints imposed on the corresponding design implementation. Those constraints may refer to a particular architectural style such as a RISC architecture, to a particular layout methodology such as standard cells, to a particular fabrication process such as CMOS or GaAS, or to a certain component library. The

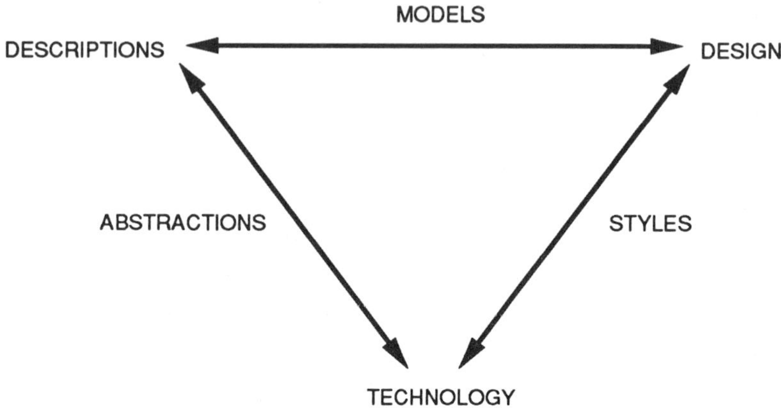

Figure 1.3: Description-design-technology dependence.

technology constraints also determine the quality of the design and the time required to finish it. Similarly, the implementation methodology determines the CAD tools needed for design. On the other hand, preferred design styles (such as pipelining) for a technology can be abstracted into language constructs to provide better understanding and simpler synthesis algorithms.

These language constructs should allow an unambiguous description for each design. Unfortunately, each design can be described or modeled using a single language in several different ways. Figure 1.4 shows two different descriptions of the same behavior and the different designs derived from each description. When asserted, the signal $ENIT$ starts a counter by setting $EN = 1$. When the counter reaches its limit, the comparator asserts its output and sets $EN = 0$, which stops the counter. The first model treats $ENIT$ as a level signal that stays asserted while the counter is counting. The second description uses the positive edge of the $ENIT$ signal to set $EN = 1$ and the output of the comparator to set $EN = 0$. As shown in Figure 1.4, this behavior will result in two different implementations. Since the modeler has chosen to use the positive edge of $ENIT$ to indicate the moment when EN becomes equal to 1, the second implementation has an extra D-type flip-flop used to store the occurrence of the positive edge of the $ENIT$ signal. The implementation shown in Figure 1.4(b) is correct but unnecessarily costly.

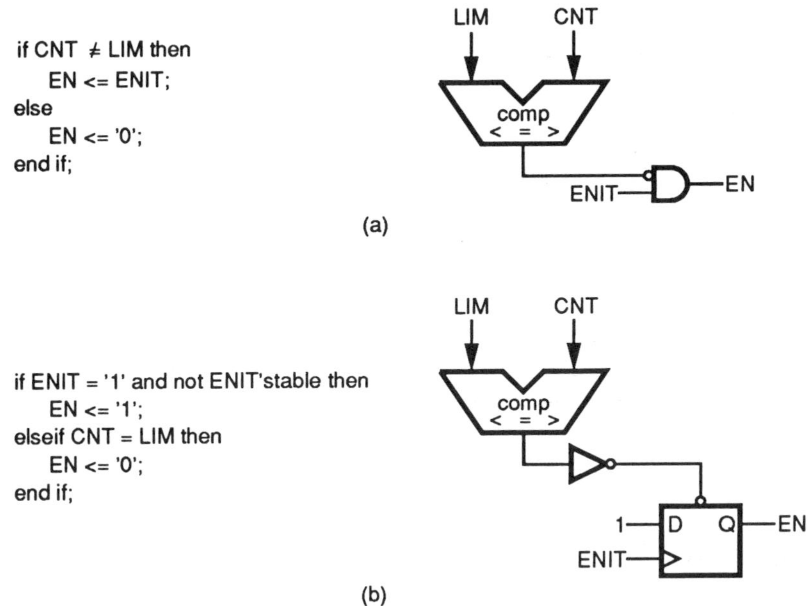

Figure 1.4: Two different descriptions and implementations of an event: (a) level sensitive, (b) edge sensitive.

This simple example shows that different modeling practices result in different designs and that complex synthesis algorithms are required for disambiguation of the design descriptions.

Similarly, a design implementation is not unique. For each function in the design, there are several design styles suitable for different design goals or constraints. For example, Figures 1.5(a) and 1.5(b) illustrate two different implementations of the EXOR function. The design in Figure 1.5(a) uses twelve transistors: two for each inverter and additional four transistors for pass-transistor selector. The design in Figure 1.5(b) uses only ten transistors: four for the 2-input NOR gate and six for the 2-input AND-NOR gate. The 12-transistor design is better suited for large loads since only 2 output transistors must be oversized. On the other hand, the 10-transistor implementation is better suited for small loads since it will utilize a smaller layout area than the 12-transistor circuit. In the case of large loads, six large transistors at the output will take a

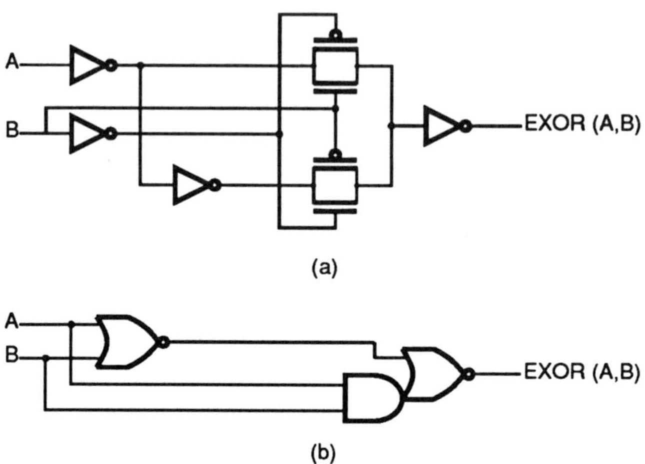

Figure 1.5: Two styles of EXOR implementation using: (a) transmission gates, (b) AND-OR-INVERT gate.

much larger area than a 12-transistor circuit.

Thus, future synthesis systems must allow for making design tradeoffs using different design styles for different technologies and design goals. High-level synthesis systems will not be accepted by designers until this capability is available.

Design styles are also reflected in design descriptions. The more detailed the design model, the more it suggests a design style. Figure 1.6 suggests two different styles of datapath implementation of the statement:

$$\textbf{if } x = 0 \textbf{ then } y = a + b \textbf{ else } y = a - b.$$

The condition $x = 0$ can be directly fed into the control logic or latched in the status register before being tested in the control logic. In the former case, the clock cycle is longer, but the statement can be executed in one clock cycle. In the latter case, the clock cycle is shorter, and the statement's execution requires two clock periods. If the design description allows arbitrary expressions as conditions in each control step, the design style without the status register is more suitable; but if only single

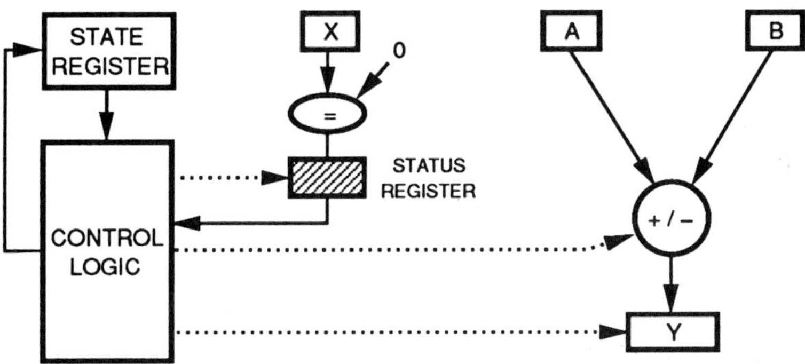

Figure 1.6: Two different design styles: without status register and with status register (shaded box).

bits are allowed as conditions, the style with the status register can be used.

When synthesis tools are forced to deal with description models and design styles that do not match, they must incorporate complicated model analysis and sophisticated style-transformation techniques. Thus, extension of synthesis to the system level requires a matching of models, abstractions and design styles.

1.5 Essential Issues in Synthesis

1.5.1 High-level Formulation

High-level synthesis (HLS) still lacks a formalism of the type already developed for layout and logic synthesis. Layout models have a good formal foundation in set and graph theory since the algorithms deal with placement and connectivity of two-dimensional objects. Logic synthesis is based on well-known formalisms of Boolean algebra because the algorithms deal predominantly with minimization and transformation of Boolean expressions. HLS uses formalisms based on several different areas, varying from the programming-language paradigm for the behavioral description to layout models for cost and performance estimation. However, it does not have an identifiable formal theory of its own. Most frequently, a mixture of finite-state machine and register-transfer notations servers as an HLS formalism.

The lack of formalism in HLS can be attributed to, first, a large number of different objects and, second, the many different types of information needed to describe each object. Circuit theory deals with only three objects: transistors, capacitors and resistors. Physical design adds to those objects wire segments and vias between two wire segments on different layers. Logic synthesis deals with only three Boolean operators: AND, OR and NOT. Even when sequential logic is considered, one storage element, a D-type flip-flop, is sufficient to extend the Boolean algebra formalism to sequential logic. On the other hand, HLS deals with many operators (e.g., ALUs, shifters, multipliers), storage elements (e.g., registers, counters, register files, RAMs, associative memories) and interconnection units (e.g., selectors, buses, signals). These are called RT components. They are connected in different configurations to form basic system components such as memories, processors, controllers and arbiters. In short, system design deals with a large number of different system component types. Even for one type of component, such as a processor, there are many versions, each with varieties of options and I/O formats.

Among many different types of information needed to describe system components, are:

(a) the specification of RT elements,

(b) the connectivity of declared RT elements,

(c) the set of register-transfers for each state,

(d) the sequencing of states for each combination of control inputs,

(e) the physical characteristics of the components, such as area, power, performance and I/O pin configuration, and

(f) a definition of the external environment, including timing relationships among input/output signals.

1.5.2 Languages and Representations

The main problem with a design description is its change with the design over time. At the beginning, the description is vague, with little or no implementation detail. More detail is added as the design evolves, and the level of abstraction is lowered until the design is ready for manufacturing. A language that spans all the levels of abstraction would be desirable from a management point of view but would be too cumbersome to be mastered by designers, who work only on one or two levels of abstractions.

Furthermore, different members of the design team require different aspects of a design, for reasons of verifiability, testability, reliability, manufacturability, maintainability and manageability. Adding all the information needed by those experts to one language known to everybody would produce a very rich language in which information for any particular aspect of the design would be difficult to find. Designers need different "views" of the description to suit the application at hand. Thus, the high-level synthesis community needs the capability to describe design views for capture, optimization and verification of different design aspects. These languages may be graphical, tabular or textual. For example, if designers want to balance resource utilization, they must be able to view or specify the usage of all available resources, such as memories, registers, ALUs or buses, on a clock-cycle basis. Similarly, designers performing floorplanning would be interested in the shape and the size of the same resources. A graphical language for floorplanning that would allow viewing of resource footprints and assignment of their positions on the chip would be necessary for manual optimization of the silicon area. Thus, the development of languages for specifying different aspects of a design as well as different formats for manual refinement or modification of partially synthesized designs is essential in order for high-level synthesis to be successfully accepted by designers.

Similarly, a design representation that would allow linking, manipulation and extraction of different design information must be available. Such a representation could be used for communication between different design tools and for entry and modification of design information.

1.5.3   Design Modeling

The existence of a multitude of different descriptions for the same design also hinders the success of HLS. Multiple descriptions are a consequence of the multitude of abstraction levels and language constructs available in the design description languages.

Every design description makes a tradeoff between behavioral and structural information. We can describe a design with a single process or break it into several processes. We can continue this decomposition hierarchically and generate a hierarchical structure of smaller and smaller components, each in turn described by one process. However, there is an essential difference between behavioral and structural descriptions. Operations in a behavioral description are written in the order of their occurrence in time, but operations in a structural description are grouped together according to the place of execution. For example, a shift operation may be used in the lines 3, 53 and 97 of a one-hundred line description. All three shift operations are performed by the same shift component in the hardware design. Consequently, the structural description will contain a description of the shift component that is activated at the beginning, in the middle and at the end of the execution sequence. Note that a behavioral description does not contain explicit information about the structure of the implementation; nor does a purely structural description contain explicit information about the behavior of the design over time. Obviously, in some intermediate descriptions behavior and structure are intermingled. The number of possible descriptions is even higher in these cases since different language constructs can be used in a design description with a fixed ratio of behavior and structure.

Structured modeling that provides modeling guidelines for different design styles or levels of abstraction can solve the problem of description diversity. This structured modeling is similar to structured programming in which a programmer is forced to use only well-defined constructs satisfying certain verifiable properties. HLS needs modeling guidelines because it is impossible to develop synthesis algorithms that will generate the same design quality from all possible design descriptions.

1.5.4 Design Quality-Measures

In order to assess the quality of a synthesized design, HLS needs good quality-measures. Counting the number of adders, multipliers and registers, although a crude cost measure, is not adequate. A good quality-measure must be well focused and point out possible inefficiencies or critical spots in the design. Quality-measures are computed from the final synthesized design description, and are used to compare two designs produced from the same specification. However, it is time-consuming to generate completely synthesized designs for comparison of their quality. Hence, more important than the final quality-measures are estimates of those measures early during the synthesis process when the final design is not available.

Estimates must satisfy three criteria: accuracy, fidelity, and simplicity. We define accuracy in terms of the error between an estimated quality value and the real quality value. Fidelity is the deviation from the average error over all design points. If an error over all design points is always of the same magnitude, we say that fidelity is high. Estimates with low accuracy but high fidelity are very useful in pruning the design space since they can predict the superiority of one design over another. Finally, estimates are evaluated by the simplicity of the estimation algorithm that determines the run time of the estimator. Estimation algorithms must be efficient to be useful during the design-space exploration. If the complexity of an estimation algorithms approaches the complexities of the synthesis algorithms, then estimation is useless, since real designs can be generated as quickly as estimates.

To be useful, high-level estimates must predict silicon area, wire delays and performance based on the transistor and wire layout. Furthermore, estimates must predict manufacturing, testing and maintenance costs, and must reflect the influence of manufacturing process variations on product cost and performance.

1.5.5 High-Level Synthesis Algorithms

As was defined earlier, synthesis is a transformation of a behavioral description into a set of connected storage and functional units. General types of algorithms used in HLS are partitioning, scheduling and allocation. Partitioning algorithms deal with the division of a behavioral description or design structure into subdescriptions or substructures in order to reduce the size of the problem or to satisfy some external constraints such as size of the chip, number of pins on the package, power dissipation or maximal wire length. Scheduling and allocation can be thought of as partitioning problems. Scheduling algorithms partition the variable assignments and operations into time intervals and allocation partitions them into storage and functional units.

Scheduling and allocation are orthogonal to each other, yet are closely intertwined. Scheduling partitions with respect to time, but allocation partitions with respect to hardware resources. If scheduling is performed before allocation, it imposes additional constraints on scheduled operations with respect to allocation. For example, two operations scheduled in the same time interval cannot be executed by the same functional unit. Similarly, if allocation is performed before scheduling, it restricts the scheduling. For example, two operations assigned to the same functional unit can not be executed in the same time interval. Scheduling and allocation are not the only algorithms needed in high-level synthesis. After operations are assigned to time intervals and units, additional algorithms for generation and optimization of the control, storage and functional units are needed. Traditionally, synthesis of control units is called sequential synthesis and synthesis of functional units called logic synthesis. Storage units can be synthesized with sequential and combinational synthesis methods.

HLS algorithms have been the focus of synthesis research for the last decade. However, most of them use oversimplified models of the target architecture and cost functions. More work is needed on algorithms that use more realistic models.

1.5.6 Component Sets and Technology Mapping

HLS algorithms generate a structure of RT components such as ALUs, multipliers, registers, counters, and memories. These components are given initially as a library of RTL components. Since components of different size, functionality and style (such as carry-ripple vs. carry-look-ahead) are needed, a library of component generators to generate component instances on demand would be more appropriate. The generators would produce a logic or transistor-level netlist, or layout, or both for each component. Generators are more difficult to develop than designing particular component instances.

Every designer or design team usually develops their own set of commonly used components. As technology and projects change, each team repeats this task over and over again. Since different component sets may be used from project to project, the problem of technology mapping of components from one set to another becomes a critical issue. However, technology mapping in high-level synthesis is much harder than at the logic level because of the number and variety of components. For example, there may be several different ALUs performing the same function but with different control encoding, polarity of inputs and outputs, and availability of intermediate carry signals.

Technology mapping from one library to another on the logic level is accomplished by deriving the behavioral description in terms of Boolean expressions and resynthesizing it with a new library. Deriving a behavioral description from a structural one may be difficult, if not impossible, in HLS since a complete behavior of a structural design is not explicitly given.

Extending technology mapping to system components such as DMA controllers, I/O interfaces and bus arbiters is even more difficult since no accurate behavioral or structural descriptions exist. Furthermore, the problem of mapping one component onto another when they match only in 90% of the functions remains unsolved; in fact, it has yet to be formulated.

1.5.7 Databases and Environments

Two types of databases are needed for HLS. The first type is a component database that stores the RT components to be used in high-level synthesis. It interfaces to tools and answers queries about component instances on three levels of abstraction: RT, logic and layout levels. On the RT level, it provides component types and delay area estimates for each component to HLS tools for partitioning, scheduling and allocation. On the logic level, it provides behavioral and structural descriptions of each component for logic synthesis, and on the layout level, it provides the height and width of a component's layout and the positions of I/O ports on its boundary for the purpose of floorplanning. In addition to queries about specific instances, the component database provides a range of component parameters, such as aspect ratios for floorplanning and area-delay ratios for scheduling. The general queries are answered during unit selection, while queries about particular instances are answered during scheduling and datapath allocation.

Once a structural description is synthesized, the database must provide behavioral descriptions of each component for simulation, and structural descriptions in terms of standard cells for manufacturing with semi-custom or custom technologies.

At a higher level, a system database stores all the information about the system as a hierarchy of communicating processes or finite-state machines. It also provides all other information for testing, manufacturing, documentation and management of the complete design. As mentioned previously, a designer does not want to see all the aspects of the design at one time. A central database will store all aspects of the design and generate different design views on demand. The type and the format of a view should be specified through some format schema. The database information is initially obtained from the original behavioral description and later augmented by synthesis tools or by a human designer through a graphical interface. The database should also check the consistency of upgrades and provide estimates in the case of incomplete information.

Since we cannot expect a completely automated synthesis system and, even if automated, a completely satisfactory design quality, high-level synthesis environments are needed to allow interactive synthesis. Such a synthesis environment must provide:

(a) interactive mapping of behavior to structure,

(b) generation of quality-measures or estimates, and

(c) verification of changes in behavioral or structural descriptions.

Interactive synthesis consists of menu-driven design changes or mappings from one representation to another. Quality-measures provide a comparative capability for an intended design change, and design change verification validates the consistency of that change. For example, a variable assignment such as $x = a+b+c$ executed in a single state can be split into two assignments using a temporary variable $temp$, $temp = a + b$ and $x = temp + c$, each executed in one state. In the original case, the implementation needed one state and two adders working in parallel, but in the modified implementation only one adder and two states were required. The quality-measures would indicate that the original implementation was more expensive but faster then the modified implementation, and change verification would prove that both implementations generate the same result. Quality-measures and verification checks can run in the background concurrently with changes performed by the designer.

1.6 Status and Future of High-Level Synthesis

CAD technology has been very successful in the last ten years. CAD tools for layout and logic design have been successful to the point that they dominate design methodologies throughout the industry in the world. This widespread methodology consists of manually refining product specifications through system and chip architectures until the design is finally captured on the logic level and simulated. Standard-cell methodology and tools have been developed for easy mapping of logic-level design into IC layout. Because of the huge investment in CAD tools, equipment and training, many people believe that this trend will continue to provide more sophisticated CAD tools for capture, simulation and synthesis of logic-level designs.

However, the logic level is not a natural level for system designers. For example, when we want to indicate that the 32-bit values of two variables, a and b, should be added and stored in a third variable, c, we simply write the expression $c = a + b$. We do not write 32 Boolean expressions with up to 64 variables each to indicate this simple operation. It would be very difficult to specify and understand a complex multi-chip system if it had to be described in terms of one million or more Boolean equations.

If we equate the layout-level of abstraction (i.e., transistors, wires and contacts) with machine-level programming, then the logic-level (i.e., gates, flip-flops and finite-state machines) can be equated with assembly-level programming. We know that complex software systems consisting of more than a million lines code are not written in assembly language.

Similarly, a complex hardware system of more than a million gates should not be captured, simulated or tested on the logic level of abstraction. System designers think in terms of states and actions triggered by external or internal events, and in terms of computations and communications. Thus, we have to develop tools to capture, simulate and synthesize designs on higher abstraction levels that are close to the human level of reasoning in order to design large complex systems.

Register-transfer (RT) and, particularly, system levels are the proper abstraction levels for efficient design exploration, testing, verification and maintenance. RT and system levels are also natural for documentation and product lifetime management. Unfortunately, the systems industry has been dependent on logic-level entry and simulation methodology for the last twenty years and has invested heavily in tools, equipment and training. But moving to higher abstraction levels may not prove to be so expensive since most of the designers already work on that level. Furthermore, CAD tools on higher abstraction levels are simpler because the problem size on higher abstraction levels is reduced by one or two orders of magnitude.

On the other hand, HLS research has been focused in the past on partitioning, scheduling and allocation algorithms. Real designs, including some DSP architectures, are very simple. Since most complex chips contain no more than one multiplier and one adder, trivial scheduling and allocation algorithms are adequate for synthesis. However, HLS does not consist only of scheduling and allocation algorithms. It involves converting a system specification or a description in terms of computations and communications into a set of available system components and synthesizing these components using custom, or semicustom technology. In order to support the design process on these higher levels of abstraction, HLS requires a synthesis infrastructure.

Little or no work has been done on theoretical foundations for high-level synthesis. There is an acute need for theoretical formulation of high-level synthesis problems, development of canonical forms, and optimization algorithms for different application architectures. Although the standard description language VHDL has gained popularity all over the world, it is basically a simulation language geared towards logic level capture and simulation. In order to move upwards effectively, we need to build other languages on top of VHDL to represent familiar concepts used by system designers for particular applications. The VHDL language is too cumbersome for conceptualization and "back-of-the-envelope" calculations. Similarly, additional work is needed on more general representations for HLS. Some initial work for specific algorithms or design styles has been published, but a general representation used by a variety of synthesis algorithms and design tools has not yet emerged. With the introduction of the VHDL language, many designers will be writing system and chip modules for standard and custom designs. Research is needed to taxonomize target architectures and provide modeling guidelines for different abstraction levels for each target architecture.

Automating the design process will not succeed without good quality-measures. Presently, every designer has simple quality-measures. More comparative work is needed to establish good quality-measures. The high-level synthesis community also needs good estimation algorithms for quality-measures that can quickly reduce the search through the design space. Algorithms have been a strong point in HLS research. However, most algorithms make oversimplified assumptions regarding target architectures. More work is needed on algorithms for realistic design models.

There is a lack of standard component sets for HLS and algorithms for technology mapping between arbitrary component sets. Technology mapping should be extended to higher-level system components, which in turn will require proper formalization of component descriptions. The success of HLS depends on a good support framework, including databases, environments and graphical front-ends, to sustain system design at higher levels. Research on databases is sporadic and work on environments is almost nonexistent. The main problem is that of higher complexities of environments, since higher levels of abstraction require coordination of tools on several abstract levels and information to be fed forward and back over more than one level of abstraction.

The two principal problems in HLS by any measure are the lack of education and training in problems and solutions of HLS, and the insertion of HLS into industrial and academic design methodologies. This book is written for the purpose of alleviating this problem. It defines the basic problems in HLS, presents known solutions to some of the problems, discusses possible methodologies using HLS, and points to the trends leading to higher levels of abstraction, software/hardware co-design and concurrent product engineering.

1.7   Summary

In this chapter we discussed the basis trends in design automation of electronic systems, and concluded that a move to automation on higher abstraction levels is needed in order to efficiently build and maintain complex systems. We also discussed the present status of high-level synthesis and pointed out future problems to be solved. Solutions to these problems are necessary for insertion of high-level synthesis into standard design methodologies.

1.8   Exercises

1. Discuss the advantages and disadvantages of:
   (a) layout synthesis,
   (b) logic synthesis,
   (c) RT synthesis,
   (d) system synthesis.

2. Define a known design methodology by drawing arcs in the Y-chart.

3. Prove or disprove the existence of an abstraction level called chip synthesis between RT and system levels of abstraction.

4. Define five different design features that would make useful constructs in a hardware description language such as VHDL.

5. **Define a VHDL subset for high-level synthesis.

6. **Define VHDL modeling guidelines for high-level synthesis.

7. Specify five quality-measures for system design. Discuss the power of each measures in terms of selecting superior design implementations.

8. *Specify the smallest possible complete set of RT components. The set is complete if all possible digital chips can be built using the

set. (For example, a 2-input NAND gate and D-type flip-flops are a complete set on the logic level.)

9. Define at least three different tasks performed by designers during interactive system synthesis. Discuss a set of graphic displays to help designers during synthesis.

10. Define at least three design views for a RISC processor.

Chapter 2

Architectural Models in Synthesis

2.1   Design Styles and Target Architectures

During the refinement of a given specification into a structural design
of standard components, a designer first selects a design style and then
defines a target architecture. We use the term design style to refer to the
principal qualitative features of a design, such as prioritized interrupt,
instruction buffer, snooping data cache, bus-oriented datapath, serial
I/O, direct memory access, and others. A target architecture defines
a design more precisely in terms of particular units, their parameters,
and the connections among units. For example, a processor architecture
would include the number of registers in the register file, the number
of buses in the datapath, the number of pipeline stages, the number of
status bits, the number of ways branching can occur, and so on.

Each synthesis program or algorithm assumes a certain target architecture. When the target architecture is more realistic, the quality of
design generated at the end of the synthesis process is higher. However,
more realistic target architectures require more sophisticated algorithms,
so there is an obvious tradeoff between design quality and the complexity
of synthesis algorithms. For instance, consider the 3-bus architecture in
Figure 2.1(a). Two operands can be fetched concurrently from the register file (RF), operated upon in the arithmetic logic unit (ALU), and the

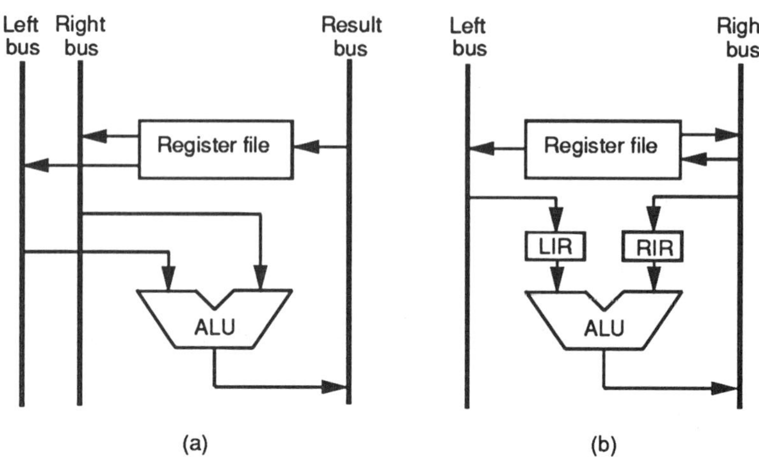

Figure 2.1: Two different datapath styles: (a) 3-bus nonpipelined design, (b) 2-bus pipelined design.

result stored back in the RF in one clock cycle. Therefore, if variables a, b, c, d, x, and y are in arbitrary registers in the RF and each cycle takes 100 ns, then the two binary operations:

$$1. \; x \Leftarrow a + b \qquad (100 \text{ ns})$$
$$2. \; y \Leftarrow c - d \qquad (100 \text{ ns})$$

can be executed in two clock cycles or 200 ns.

The datapath model shown in Figure 2.1(a) is actually oversimplified and unrealistic. Since buses consume a large portion of a chip's area, three buses may not be the most efficient use of silicon. In addition, the RF and the ALU are used only during a portion of the clock cycle, and, more importantly, they are used exclusively. When signal changes are propagated through the ALU, no signal changes are propagated through the RF and vice versa.

A significant performance gain can be obtained by inserting two registers, LIR and RIR, before the inputs of the ALU, by halving the clock cycle, and by increasing the usage of the RF and the ALU in each clock cycle. An datapath model reflecting this architecture is shown in Figure 2.1(b). Within this model, the earlier two operations can be written as

$$
\begin{aligned}
&\text{1. LIR} \Leftarrow a; \text{ RIR} \Leftarrow b; &&(50 \text{ ns})\\
&\text{2. } x \Leftarrow \text{LIR} + \text{RIR}; \text{ LIR} \Leftarrow c; \text{ RIR} \Leftarrow d; &&(50 \text{ ns})\\
&\text{3. } y \Leftarrow \text{LIR} - \text{RIR}; &&(50 \text{ ns})
\end{aligned}
$$

and can be executed in three clock cycles or 150 ns, resulting in a 25% improvement over the previous architecture. Note, however, that we obtained this improvement because the two binary operations were independent of one another. There would be no gain if we had to execute the operation $x \Leftarrow a + b + c$. It would require two clock cycles on the first architecture:

$$
\begin{aligned}
&\text{1. temp} \Leftarrow a + b; &&(100 \text{ ns})\\
&\text{2. } x \Leftarrow \text{temp} + c; &&(100 \text{ ns})
\end{aligned}
$$

and four clock cycles on the second:

$$
\begin{aligned}
&\text{1. LIR} \Leftarrow a; \text{ RIR} \Leftarrow b; &&(50 \text{ ns})\\
&\text{2. temp} \Leftarrow \text{LIR} + \text{RIR}; &&(50 \text{ ns})\\
&\text{3. LIR} \Leftarrow \text{temp}; \text{ RIR} \Leftarrow c; &&(50 \text{ ns})\\
&\text{4. } x \Leftarrow \text{LIR} + \text{RIR}; &&(50 \text{ ns})
\end{aligned}
$$

In other words, it would execute in 200 ns in either case. No gain is possible since the second operation is dependent on the result of the first operation.

The architectural model depicted in Figure 2.1(a) is simple and requires only a simple scheduling algorithm for assigning operations to clock cycles. The model in Figure 2.1(b) is more realistic but requires a synthesis procedure that must include both a construction of the dependence relations among operations and a scheduling algorithm that attempts to improve performance by reordering operation execution according to these dependencies. This example shows that an oversimplified model of the target design may require a simple synthesis algorithm, but it may generate an inferior design unacceptable to real designers.

In this chapter, we will describe the basic target architectural models used in logic, register-transfer, and system designs. We will also briefly describe the design process used to generate these architectures from input descriptions.

2.2 Combinatorial Logic

Microarchitectural components are composed of functional units and storage units. Functional units perform transformations on data values; storage units preserve those values over time. Basic storage units include registers, register files, and memories, whereas basic functional units include adders, comparators, ALUs, multipliers, shifters, incrementers, and selectors. Functional units can be further divided into unsliceable logic (such as encoders, decoders and carry look-ahead generators), sliceable logic (such as logic units, selectors, and shifters), and mixed units that can be implemented both as sliceable and unsliceable logic (such as adders, subtracters, parity generators, and comparators). For example, an adder can be implemented as a sliceable component using the ripple-carry style, or as an unsliceable unit using a carry look-ahead style.

All functional units are blocks of combinatorial logic whose outputs can be described by Boolean expressions of their inputs. For example, a 16-bit adder with data inputs $A = \{a_i \mid 0 \leq i \leq 15\}$, $B = \{b_i \mid 0 \leq i \leq 15\}$, and input carry C_{in}, and data output $SUM = \{sum_i \mid 0 \leq i \leq 15\}$ and output carry C_{out}, can be described by the loop:

$$
\begin{aligned}
c_0 &= C_{in} \\
c_{i+1} &= (a_i \text{ AND } b_i) \text{ OR } (c_i \text{ AND } (a_i \text{ OR } b_i)) \\
sum_i &= a_i \text{ EXOR } b_i \text{ EXOR } c_i \\
C_{out} &= c_{16}
\end{aligned}
$$

where the carries c_i are defined with a recurrence in which each carry bit c_{i+1} depends on the previous carry c_i. These equations describe a ripple-carry style adder. By a backward substitution, we can make each carry bit c_i independent of all carry bits c_j, for $j \leq i - 1$. Our new equations roughly describe a carry look-ahead style adder. Although the carry look-ahead style speeds up carry propagation, it makes the computation of each carry bit nonuniform, since each carry bit c_i depends on all inputs a_j and b_j, for $j < i$. Consequently, the implementation of a carry look-ahead adder is unsliceable and, as a result, difficult to layout on silicon.

Another format for describing a functional unit is with a table that designates output values for each combination of input values. Obviously, this form of description is limited to functional units with a small number

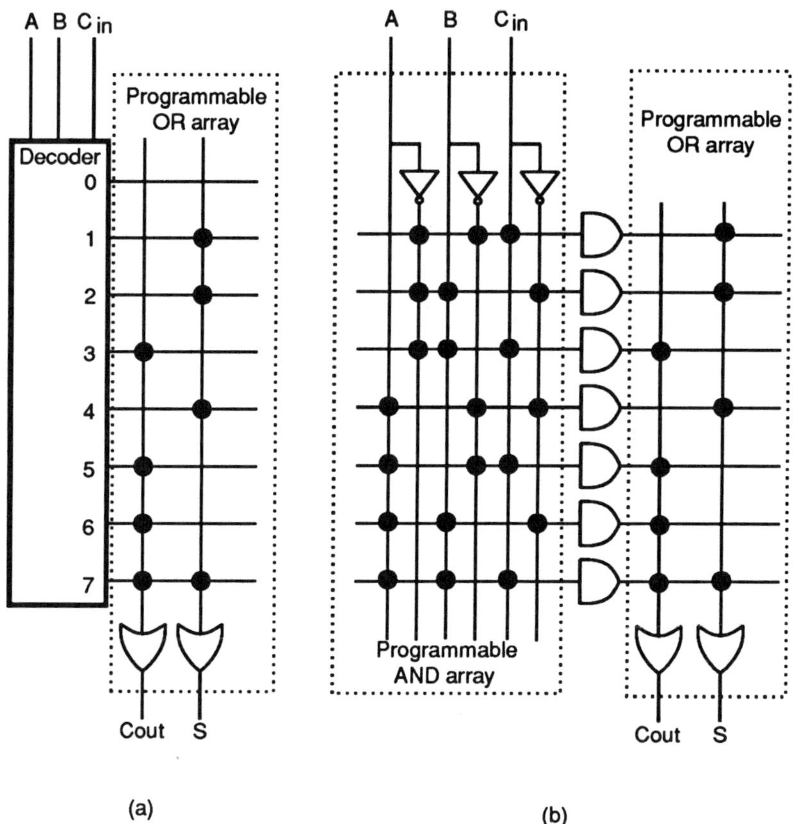

Figure 2.2: Full adder implementation with: (a) a ROM, (b) a PLA.

of inputs. The tabular form can be extended to functional units with a larger number of inputs if it has output values of 0(1) for a very small number of its input values. In this case, we can tabulate only the input combinations for those output values of 0(1).

Functional units can be implemented in several different ways. One simple implementation uses a read-only memory (ROM), as shown in Figure 2.2(a). The output value is programmed into a ROM for each combination of input values. The existence of a dark dot in the OR array indicates a connection between the input decoder and an OR gate whose output is equal to 1 when a proper combination of input values is decoded. The function programmed into the ROM in Figure 2.2(a) outputs the binary representation of the number of inputs equal to 1. For three inputs A, B, and C_{in}, this unit, generating sum and carry outputs S and C_{out}, is called a full adder.

Combinatorial functions also can be implemented with a programmable logic array (PLA), as shown in Figure 2.2(b). Whereas a ROM architecture requires a full decoder, a PLA architecture allows a designer to specify only those input combinations in the programmable AND array that make outputs equal to 1(0). In the case of the full adder, only seven of the eight input combinations must be decoded.

Small combinatorial functions can be implemented using a selector (multiplexer) circuit, as shown in Figure 2.3(a). A selector can be thought of as a single-output dynamically programmable ROM. The control inputs correspond to ROM inputs, and the data inputs serve the role of connections in the OR array. The 4-input selector in Figure 2.3(a) implements the well-known exclusive-OR (EXOR) function of two variables, A and B. Obviously, the same selector can implement any combinatorial function of two variables.

Finally, the most popular implementation of combinatorial functions today uses Boolean gates, such as NAND, NOR, INVERTER, AND-OR-INVERT, and OR-AND-INVERT. Figure 2.3(b) shows the EXOR circuit implemented using a 2-input NOR gate and a 2-input AND-OR-INVERT gate.

Comparing these different implementations, we can see that a ROM has a full decoder, while a PLA has a programmable AND array. Thus, a ROM implementation requires more silicon area than a PLA implementation for functions whose output is 0(1) for only a few input combinations. Furthermore, adding an extra input to ROM doubles the silicon area, while adding an extra input to a PLA increases the area only slightly. Similarly, the selector implementation takes more silicon area than logic gates. For that reason, the selector implementation is rarely used. With regard to timing, ROMs, PLAs, and selectors provide two-level logic implementations of combinatorial functions. When the number of variables increases, so does the number of inputs to the AND and OR gates. A large number of inputs results in long charging and discharging times, which translates into long delays. In addition, the regular structure of ROM, PLA and selector circuits introduce long wire

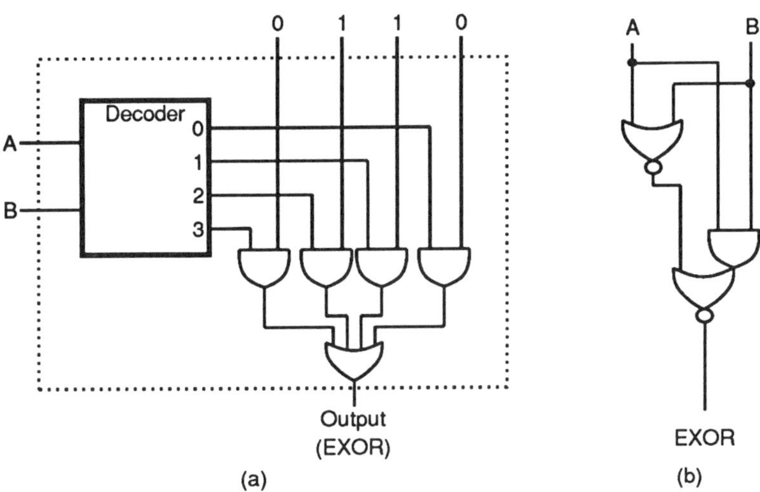

Figure 2.3: EXOR implementation with: (a) a selector, (b) logic gates.

connections, increasing the load capacitance and further encumbering circuit delay. In contrast, Boolean gates allow multi-level implementations using gates with only a few inputs each. Despite increasing the levels of gating, the total performance is much better than equivalent ROM, PLA, and selector implementations because shorter wires and fewer transistors in series allow much faster switching times. In general, Boolean gates are preferred when performance is the main design goal, but ROMs and PLAs may be preferred for reasons of manufacturability, maintainability, verifiability, and testability.

The synthesis process for combinatorial functions consists of the following tasks (for ROM, PLA and selector implementations, only steps 1 and 2 are needed):

1. Compilation
2. Minimization
3. Technology mapping
4. Optimization
5. Transistor sizing

A functional unit can be described in a variety of languages and in many different ways using the same language. Compilation provides a translation from the input description to a standard intermediate form used by minimization and optimization algorithms. Minimization reduces the number of terms and the number of variables in each term, which in turn results in a smaller number of transistors needed for implementation. Technology mapping replaces Boolean operators with Boolean gates from the given library.

Although minimization reduces the number of operations in the Boolean expression, it does not necessarily produce the most efficient implementation. The efficiency of the implementation depends on the library of gates available to the designer. Optimization fine-tunes the gate-level implementation by reducing critical paths. Critical paths through the design can be reduced by factoring variables that are not on the critical path. Optimization procedures work on the principle that maximal factoring produces a maximal level of gating, which minimizes the number of transistors needed for an implementation but introduces longer delays than necessary. Less factoring produces faster, more parallel versions of the same design that require more transistors. Optimization algorithms trade off area cost for improved delay and vice versa.

The final task in the design process defines proper transistor sizes in each gate. Each gate and each wire in the implementation represents a certain capacitance that has to be charged and discharged when signals switch from 1 to 0 and 0 to 1. Large transistors have more drive capability, reducing the switching times on the loads that they drive. On the other hand, they represent larger capacitive loads for their drivers. Transistor-sizing algorithms select proper transistor sizes in order to improve overall performance.

2.3 Finite State Machines

A finite-state machine (FSM) is the most popular design model in computer science and engineering. The model consists of a set of states, a set of transitions between states, and a set of actions associated with states, transitions or both states and transitions. More formally, a FSM is a quintuple

$$< S, I, O, f : S \times I \rightarrow S, h : S \times I \rightarrow O > \qquad (2.1)$$

where $S = \{s_i\}$ is a set of states, $I = \{i_j\}$ is a set of input values, and $O = \{o_k\}$ is a set of output values; f and h are next-state and output functions that map a cross product of S and I into S and O, respectively. The basic FSM model can be limited or extended to represent different target architectures.

2.3.1 Autonomous FSMs

An autonomous FSM is obtained if the input set I is empty. Therefore, next-state and output functions f and h are defined as mappings $S \rightarrow S$ and $S \rightarrow O$. This model is used for free running components such as the modulo-3 counter shown in Figure 2.4.

The function $f : S \rightarrow S$ is shown in Figure 2.4(a) as a state diagram in which each node represents a state; transitions between states are represented by unidirectional edges. In Figure 2.4(b), the same state diagram is given in tabular form. In this figure, we assign an encoding of two bits to each state. Each bit of information is stored in a storage element such as a D-type flip-flop. The Boolean expressions for the inputs to the flip-flops can be derived from the next-state table. In Figure 2.4(c), an implementation of the FSM is given using two D-type flip-flops, FF_1 and FF_0. D_1 must be 1 in state s_1 and D_0 must be 1 in state s_0. Thus,

$$D_1 = s_1 = Q_1' \text{ AND } Q_0. \tag{2.2}$$

Since the state encoding 11 is not used, $Q_1 Q_0 = 11$ will never occur. This means that we can rewrite Equation 2.2 as

$$\begin{aligned} D_1 &= (Q_1' \text{ AND } Q_0) \text{ OR } (Q_1 \text{ AND } Q_0) \\ &= (Q_1' \text{ OR } Q_1) \text{ AND } Q_0 \\ &= Q_0. \end{aligned} \tag{2.3}$$

Similarly,

$$D_0 = Q_1' \text{ AND } Q_0'. \tag{2.4}$$

The timing behavior of the modulo-3 counter is shown in Figure 2.4(d) with a waveform diagram. The state is represented by the pair of flip-flop outputs Q_1 and Q_0.

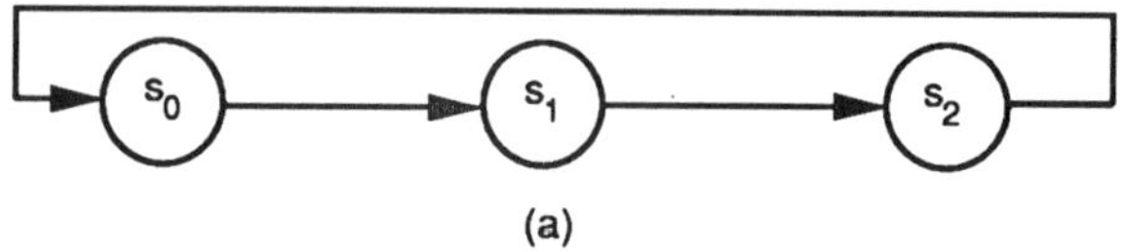

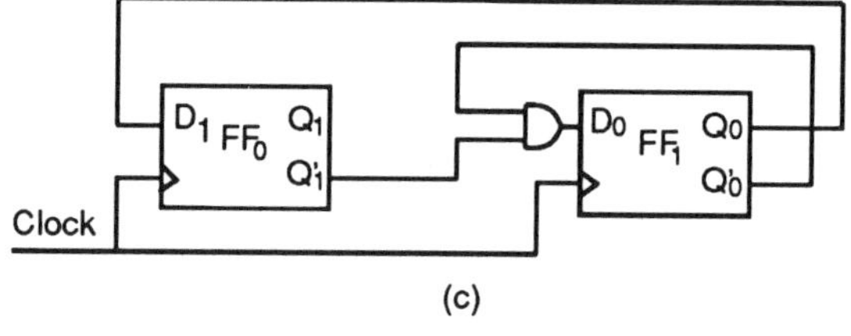

Present state	Next state
$Q_1\,Q_0$	$Q_1\,Q_0$
s_0 = 0 0	s_1 = 0 1
s_1 = 0 1	s_2 = 1 0
s_2 = 1 0	s_0 = 0 0

(b)

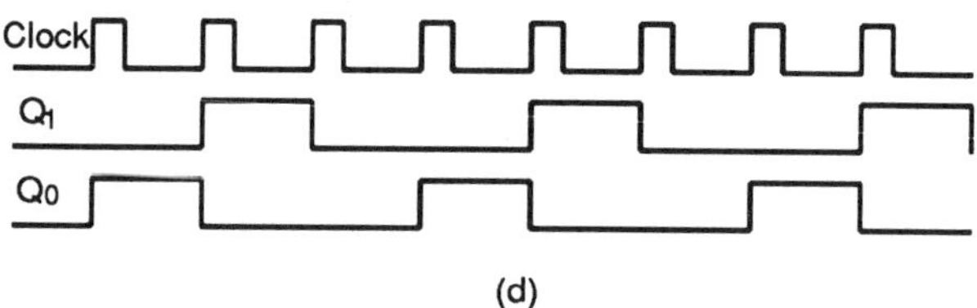

Figure 2.4: Modulo-3 counter: (a) state diagram, (b) next-state table, (c) implementation with D-type flip-flops, (d) state waveforms.

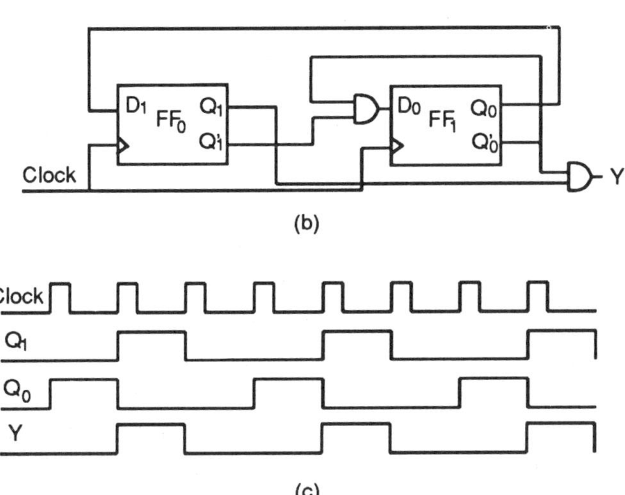

Figure 2.5: Modulo-3 divider: (a) next-state and output table, (b) implementation with D-type flip-flops, (c) state and output waveforms.

The example in Figure 2.4 demonstrated an FSM without inputs and outputs. By adding the output signal Y, which is equal to 1 in state s_2 only, we obtain a modulo-3 divider. If Y is ANDed with the clock signal, the result is a clock sequence in which only every third clock pulse appears at the output.

Figure 2.5(a) shows the next-state table for the modulo-3 divider, and Figure 2.5(b) shows the implementation using D-type flip-flops. The Boolean expression for the output Y detects the modulo-3 divider being in state s_2. Thus,

$$Y = Q_1 \text{ AND } Q_0'. \tag{2.5}$$

The state and output waveforms for a circuit starting in state s_0 are shown in Figure 2.5(c).

2.3.2   State-Based and Transition-Based FSM

Both state-based and transition-based FSMs have a nonempty input set I. They differ in the specification of the output function h. In a state-based (Moore) FSM, the output value depends only on the state of the FSM ($h : S \rightarrow O$), but in a transition-based (Mealy) FSM, the output value depends on the state and input values ($h : S \times I \rightarrow O$). In a transition-based FSM, the output will change when input value changes, while the state will change only on the next clock pulse. In a state-based FSM, the output will persist until the state changes, no matter when the input value changes.

To illustrate the difference between state-based and transition-based FSMs, we extended the specification of the modulo-3 divider by adding the control input called *Count*. When *Count* = 1, the modulo-3 divider behaves as previously described. When *Count* = 0, however, the modulo-3 divider returns to state s_0, in which it persists until *Count* = 1. The state-based modulo-3 divider is shown in Figure 2.6. The next-state and output functions are described by the tables in Figure 2.6(a). These tables cannot be combined into a single table because the next-state function depends on the input signal *Count* and the present state of the FSM, while the output function depends on the present state, only. The implementation using D-type flip-flops is shown in Figure 2.6(b). The input and output waveforms for an arbitrary input are shown in Figure 2.6(c).

In the state-based implementation shown in Figure 2.6, $Y = 1$ during the complete clock cycle in which the modulo-3 divider is in state s_2. Even if *Count* goes to 0 during that clock cycle, Y will retain its output value. Such is not the case for the transition-based implementation shown in Figure 2.7. Note that the output signal Y follows the input signal *Count* when it goes to 0 in state s_2.

Present state	Input	Next state
$Q_1 Q_0$	Count	$Q_1 Q_0$
s_0 = 0 0	1	s_0 = 0 1
s_1 = 0 1	1	s_2 = 1 0
s_2 = 1 0	1	s_0 = 0 0
don't care	0	s_0 = 0 0

Present state	Output
$Q_1 Q_0$	Y
s_0 = 0 0	0
s_1 = 0 1	0
s_2 = 1 0	1

(a)

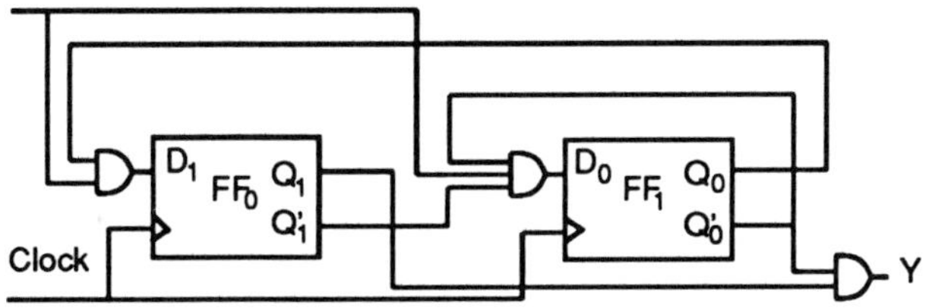

(b)

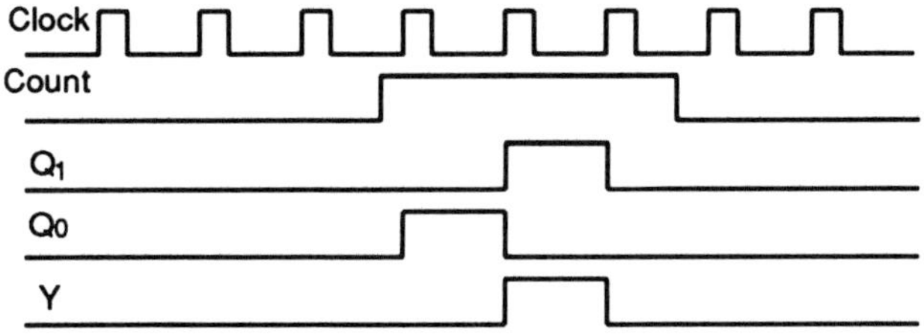

(c)

Figure 2.6: State-based modulo-3 divider: (a) next-state and output tables, (b) implementation with D-type flip-flops, (c) input and output waveforms.

Present state	Input	Next state	Output
$Q_1 Q_0$	Count	$Q_1 Q_0$	Y
s_0 = 0 0	1	s_0 = 0 1	0
s_1 = 0 1	1	s_2 = 1 0	0
s_2 = 1 0	1	s_0 = 0 0	1
don't care	0	s_0 = 0 0	0

(a)

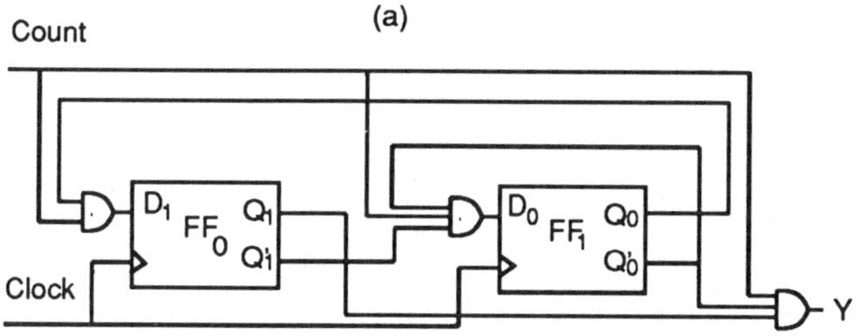

(b)

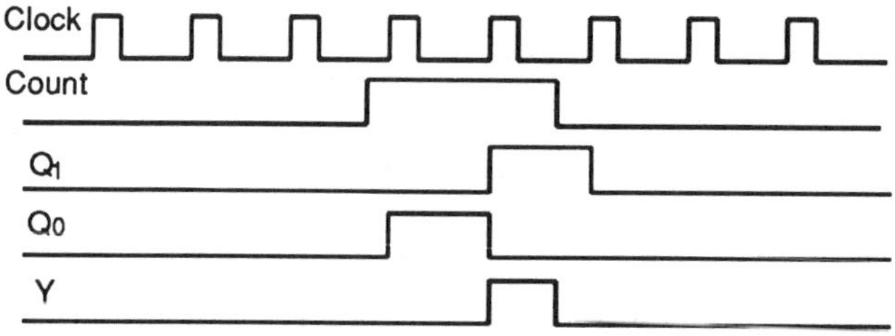

(c)

Figure 2.7: Transition-based modulo-3 divider: (a) next-state and output table, (b) implementation with D-type flip-flops, (c) input and output waveforms.

The synthesis process for state-based and transition-based FSMs consists of the following tasks:

1. Compilation
2. State minimization
3. State encoding
4. Synthesis of next-state and output functions

Compilation translates an input description from a particular language into a standard FSM description. The initial description may have redundant states. Two states are redundant when every input sequence applied to the FSM in either state generates the same output sequence. In such instances, one of the two redundant states can be eliminated during state minimization. After state minimization, each state must be assigned a binary code.

One goal of state encoding is the minimization of the number of bits used to encode each state, the result of which is to minimize the number of flip-flops used to store the state encoding. The other goal of state encoding is to minimize the implementation of the next-state and output functions. This can be done by using a redundant encoding. An encoding is redundant if it uses more than the minimal $\log_2(n)$ bits needed to encode n states. Although a redundant encoding requires more bits and, thus, more flip-flops, the decoding operations performed by the next-state and output functions can be greatly simplified. Thus, the savings in gates for the next-state and output functions may outweigh the additional cost in flip-flops. Synthesis of next-state and output combinatorial functions is performed with the techniques described earlier in Section 2.2.

2.4   Finite State Machine with a Datapath

The FSM model works well for a few to several hundred states. Beyond several hundred states, the model becomes incomprehensible to human designers. Even low-complexity components, such as I/O interfaces, and bus controllers, can have several thousand states when all the storage elements are counted. In order to make the FSM model usable for more complex designs, we introduce a set of integer and floating-point variables stored in registers, register files and memories. Each variable replaces thousands of different states. For example, a 16-bit integer variable represents 2^{16} or 65536 different states; thus, the introduction of a 16-bit variable reduces the number of states in the FSM model by 65536. The use of variables leads us to the concept of an FSM with a datapath (FSMD).

Let us define a set of storage variables VAR, a set of expressions, $EXP = \{f(x, y, z, ...) \mid x, y, z, ... \in VAR\}$, and a set of storage assignments, $A = \{X \Leftarrow e \mid X \in VAR, e \in EXP\}$. Let us further define a set of status signals as logical relations between two expressions from the set EXP, $STAT = \{Rel(a, b) \mid a, b \in EXP\}$. Given these definitions, an FSMD can be defined as the quintuple

$$< S, I \cup STAT, O \cup A, f, h > \qquad (2.6)$$

where S, f, and h are defined as before, the set of input values is extended to include a combination of status values, and the output set is extended to include storage variable assignments.

A modulo-3 divider can be modeled as a 1-state FSMD, as shown in Figure 2.8. In this implementation, we introduce a 2-bit storage variable x that represents the counter. When x is not equal to 2, x is incremented by 1, and when $x = 2$, x is set to 0. The next-state and output functions are shown in Figure 2.8(a), and the FSMD implementation is shown in Figure 2.8(b). The datapath consists of a register for storing variable x, a selector to supply 0 or $x + 1$ to the register, and an adder to increment the value of x. The register is loaded with 0 or $x+1$ as long as $Count = 1$. When $Count = 0$, the value of the register is set to 0.

A modulo-3 divider also can be modeled as a state-based FSMD, as shown in Figure 2.9, where a different state is used for each distinct combination of output values and storage assignments. For this model, we need three different states: s_0, when $Y = 0$ and $x \Leftarrow 0$; s_1, when $Y = 0$ and $x \Leftarrow x + 1$; and s_2, when $Y = 1$ and $x \Leftarrow 0$. The next-state and output tables for a state-based architecture is shown in Figure 2.9(a), and its implementation is shown in Figure 2.9(b). The states are implemented with a 2-bit variable called $State : Exp$ and the next-state logic is implemented using two 2-input selectors, a 2-bit decoder, and an AND-OR gate. The output logic is implemented using one 2-input AND gate. Although the implementation of the next-state and

Present State	Input	Next State	Output
s_0	(Count = 1) AND (x ≠ 2) (Count = 1) AND (x = 2) Count = 0	s_0	x = x + 1, Y = 0 x = 0, Y = 1 x = 0, Y = 0

(a)

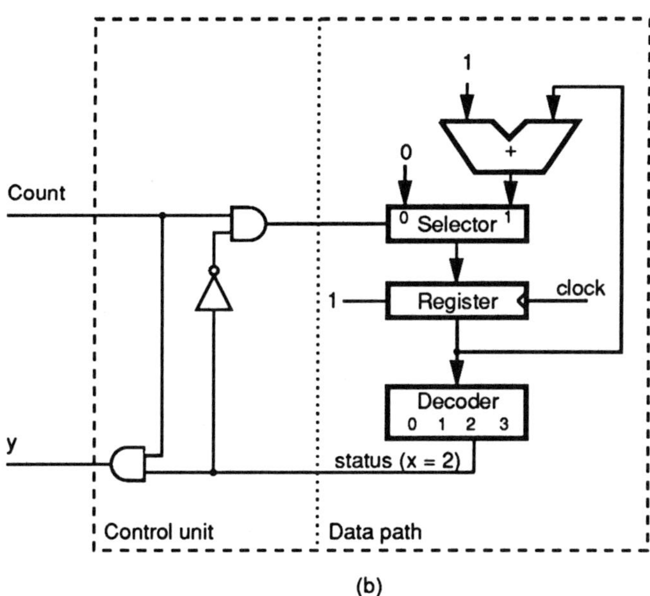

(b)

Figure 2.8: Modulo-3 divider modeled as a transition-based FSMD: (a) next-state and input tables, (b) datapath implementation.

Present State	Input	Next State	Output
s_0	Count = 0 (Count = 1) AND (x ≠ 2) (Count = 1) AND (x = 2)	s_0 s_1 s_0	x = 0, Y = 0
s_1	Count = 0 (Count = 1) AND (x = 0) (Count = 1) AND (x = 1)	s_0 s_1 s_2	x = x + 1, Y = 0
s_2	Count = 0 Count = 1	s_0 s_1	x = 0, Y = 1

(a)

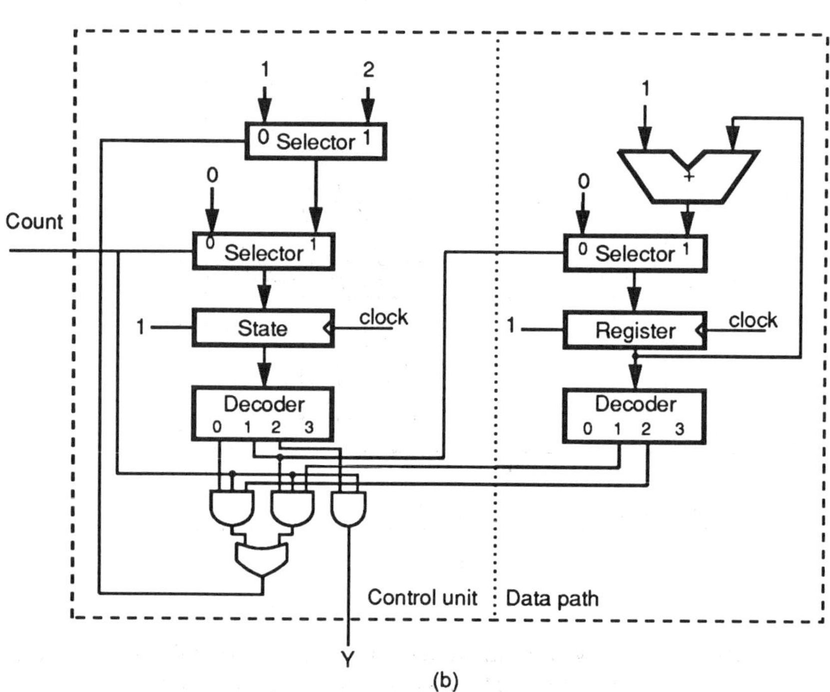

(b)

Figure 2.9: Modulo-3 divider modeled as a state-based FSMD: (a) next-state and output table, (b) datapath implementation.

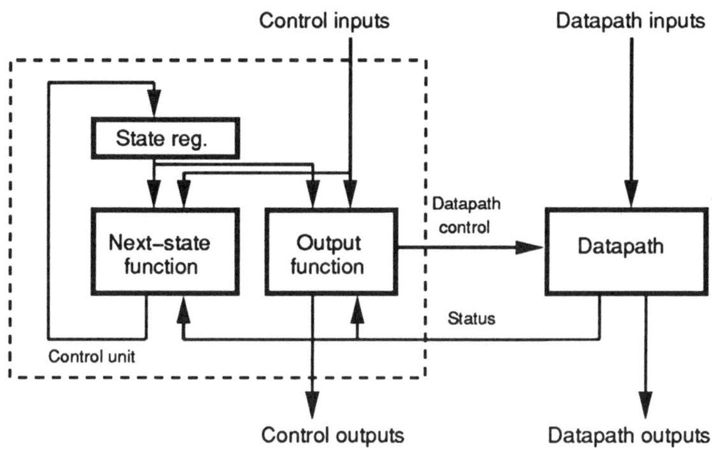

Figure 2.10: Generic FSMD block diagram.

output functions using selectors, decoders, and gates is nonoptimal, it is straightforward to understand and to use for deriving control expressions.

FSMD models are used to describe digital systems on the register-transfer level. A block diagram of a generic FSMD implementation is shown in Figure 2.10; it consists of an FSM, called a control unit, and a datapath. The control unit contains a state register and two combinatorial blocks computing the next-state and output functions. The inputs to the control unit can be categorized as control inputs and status bits; the outputs can be categorized as control outputs and datapath control signals. The datapath also has inputs and outputs. The datapath inputs and outputs are typically words, while control inputs and outputs are more often single bits. The FSMD is state based if the output function does not depend on control inputs and status bits.

We present a typical implementation of the next-state function in Figure 2.11. The next state is selected between the incremented value of the present state and a branch state stored in a ROM or PLA. The selection is made through the address selector by the test bit, which is itself selected through the status selector among control inputs and status bits. The control inputs represent external conditions, such as interrupts, and status bits represent internal conditions, such as a zero

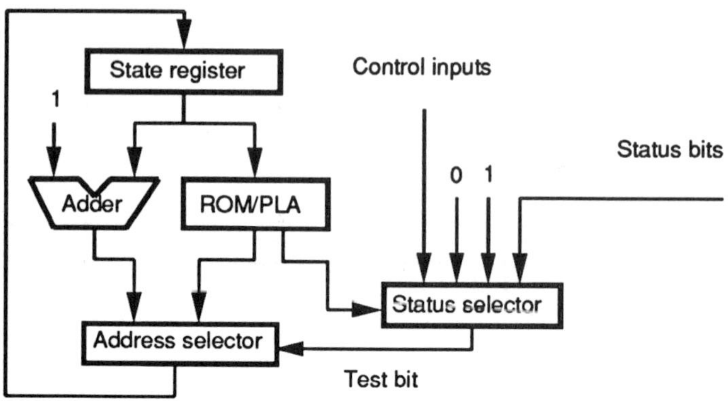

Figure 2.11: A typical processor implementation of the next-state function.

result, overflow, or sign bit. The selection code for the test bit is stored with the branch state in the ROM or PLA. The implementation in Figure 2.11 resembles the control unit of standard processor chips.

The design process for FSMDs has some additional tasks in comparison to FSM synthesis. There are eight basic design synthesis steps:

1. Compilation
2. Unit selection
3. Storage binding
4. Unit binding
5. Interconnection binding
6. Control definition
7. Control-unit synthesis
8. Functional-unit synthesis

Compilation translates the original description, written in a hardware description language (HDL), into a tabular form for the next-state and output functions. Unit selection determines the number and type of storage, functional and interconnection units to be used in the design. Since the output contains variable assignments, the declared variables are stored in registers and memories. Variables with nonoverlapping lifetimes may share storage elements. Two variables x and y have nonoverlapping lifetimes if variable x is assigned a value whose usage ends before variable

y is assigned a value and vice versa. Storage binding assigns variables to storage elements. Similarly, operations executed in different states are independent of each other and can be executed by the same functional unit. Unit binding assigns operations to functional units. Selectors or buses are interconnection units that are required to connect storage and functional units. Interconnection binding assigns interconnection units to datapaths between units. Many HLS algorithms perform selection and binding together in a process called allocation. We describe algorithms for storage, unit and interconnection selection and binding in Chapter 8.

Control definition involves generating Boolean expressions for every datapath control signal that controls selectors, storage units, and functional units. Finally, control-unit synthesis and functional-unit synthesis generate detailed designs for the control unit and datapath using the previously defined FSM and FU design processes. All of these eight tasks are interdependent in the sense that optimally solving each will not necessarily guarantee overall optimality of the design.

2.5 System Architecture

At a higher level of abstraction, an FSMD can be viewed as a process that consumes inputs and produces outputs. A system can be described as a set of communicating processes. Each process can be described using standard programming language constructs, such as loops, if and case statements, and assignment statements. Such a behavioral description of the system assumes the existence of storage elements for all global and local variables, only. It also assumes that operations, the reading and storing values, and the transporting of values from storage to functional units and back take no time. This model does not include the concepts of time delay, time interval, or state. The concept of time is reduced to the order in which statements are executed.

A typical process description of the modulo-3 divider is shown in Figure 2.12. The loop can be thought of as a one-state FSMD, as in Figure 2.8. As a one-state FSMD, the body of the loop will be computed by the datapath on each clock cycle. On the other hand, we can assume that the conditional expression in each "if" statement and each branch of each "if" statement are executed in different states. Thus, every

```
Loop forever
  if  count=1
    then
      if  x=2
        then
          begin
            x=0
            y=1
          end
        else
          begin
            x=x+1
            y=0
          end
      endif
    else
      begin
        x=0
        y=0
      end
  endif
endloop
```

Figure 2.12: Behavioral description of modulo-3 divider.

behavioral description or, in that respect, with every computer program can be trivially implemented with a FSMD architecture.

In reality each system is described by a set of processes, procedures and functions. Such a system description is partitioned into subdescriptions, each of which is realized with a system component. Thus, a system consists of a set of interconnected components, such as processors, memories, controllers, bus arbiters, DMA controllers, and interface logic. Each component can be implemented with one or more communicating FSMDs. Communication is between control units, or between datapaths, or between both. Control signals are rarely used as inputs to the datapaths of other processes, and, likewise, datapath values are rarely used as inputs to other control units. The number of signals and the timing relationship between signals during communication is called a protocol. If all processes are clocked by the same clock signal, the system is called synchronous; if clock rates are different, communication is called asynchronous.

In the case of both synchronous and asynchronous communications, the most frequently used protocol is a request-acknowledge handshaking

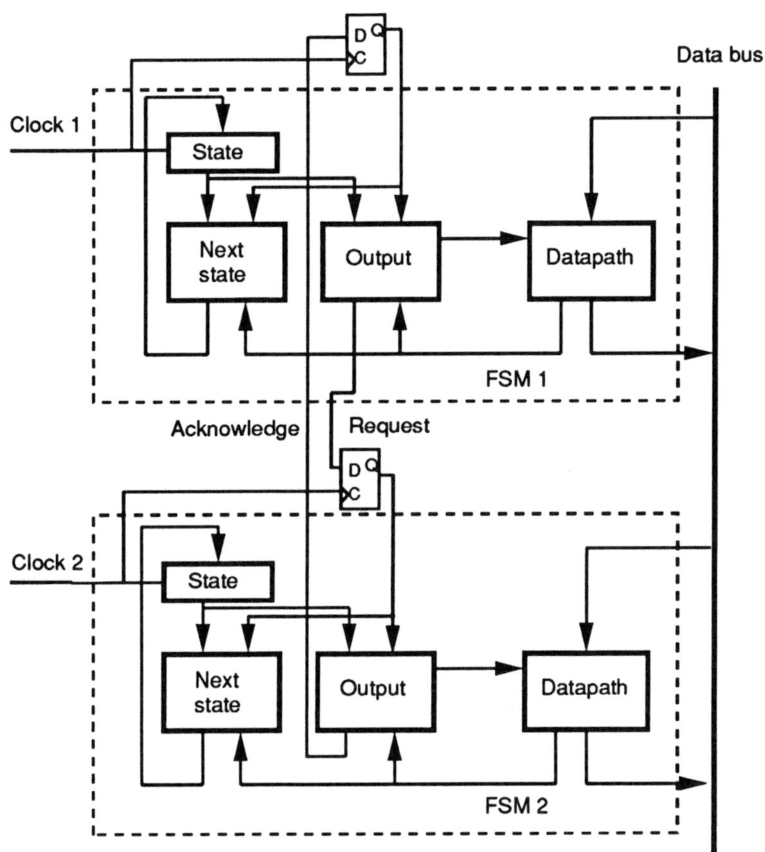

Figure 2.13: Communication between FSMDs.

protocol, which is depicted in Figure 2.13. The first FSMD interrupts the second FSMD by setting the request signal to 1. It stays in the request state until the acknowledge signal is set to 1 by the second FSMD, at which time it leaves the request state, setting the request signal back to 0. The second FSMD jumps to the acknowledge state when interrupted. It sets the acknowledge signal to 1 and stays in the acknowledge state until the request is removed, at which time it leaves the acknowledge state. This protocol can be implemented by two simple D-type flip-flops clocked by the corresponding FSMD clock signals shown in Figure 2.13. Data can be exchanged between corresponding datapaths in a similar manner. The request-acknowledge sequence can also be used to free the bus for data transfer. After the acknowledge signal is received, the first FSMD sets the data-ready signal to 1 and puts the data on the bus. The second FSMD reads the data from the bus and sets the data-received signal to 1, after which the first FSMD takes the data off the bus and resets the data-ready signal. The second FSMD responds by resetting its data received signal.

The request-acknowledge handshaking protocol can be simplified if the clock rates are known and the response time of each FSMD can be accurately predicted. Under these conditions, the first FSMD keeps the data on the bus for a predetermined period of time and does not wait for the data-received signal, since the time needed by the second FSMD to read the data is known. Such is normally the case when the second FSMD is a memory. The design process system synthesis from a behavioral description consists of five steps:

1. Compilation
2. Partitioning
3. Interface Synthesis
4. Scheduling
5. FSMD synthesis

Compilation translates the behavioral description, written either in an HDL or a standard programming language, into a standard representation, such as a control/data flow graph (CDFG), that explicitly shows control and data dependences within the input description. Partitioning divides this representation into local groups, i.e., where each group will be implemented on a distinct chip or some other physical medium. The groups are further partitioned into subgroups, where each subgroup is implemented using a FSMD architecture; a subgroup can be as large as a processor or as small a counter. Partitioning into chips or between FSMDs requires well-defined communications. Interface synthesis determines the communication channels and protocols and generates the hardware to execute these protocols. Scheduling divides the CDFG into states or control steps. Variable assignments in the same control step are executed concurrently. Such a scheduled behavioral description can be then synthesized as an FSMD.

2.6 Engineering Considerations

The previous sections described various target architectural models in terms of control units and datapaths, and their decomposition into combinatorial and sequential components. Each target-architectural model requires further characterizations in terms of engineering considerations, such as clocking, busing, and pipelining.

2.6.1 Clocking

The most basic storage element is a set-reset (SR) latch, which can be realized by cross coupling two NAND or NOR gates. Such a latch is set into state 1 when the *set* input is equal to 1, and reset to 0 when the *reset* input is equal to 1. When both inputs are equal to 0, the latch persists in its previous state. The case where both inputs are equal to 1 is forbidden because it is difficult to predict which state the latch will enter if both inputs become equal to 0 simultaneously. The SR latch is an asynchronous element since its output depends only on its inputs. Asynchronous circuits that are built from such asynchronous elements are difficult to debug because each component latch in the design changes its state at a different time.

In order to simplify the debugging, testing, maintenance, and manufacturing of designs, time is divided into intervals, and all storage elements are set into their respective states at the beginning of each time interval by a clock signal. Using a clock signal, an SR latch can be converted into the commonly used D-type latch shown in Figure 2.14. The behavior of a D-type latch is very simple. The output of the latch follows,

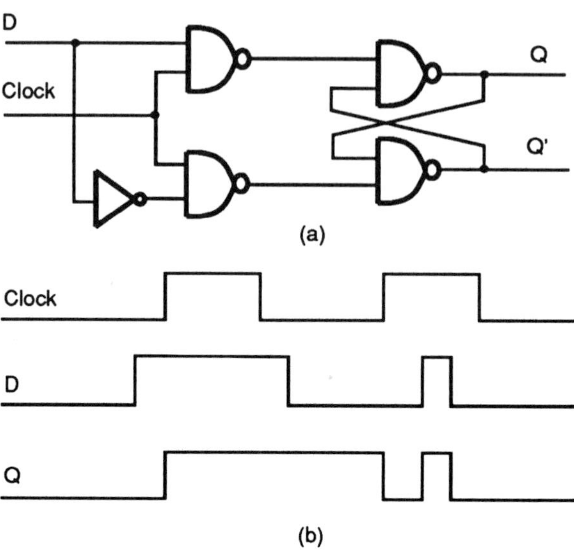

Figure 2.14: D-type latch: (a) NAND gate implementation, (b) a sample of input and output waveforms.

with some delay, the input data signal D, when clock = 1. When clock becomes 0, the latch retains the last value of the D input. To prevent any ambiguity, the D input must be stable during a short interval before and after changes in the clock.

Clocking has the advantage of all latches being triggered at the same time. However, there is a risk of data leakage from serially connected latches. If the delay through a latch is less than the duration the clock signal remains at 1, data can propagate through several latches connected in series, disrupting the orderly transfer of data from latch to latch. For example, consider a simple shift register consisting of three D-type latches, as shown in Figure 2.15(a). If the input-output delay through the latch is t_{DQ} and the clock signal is equal to 1 longer than $3 \times t_{DQ}$, all three latches in the shift register will contain the same input data by the time the clock becomes equal to 0. This is obviously erroneous behavior for a shift register.

Although we can solve the leakage problem by making clock width

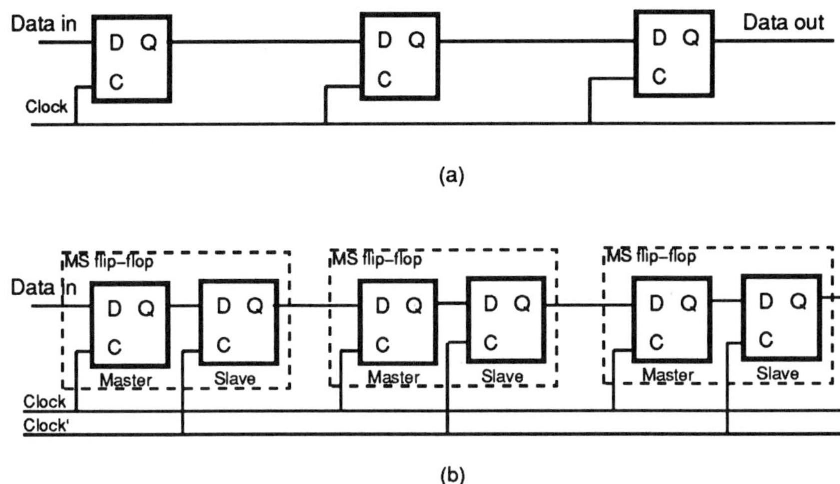

Figure 2.15: A shift register implemented with: (a) latches, (b) master-slave flip-flops.

$t_w = t_{DQ}$, this solution requires zero manufacturing tolerances, which is not possible. A more practical solution is to use master-slave (MS) flip-flops. A MS flip-flop consists of two latches operating in master-slave mode. The master latch is triggered by the clock signal, while the slave latch is triggered by the complement of the clock signal. This ensure that data only propagates through the master latch when $Clock = 1$ and through the slave latch when $Clock = 0$, as shown in Figure 2.15(b). Yet another solution is to use an an edge-triggered (ET) flip-flop, which can be used equally effectively as a MS flip-flop and at the same cost. An ET flip-flop changes its state only during the clock transition from 0 to 1 (or 1 to 0).

As shown in Figure 2.16(a), a clock signal is characterized by a clock period and a clock width. The data is transferred to the output of the flip-flop only during the clock width or during one of the clock signal transitions from 0 to 1 (or 1 to 0). Since signal transitions are not instantaneous, there is still the possibility that both master and slave latches (or two connected ET flip-flops) will be open at the same time. To avoid such a hazard, many systems employ a multi-phase clock, that

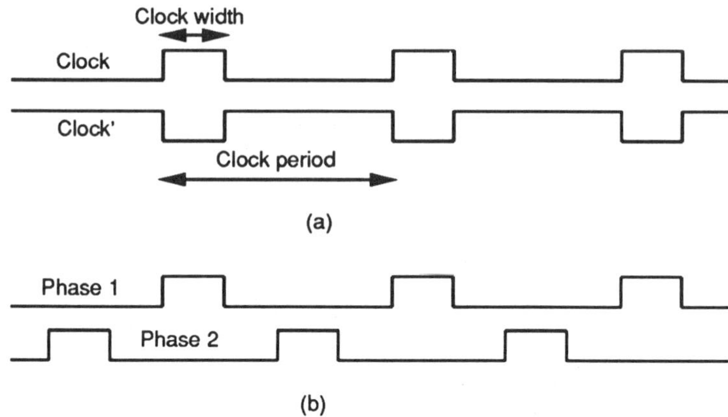

Figure 2.16: Clocking styles: (a) single clock, (b) 2-phase clock.

is, a set of clock signals that have the same period and are phased in
time, such as the 2-phase clock shown in Figure 2.16(b). One phase
serves the role of the clock signal, while the other serves the role of its
complement. Note that a clocking strategy with more than two phases
does not add any new features or improve the performance beyond the
capability of two phase clocking, although it does add more flexibility to
the design phase. Generally, any design can be constructed using only
a 2-phase clocking strategy. In such a design, latches are partitioned
into odd and even stages with even latches driving odd latches and odd
latches driving even latches.

2.6.2 Busing

The most commonly used interconnection unit for chip and system design
is a bus. The popularity of buses can be attributed to their simple layout
and to the ease with which extra inputs or outputs can be added. The
main component in a bus design is a tri-state driver, as shown in Figure
2.17(a). The output Y follows its input data as long as *Control* is equal
to 1.

A bus is used for communication between different FSMDs in a system. Each FSMD has input and output latches or registers that receive

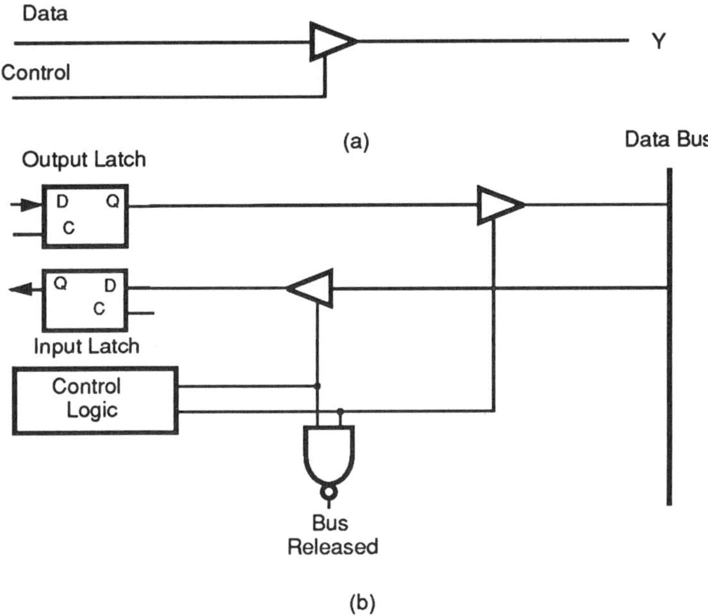

Figure 2.17: Bus communication: (a) tri-state driver symbol, (b) FSMD interface to a bus.

data from a bus or drive the bus using a tri-state driver. Tri-state drivers are controlled by the signals that are generated by the output-function logic of the controlling FSM.

A bus controller is another FSMD. It receives requests for bus use and grants requests using a prioritized or fair-distribution scheme. When an FSMD is not driving or receiving data, it signals to the bus controller that it has released the bus. The bus controller communicates with each FSMD through several control signals. Each bus user usually requests the bus through a bus-request signal, and the bus controller grants the bus through a bus-granted signal. When a bus user does not need the bus anymore, it cancels the request and the bus controller cancels the grant. More sophisticated protocols than this one described may be used for controlling the bus. Usually the bus controller is a part of one FSMD, such as a processor. The processor grants the bus to other controllers, such as the direct-memory access (DMA) controller, when the processor

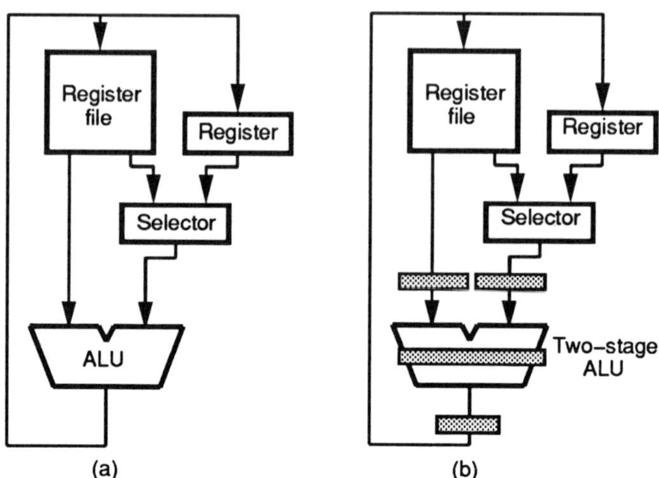

Figure 2.18: Datapath pipelining: (a) nonpipelined datapath, (b) pipelined datapath with 2-stage adder.

does not need it for its own use.

2.6.3 Pipelining

Each functional unit takes some nominal amount of time, called unit delay, to compute a result from the values of its operands. If the clock period is long and unit delays are short, it is possible to chain functional units and perform two or more operations on the same data in a single clock cycle. On the other hand, if the unit delay for a particular functional unit is long compared to the clock cycle, it may take several clock cycles to obtain a result. A control unit must account for such multicycle delays, so that the result from a multicycle unit is not needed for as many clock cycles as it takes to generate that result. In either case, the functional units are not highly utilized. In the case of chaining, each chained unit is used only during a fraction of the cycle and is idle otherwise. Similarly, in the case of multicycling, the entire unit is blocked during several clock cycles, even though the result is computed only in a small portion of the unit.

Pipelining increases the utilization of functional units. Each unit is divided into stages and the partial results are latched after each stage. The clock cycle can be shortened to equal the delay of the longest stage. This method allows concurrent operation on several pairs of operands, each getting partially computed in one stage of the pipelined unit. The total operation delay may even increase slightly because of the extra time needed to latch partial results, but the total throughput on an independent stream of operands is proportional to the number of stages.

For example, Figure 2.18 illustrates the difference between nonpipelined and pipelined datapaths. The ALU is divided into two stages, but the entire register-to-register transfer is divided into four stages by providing input and output latches for the ALU. Thus, any value can be read out of the register file only four clock cycles after its computation is started.

Similarly, a control unit can be pipelined by dividing its next-state and output logic into stages and inserting latches between each stage. Since the control logic does not have many levels of gating, designers usually insert latches only at the input and output of the control logic, as shown in Figure 2.19.

2.7 Summary and Future Directions

In this chapter, we introduced the basic target architectures for high-level synthesis. Designers frequently use FSMs, FSMDs, and systems of communicating FSMDs. However, these high-level models as well as many synthesis algorithms do not take into account engineering consideration such as clocking, busing, and pipelining. Other considerations, such as testability, power distribution, maintainability, manufacturability, and verifiability, were not discussed in this chapter.

Future research is needed to extend our existing models and synthesis algorithms to include important engineering considerations. Of particular importance for system CAD is the definition of protocols and synthesis algorithms for communication interfaces between synchronous systems, asynchronous systems, and, in particular, mixed synchronous/asynchronous systems. Yet perhaps the most important need in high-level synthesis is for comprehensive models that will include all engineering considerations for a particular application as well as classification of target architectures used in different applications, so that high-quality algorithms can be developed for each specific application.

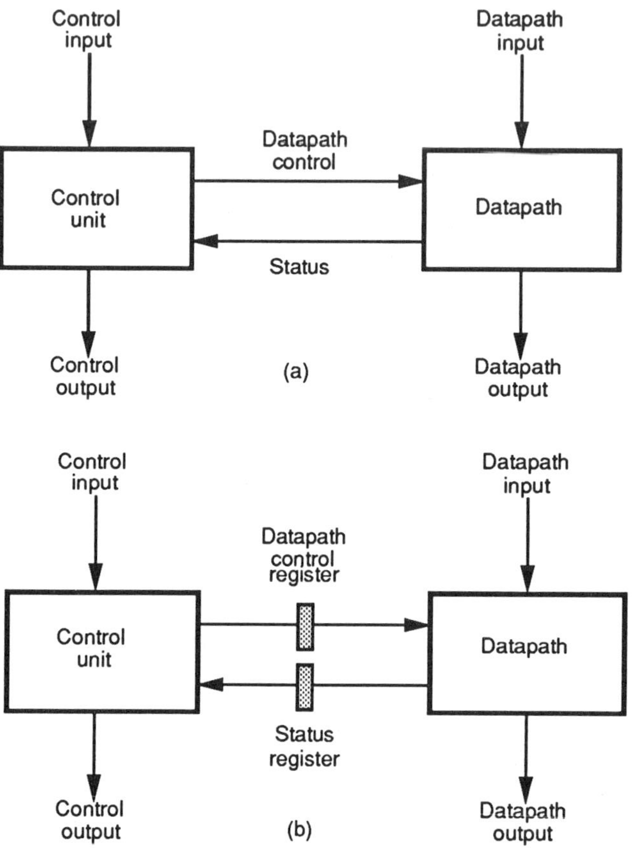

Figure 2.19: Control unit pipelining: (a) nonpipelined control unit, (b) pipelined control unit.

2.8   Exercises

1. The datapath in Figure 2.1(a) consists of a register file (RF) and an arithmetic logic unit (ALU) connected by three buses. To minimize the wiring on the silicon chip, we can combine: (a) the left bus and right bus; (b) the left bus and result bus; and (c) all three buses. Insert a minimal number of registers to maintain proper operation of the datapath while maximizing its throughput.

2. It is possible to add registers at the (a) input, (b) output, and (c) input and output of the ALU in Figure 2.1(a). Which of the above alternatives gives any improvement in performance for a 1-bus, 2-bus, and 3-bus datapath style? Assume that the ALU delay and RF read time are 50 ns each and that the RF write time is equal to 0 ns.

3. Implement a 2-bit counter that counts up when the count signal is equal to 1 and down when the count is equal to 0, by programming (a) the OR array of a 3-input, 8-word ROM, and (b) the AND and OR arrays of a 3-input, 8-word PLA.

4. Discuss the advantages and disadvantages of using a ROM versus a PLA in your design.

5. Design an 8-bit incrementer whose output is equal to its input plus 1 using (a) a ripple-carry style, and (b) carry look-ahead style. You may use only two types of components: a 2-input EXOR and a 2-input AND gate.

6. Model a modulo-3 divider circuit as a 2-state transition-based FSMD with one input signal, count, and one output signal, Y. When count = 0, the FSMD goes to state s_1, in which it is counting up modulo-3. The output signal Y is equal to 1 only when the counter reaches value 2; otherwise it is equal to 0. Give (a) the next-state and output tables, (b) a block diagram, and (c) input, state and output waveforms.

7. Redo exercise six, modeling a modulo-3 divider as a state-based FSMD. Compare the input and the output waveforms for the state-based and transition-based models.

8. Figure 2.11 shows a typical processor control unit using a one-way branch in which the next control word to be read from a ROM/PLA is chosen between the next address and a branch address. Explain performance and silicon area tradeoffs for introducing (a) a 2-way and (b) a 4-way branching capability.

9. Give a behavioral description of (a) a 4-bit ALU, (b) a 4K memory, and (c) a 4-instruction processor.

10. Design the communication interface between two FSMDs driven by clock signals with a 4:1 ratio in their clock periods. Each FSMD issues a request and an acknowledge for the duration of only one clock period.

11. Design a D-type latch using only 2-input NOR gates.

12. Design an asynchronous counter that counts in modulo 16. Use only D-type MS flip-flops and AND, OR, and INVERT gates.

13. Design a bus controller using a fixed-priority distribution scheme. The bus controller serves four different devices using a simple 2-signal request-acknowledge protocol per device.

14. Define the vectored-interrupt scheme used by a CPU to control four devices; use only two signals, interrupt-request and interrupt-acknowledge.

15. Define device logic for handling vectored interrupts with daisy chaining. The device logic has one input (daisy-chain-in), two outputs (interrupt-request and daisy-chain-out), and a bus output for the vector address.

16. Assume that reading two operands from a register file and storing the result back into the register file takes 100 ns (Figure 2.18(a)). Assume also that a 3-stage pipeline reduces the clock to 40 ns. Show the operations executed in each clock period and compare execution times for the following loop from [PaKn89]:

$$\text{while } (x \; < \; a) \text{ repeat:}$$
$$x1 \; := \; x \; + \; dx;$$

$$u1 := u - 3xudx - 3ydx;$$
$$y1 := 0 + udx;$$
$$x := x1;$$
$$u := u1;$$
$$y := y1;$$

end;

17. Assume a state-based FSMD operates with a 100 ns clock period (Figure 2.19(a)); also, assume that a 3-stage control pipeline reduces the clock to 40 ns (Figure 2.19(b)). Show all the operations executed in each clock period for the 100-ns and 40-ns designs for the loop in Exercise 16.

Chapter 3

Quality Measures

3.1 Introduction

High-level synthesis generates a structural design that implements the given behavior and satisfies design constraints for area, performance, power consumption, packaging, testing and other criteria. Quality measures are needed to support HLS in two ways. First, accurate measures are needed to determine the quality of the final synthesized design. The quality measures of the final design allow comparison with given constraints and identify the critical spots in the design methodology, tool set or management structure. For example, the total layout area of a chip determines the manufacturing yield, which determines the cost. Similarly, a high ratio of metal1 versus metal2 wires may point to inferior placement or routing algorithms. Second, good estimates of design quality are needed to guide HLS tools and drive the selection of design style, target architecture and other architectural properties. For example, if the clock period is too long a nonpipelined multiplier must be replaced with a pipelined multiplier. Similarly, a design with three buses requires more area than a design with two buses. Thus, different quality measures are used to evaluate the quality of different design alternatives during partitioning, scheduling and allocation in HLS as described in Chapters 6, 7 and 8.

There are typically three major measures used to support design decisions: size, performance and power consumption. Size measures deal

with the physical size of a design. Performance measures are related to
the propagation delay through a design and the total execution time.
Power measures represent the power dissipation across a design and indicate hot spots in the implementation. There are also other measures
used to evaluate the design quality with respect to verifiability, testability, reliability and manufacturability. Since all design decisions depend
on these measures, their accuracy and fidelity are essential for generation
of high-quality designs.

In this chapter, we focus on some simple quality measures for area and
performance. First, we review the relationship between structural and
physical designs. Then we describe the computation of area and performance measures for FSMD architectures in custom CMOS technology.
Finally, we provide an intuitive discussion of other quality measures.

3.2   The Relationship between Structural and Physical Designs

High-level synthesis transforms an input behavioral description into a
structural design composed of a datapath netlist and a control unit. Typically, the datapath netlist consists of a set of generic register-transfer
(RT) components (e.g., functional, interconnected and storage units).
The control unit specifies the control signals for executing register transfers in each state, as well as the sequencing for the design's next state
(Figure 3.1).

The process of generating manufacturing data for custom or semicustom technologies from a structural design consists of several steps (Figure 3.1), including technology mapping, module generation, floorplanning, placement and routing. Technology mapping assigns real components from a physical library to the generic components in the structural
design. It is usually followed by some optimization procedures to reduce
the total area and delay of a design. We also assume that the technology
mapper selects different layout styles for different parts of the design. For
example, regularly structured units, such as adders, subtracters, ALUs,
multiplexers and registers, can be mapped into bit-sliced stacks, general
cells or standard cells; the control-state table can be mapped into a PLA
or standard cells; and storage units, such as RAMs, ROMs and register

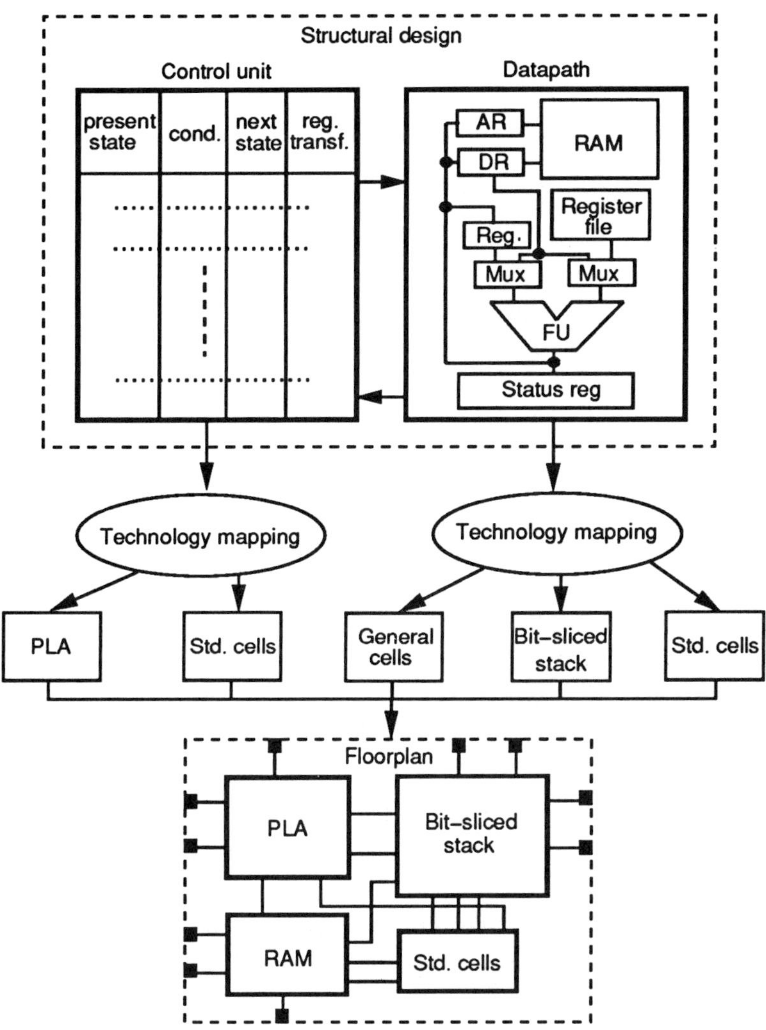

Figure 3.1: The relationship between structural and physical designs.

files, as well as functional units, such as multipliers, are mapped into macro cells. The design may contain a mixture of arbitrarily sized bit-sliced stacks, general cells, macros and standard-cell blocks, which we simply refer to as "modules". Module generators perform floorplanning and routing of each module independently. The chip floorplanner determines the positions and interconnections of modules and generates the chip layout.

The total area of a design is the sum of the area of its modules, I/O pads and pad drivers, the chip routing areas and the remaining wasted areas, as illustrated in Figure 3.1. Since macros are predesigned, their areas and shapes can be obtained directly from component libraries. However, the areas and shapes of datapath and control modules mapped into bit-sliced stacks, standard cell blocks or PLA macrocells vary greatly depending on their layout styles and architectures. In the next two sections, we will discuss area and performance measures for datapaths and control units using some sample layout styles.

3.3   Area Measures

Traditional area measures divide the physical design into active-unit area and interconnection area. Active units include functional units, such as ALUs, adders and multipliers, and storage units, such as registers, register files, ROMs and RAMs. Interconnection units include multiplexers, buses and wires.

Various methods have been used to measure active-unit areas. One method uses the number of AND, OR and NOT operators in the Boolean expressions describing a functional unit. Another measure approximates the active-unit area with the sum of cell areas where each cell implements one symbol in the schematic diagram. This cell area is obtained from cell library databooks provided by manufacturers. The area can also be approximated by the number of transistors multiplied by a transistor density-coefficient in μm^2/transistor obtained by averaging layout area per transistor over all cells available in the library or over some sample real designs.

With regard to interconnection area, most metrics assume that the layout area is directly proportional to the number and size of some inter-

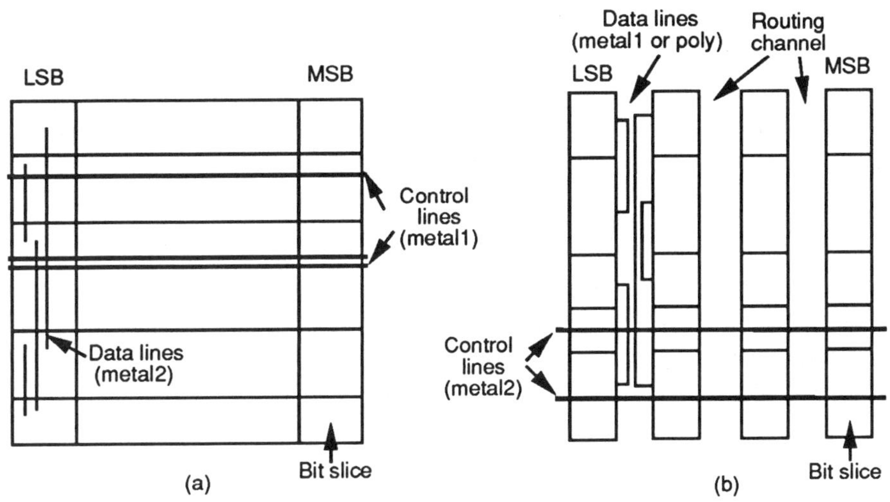

Figure 3.2: Two data path layout architectures using: (a) custom cells, (b) standard cells.

connection units. Hence, the number of multiplexers or equivalent 2-to-1 multiplexers is a common measure for the interconnection area in datapath-allocation algorithms. Likewise, the number of multiplexer inputs and the number of wires are common measures for the routing area.

In this section, we describe area measures for the datapath and control unit of the FSMD model using two different layout styles for CMOS technology.

3.3.1   Datapath

A datapath consists of a set of regularly structured RT components, such as ALUs, multiplexers, latches, drivers and shifters. Datapath layout is accomplished with a stack of functional and storage units that are placed one above the other. Each unit consists of bit slices and all units have bit slices of the same width. However, bit slices in different units may be of different height. All bit slices are aligned starting with the least-significant bit (LSB) and distinct units are stacked on top of another. Thus, the stack grows horizontally when the bit width increases, and it grows vertically when the number of units increases (Figure 3.3(a)). Each bit slice of a unit may be a handcrafted custom cell or may be implemented with one row of connected standard cells as shown in Figure 3.2(a) and (b), respectively. The difference between custom and standard-cell styles is in the layers used for routing of control and data wires, use of custom or standard cells and routing of data lines over the cells or in a separate channel.

In the first layout architecture (Figure 3.3(b)), diffusion strips for P and N transistors are placed horizontally. Power and ground wires run horizontally in the first metal layer. The control lines common to different bit slices in each unit run also horizontally in the first metal layer. Data lines connecting distinct units in each bit slice run vertically in the second metal layer. In the second layout architecture (Figure 3.3(c)), a bit slice of each unit consists of one or more standard cells. P and N diffusion strips are placed vertically. Power and ground wires run vertically in the first metal layer. Control lines run over the standard cells in the second metal layer. Data lines are placed in the routing channel and run vertically in the first metal or the polysilicon layer. The connections between standard cells inside each bit slice is also placed in the routing channel.

To compute the height (H_{dp}) of a bit slice (Figure 3.3(a)), we observe that H_{dp} is proportional to the number of transistors in the bit slice. Each bit slice of a unit in Figure 3.3(c) consists of several diffusion strips separated by gaps. The transistors on each diffusion strip are separated by metal-diffusion contacts or by the minimal poly-to-poly spacing. Thus, the width of $Unit$ (W_{unit}) in Figure 3.3(c) can be computed as a product of the number of transistors ($tr(unit)$) and the transistor-pitch coefficient (α) in μm/transistor. α is obtained by averaging the ratio of cell width and the number of transistors per cell over all units in the library. Thus,

$$W_{unit} = \alpha \times tr(unit). \qquad (3.1)$$

Consequently, the height of the bit-sliced stack of n units is

$$H_{dp} = \sum_{i=1}^{n} W_{unit_i} = \alpha \times (\sum_{i=1}^{n} tr(unit_i)). \qquad (3.2)$$

The Equation 3.2 will hold for standard-cell architecture even if each bit slice is implemented with two or more rows of standard cells. Ob-

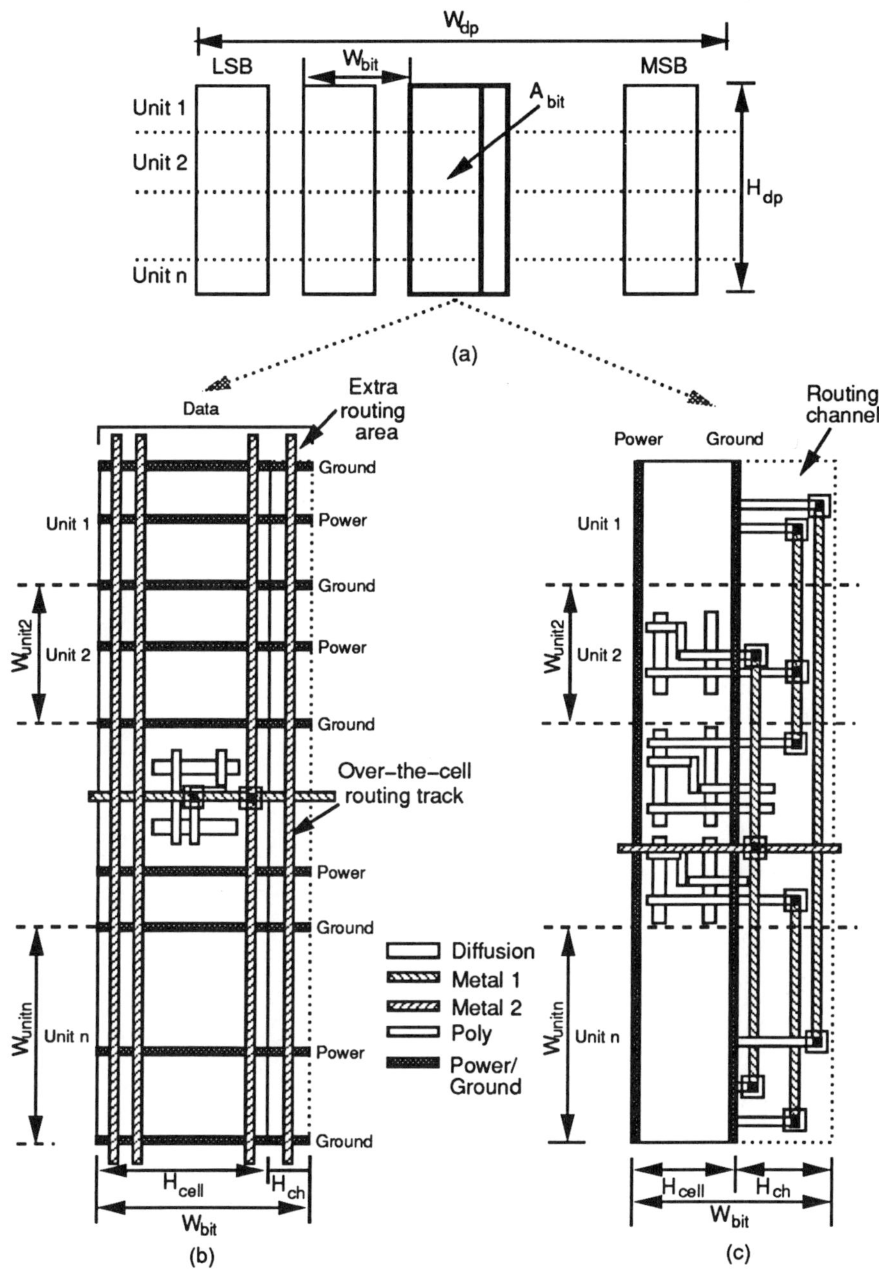

Figure 3.3: The layout models: (a) datapath stack, (b) custom cell architecture, (c) standard cell architecture.

viously, a different coefficient α' must be used in that case. Thus, the height of the bit-sliced stack of n units with an m-row implementation is

$$H_{dp} = \alpha' \times (\sum_{i=1}^{n} tr(unit_i))/m. \tag{3.3}$$

Similar assumptions can be made for custom-cell architecture shown in Figure 3.3(b). Although the P and N strips are placed horizontally in several rows, W_{unit} can be computed by Equation 3.2 using a different transistor-pitch coefficient α''. This assumption holds because the height of the unit slice (H_{cell}) is a constant and the unit width (W_{unit}) thus must reflect the size of the cell in number of transistors.

The width W_{bit} of a bit slice is equal to the sum of the height of the unit slice (H_{cell}) and the height of the routing channel (H_{ch}). For both layout architectures H_{cell} is a constant since all unit slices are predesigned to be of the same height. H_{ch} is calculated as a product of the wire pitch (β), and the difference between the number of estimated routing tracks (Trk_{est}) required to completely connect all nets in one bit slice and the number of available over-the-cell routing tracks (Trk_{top}). Thus,

$$H_{ch} = \begin{cases} 0; & \text{if } Trk_{top} \geq Trk_{est} \\ \beta \times (Trk_{est} - Trk_{top}); & \text{if } Trk_{top} < Trk_{est} \end{cases} \tag{3.4}$$

where Trk_{top} in standard-cell architecture is equal to zero, and coefficient β is equal to the sum of the minimal wire width and the minimal spacing between two metal wires.

An estimate for the required number of tracks in each bit slice can be obtained only after the position of each unit in the bit slice is determined. A fast algorithm with pseudo linear time complexity, such as the min-cut algorithm [FiMa82] can be used for this purpose. The required number of tracks can be estimated by the maximal density which is defined as the maximum number of connections across any cut perpendicular to the channel. A better estimate can be obtained by using some simple routing algorithms, such as the left-edge algorithm [HaSt71] which has $O(nlogn)$ complexity where n is number of nets (explained in Chapter 8). Thus, the datapath area (A_{dp}) can be calculated as a product of the number of bits (b_w) and the area of one bit slice, i.e.,

$$A_{dp} = b_w \times H_{dp} \times (H_{cell} + H_{ch}). \tag{3.5}$$

The Equation 3.5 gives an upper bound on the datapath area. The bound is proportional to the product of the number of transistors and the number of routing tracks. The number of transistors can be approximated from the Boolean expressions describing each unit slice or counted from its schematic. The number of tracks can be approximated by the track density after a linear placement. Better estimates can be achieved with algorithms of higher complexity. Since the number of components in the datapath is small, those more accurate estimates are not necessarily computationally intensive.

3.3.2 Control Unit

The control unit in a FSMD model can be described by the control state-table which specifies the next-state and control signals as a function of present states and conditional or status signals in a tabular form. Figure 3.4(a) shows a sample control state table with the input present-state and conditional/status signals and the output next-state and control signals. The present states are encoded as binary values $p_k...p_1 p_0$, where $k \geq \lceil log_2 m \rceil - 1$ and m is the number of states. Similarly, next states are encoded as binary values $r_k...r_1 r_0$. Each output c_i controls a functional, storage or interconnection components in the datapath, whereas the outputs r_j specify the next state.

The control unit consists of a state register and control logic. There are two commonly used techniques for implementation of control units: standard cells and programmable logic arrays (PLA) as described in Chapter 2. In this section, we will give area estimates for standard-cell and PLA implementations based on sum-of-product expressions of next-state and control signals (Figure 3.4(b)).

In reality, a number of optimization procedures, such as logic minimization or PLA folding, are applied in order to reduce the size of the control logic. We ignore the impact of optimization and give an upper bound on the control-unit area. Procedures for tighter upper bounds on area estimates (that include the impact of optimization techniques) are open problems given as exercises at the end of the chapter.

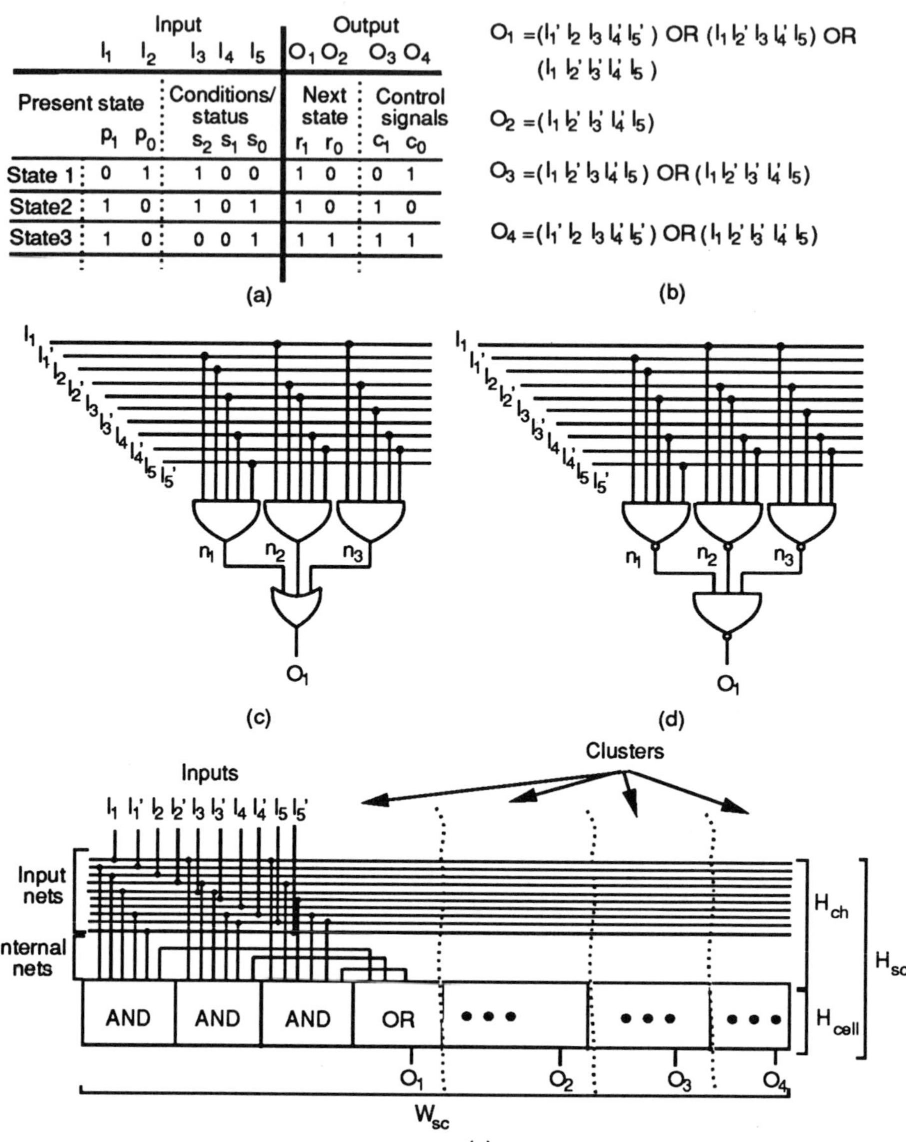

Figure 3.4: Control unit description: (a) state table, (b) Boolean equations for output signals, (c) two-level AND-OR implementation, (d) two-level NAND-NAND implementation, (e) standard cell layout style.

Standard-Cell Implementation

To simplify area estimates for standard cell implementations, we make a number of assumptions. We assume that a product term in the sum-of-product expression for each output signal includes each present-state signal and (in the worst case) each conditional/status signal; these inputs to the product term may be complemented. We also assume that each product term is implemented by an AND gate and the sum-of-products by an OR gate, as illustrated in Figure 3.4(c) for output O_1. Obviously, we can replace the AND-OR implementation with an equivalent NAND-NAND implementation as shown in Figure 3.4(d).

We assume that all the gates for implementation of the control unit are placed in a single row of cells, the inputs appear at the top, and the outputs appear at the bottom, as shown in Figure 3.4(e). We also assume that all the gates needed for the implementation of an output signal are clustered together as shown for signal O_1 in Figure 3.4(e). This requirement for strong clustering prevents sharing of an AND gate between two expressions with the same product term. The layout area of the control unit using the standard-cell implementation (A_{sc}) is then equal to the product of width (W_{sc}) and height (H_{sc}), that is,

$$A_{sc} = W_{sc} \times H_{sc}, \tag{3.6}$$

where W_{sc} is proportional to the number of transistors and H_{sc} is proportional to the number of routing tracks. The number of transistors can be computed from the sum-of-product expression for each output signal.

In CMOS technology, each n-input AND or OR gate has $2n + 2$ transistors. Note that n-input NAND and NOR gates need only $2n$ transistors. Since each product term in a sum-of-product expression is implemented with one AND gate and one OR gate (e.g., Figure 3.4(c)), we can compute the required number of transistors. Each literal in a product term contributes 2 transistors, each product term contributes 2 transistors in the AND gate and 2 transistors in the OR gate and the OR gate contributes an additional 2 transistors.

Let $occur(O_i)$ and $term(O_i)$ be the number of occurrences of literals and number of terms in the sum-of-product expression of the signal O_i, respectively. Let $tr(Reg)$ be the number of transistors in one-bit state register. Thus, for a AND-OR implementation of the control unit the width of the control unit is computed as

$$W_{sc} = \alpha \times \ \ ((\textstyle\sum_{i=1}^{\lceil log_2 m \rceil + n}(2term(O_i)occur(O_i) + 4term(O_i) + 2))$$
$$+(\lceil log_2 m \rceil \times tr(Reg))), \qquad (3.7)$$

where m is the number of states and n is the number of control signals and α is the transistor-pitch coefficient as defined in Section 3.3.1.

The height H_{sc} is computed as the sum of the standard-cell height (H_{cell}) and the channel height (H_{ch}). H_{ch} is proportional to the number of tracks used by input signals and internal nets connecting AND and OR gates. We assume that each input signal requires two tracks for the true and complemented values. We assume that there are $\lceil log_2 m \rceil$ state signals and C conditional/status signals. The maximum number of tracks required for routing internal nets in each cluster is equal to the number of terms in the sum-of-product expression for a particular output signal. Since clusters do not overlap, the maximum number of required tracks to route all internal nets is equal to the largest number of terms used in any particular sum-of-product expressions of an output signal. Thus, the height of the control unit H_{sc} is

$$H_{sc} = H_{cell} + \beta \times (2(\lceil log_2 m \rceil + C) + MAX(\forall_{i=1}^{\lceil log_2 m \rceil + n} term(O_i))), \quad (3.8)$$

where β is the wire pitch in the channel.

Although we assumed a single-row layout, control logic modules with different aspect ratios are needed in reality. Different aspect ratios can be obtained by laying the control logic module in several rows. We approximate this folding process by assuming that we can evenly partition the single-row layout of Figure 3.5(a) into three rows of equal width as shown in Figure 3.5(b). We assume that input and output pins are positioned at the top and bottom of the control unit, respectively. Further, we assume that inputs reach all rows using existing polysilicon lines in each row as feedthroughs, and output nets are routed vertically (over the cells) in metal2. We also assume that output clusters are not split across folded rows. Then, the width of such a multiple-row module is W_{sc}/R where R is the number of rows. The height of the module is the sum of all row heights. The height of each row can be computed by Equation 3.8 in which only the internal nets of clusters in

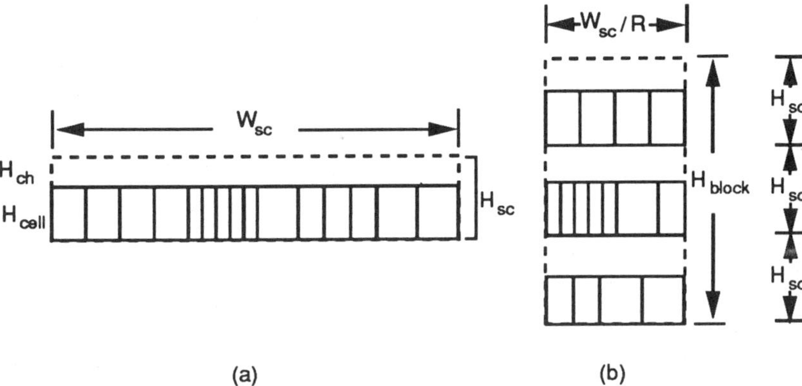

Figure 3.5: Different aspect ratios of the control logic: (a) one-row implementation, (b) three-row implementation.

that particular row contribute to the channel density of that row. Since $H_{ch} \times R \geq H_{ch1} + H_{ch2} + ... + H_{chR}$, we can use H_{ch} as an upper bound on the channel height for each row. With these assumptions, the total height of the folded implementation (H_{block}) is $H_{sc} \times R$, while its width is W_{sc}/R Hence, we can still use the Equation 3.6 as an estimate for control-logic layouts of different aspect ratios.

These simple-minded bounds may not be adequate when more accurate estimates are needed. Better estimates can be obtained by generating more accurate netlists for the control logic and by modeling placement and routing algorithms more accurately.

We can generate a large netlist for the control-state table and perform a limited technology mapping by decomposing each AND or OR gate into series of gates from the given library. Placement can be modeled by probabilistic distributions of pin positions and wire lengths. For example, Kurdahi and Parker [KuPa89] assume a uniform distribution of pins across R rows of cells and a geometric distribution of the wire length with average wire length obtained experimentally by running many examples. Routing algorithms are approximated by computing routing density across each channel.

To obtain even better estimates, more accurate models of the placement and routing algorithms must be used. Pedram and Preas [PePr89] model a placement algorithm that minimizes the sum over all nets of the half-perimeter length of the rectangle enclosing pins of each net. They also model global routing by approximating a minimal rectangular Steiner tree for connecting pins on each net and channel routing by approximating the left-edge algorithm. Instead of modeling placement and routing algorithms, some linear algorithms can be used to obtain even more accurate estimates. Zimmermann [Zimm88] uses the well-known min-cut algorithm [FiMa82] to quickly generate an acceptable floorplan from netlist of components with known area aspect ratios.

As we mentioned earlier, most estimation methods start with a netlist of components and assume a standard-cell layout architecture. Currently, there are no estimation methods for HLS that start from behavioral descriptions, state tables or Boolean expressions.

Programmable Logic Array

A programmable logic array (PLA) is frequently used to implement combinatorial and sequential logic, and in particular, the control units of FSMDs. A PLA consists of AND and OR arrays supported by input and output buffers, input and output latches and product term buffers as shown in Figure 3.6. Input buffers are needed to drive the AND array, product term buffers to drive the OR array and output buffers to drive the external logic. The input and output latches are used for the sequential logic.

The width of the PLA module (W_{PLA}) is the sum of the width of the input AND array (W_{in}), the width of product-term buffers (W_p) and the width of the OR array (W_{out}) (Figure 3.6(b)). W_{in} equals the number of inputs (n) multiplied by the the maximum of the latch width (l_w) and the buffer width (b_w). Similarly, W_{out} equals the product of $MAX(l_w, b_w)$ and the number of outputs (m).

The height of the PLA (H_{PLA}) is computed as a sum of the latch height (l_h), buffer height (b_h) and the height of the AND-OR plane. The height of the AND-OR plane is determined by the product of the number of distinct product terms (p), and the transistor-row pitch (r). Thus, the

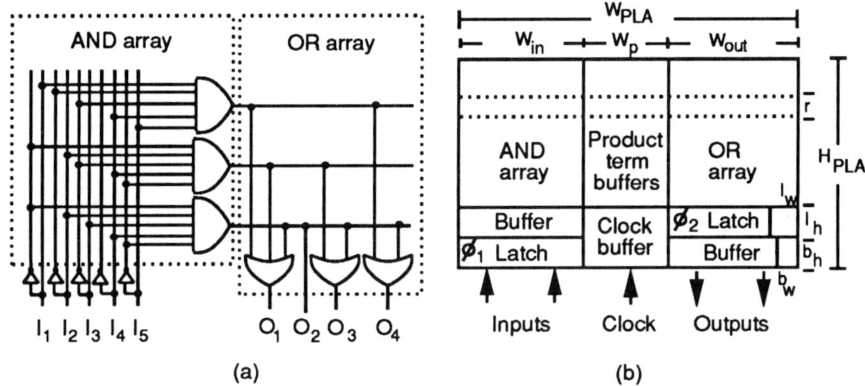

Figure 3.6: PLA layout model: (a) logic mapping, (b) layout model.

area of a PLA is

$$A_{PLA} = (((n + m) \times MAX(l_w, b_w)) + W_p) \times (l_h + b_h + r \times p). \quad (3.9)$$

3.4 Performance Measures

Traditionally, design performance for HLS has been characterized by such measures as the clock frequency, the number of instructions or the number of additions/multiplications executed in a second. These individual measures do not characterize the actual design performance with respect to a particular description. Instead, the performance of a design for a particular description is measured by the total time needed to execute that description. Since a description may contain loops without fixed bounds and "if" statements in which "then" and "else" branches require different amounts of time, the execution time of such a description will depend on the input data. Thus, the execution time for any description is equal to the product of the number of control steps needed to execute the description and the duration of the control step (usually a clock period). Hence, minimizing the clock period always improves the performance for a fixed schedule. Performance can also be improved by minimizing number of control steps on the critical path through the computation, using more resources and executing in parallel whenever

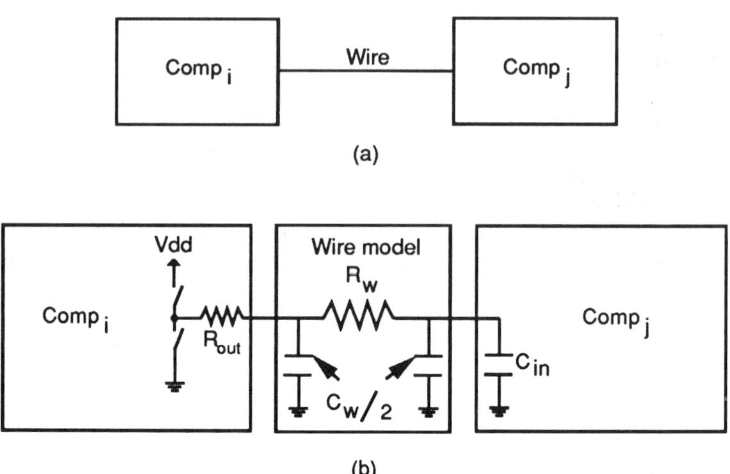

Figure 3.7: Wire: (a) RT model, (b) equivalent RC delay model.

possible.

In this section, we will describe electrical delay models, and use them to construct delay models for combinational and sequential circuits. Finally, we will describe the clock-cycle computation for the FSMD architecture.

3.4.1   Electrical Models

The lumped RC model, also called the Elmore delay model [PeRu81], is widely used for delay calculation. In the lumped RC model, the propagation delay along a path from the start point to the end point, $(t_p(\text{start,end}))$, is computed as a product of lumping all of the resistances R_j and capacitances C_k along the path, that is,

$$t_p(\text{start}, \text{end}) = \sum_j R_j \times \sum_k C_k. \qquad (3.10)$$

We can use Equation 3.10 to obtain delay of a connecting wire between two components as shown in Figure 3.7(a). In CMOS technology we model a component with its input capacitance (C_{in}) and its output resistance (R_{out}), as shown in Figure 3.7(b). For the connecting wire, we use the well known π-model with input capacitance $(C_w/2)$, wire resistance (R_w) and the output capacitance $(C_w/2)$. Since a wire is a thin sheet of metal of fixed thickness defined by the fabrication process consisting of rectangular segments, the wire resistance is equal to the product of the sheet resistance (R_s) in Ohm/square and the ratio of the wire length (L_w) and wire width (W_w), that is,

$$R_w = R_s(\frac{L_w}{W_w}). \tag{3.11}$$

The wire capacitance (C_w) is equal to the product of the wire area and the ratio of the dielectric constant (ε) and the wire thickness (t), that is,

$$C_w = (L_w W_w)(\frac{\varepsilon}{t}). \tag{3.12}$$

Using Equations 3.11 and 3.12, we can compute the propagation delay $(t_p(net_k))$ of a wire net net_k used by a component $(comp_i)$ to drive load components $(comp_j, 1 \leq j \leq n)$ as

$$t_p(net_k) = (R_{out}(comp_i) + R_w)(C_w + \sum_{j=1}^{n} C_{load}(comp_j)). \tag{3.13}$$

3.4.2 The Delay of Combinational Circuits

Each combinational circuit consists of one or more components. The delay of a combinational circuit is the propagation delay on the critical path through the circuit, that is, the longest delay from any input to any output. For example, for the circuit in Figure 3.8 the critical path is from input I_3 to the output O_2. The critical path goes through components A, C, E and F and nets n_1, n_4, n_6, n_7 and n_8. Thus, the delay on a path from an input port to an output port is equal to the sum of propagation delays of all components and all the nets on the critical path. Let $COMP(I,O)$ and $CONN(I,O)$ define all the components and all the nets on the path from an input port I to an output port O. Thus, the propagation delay $(t_p(I,O))$ from the input port I to the output port O is

$$t_p(I,O) = \sum_{comp_i \in COMP(I,O)} t_p(comp_i) +$$

$$\sum_{net_j \in CONN(I,O)} t_p(net_j). \tag{3.14}$$

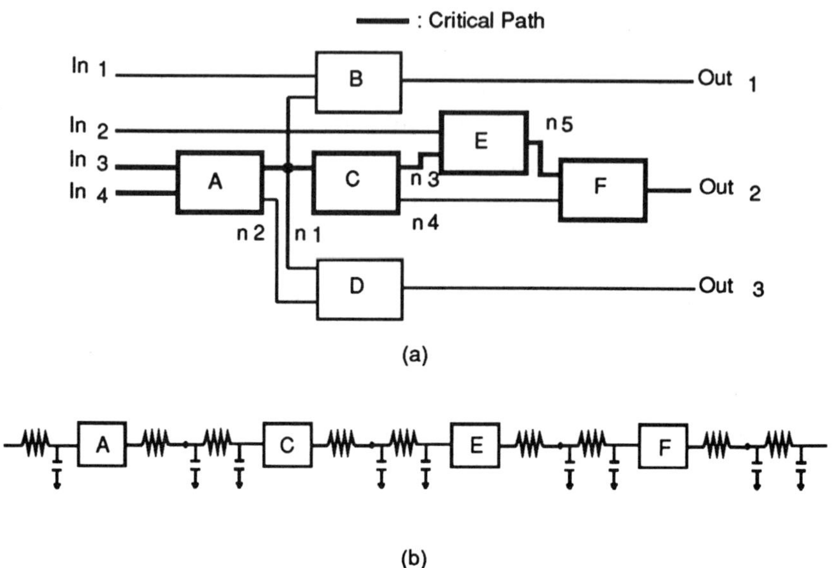

Figure 3.8: The critical path in a combinational circuit.

3.4.3   The Delay of Storage Elements

Storage elements are typically composed of latches and flip-flops. The simple D-type latch was described in Section 2.6.1 and is shown in Figure 3.9(a). We can identify two different delay paths through this D-type latch: the data-to-output delay (t_{DQ}) and the clock-to-output delay (t_{CQ}), as shown in Figure 3.9(b). From the latch's schematic in Figure 3.9(a), we can also conclude that $t_{DQ} > t_{CQ}$ since an extra inverter is on the data-to-output path.

In order to ensure proper operation of the latch, two timing constraints are imposed: the setup time (t_{setup}) and hold time (t_{hold}). The setup and hold times specify time intervals before and after clock transitions during which a data signal must be stable to ensure proper operation of the latch (Figure 3.9(b)). It is obvious from Figure 3.9 that $t_{DQ} = t_{setup} + t_{CQ}$.

Master-slave (MS) and edge-triggered (ET) flip-flops respond to the transition of the clock signal. In a MS flip-flop the master latch is active

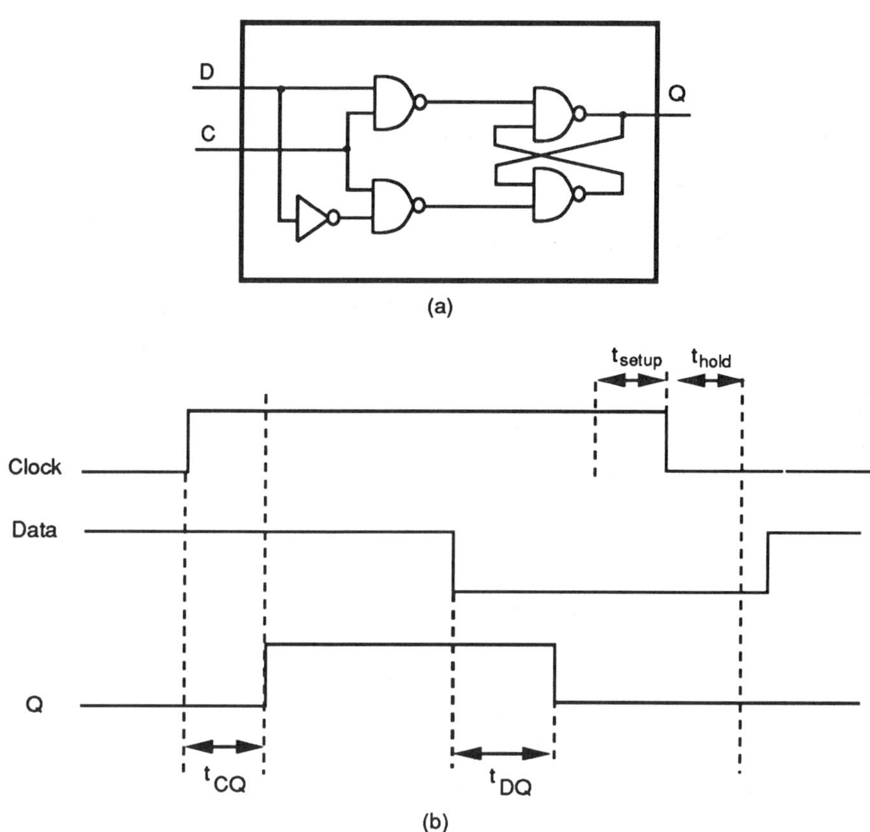

Figure 3.9: D-latch: (a) logic schematic, (b) delays and constraints.

when the clock signal is high while the slave latch is active when the clock signal is low (Figure 3.10(a)). The output of the MS flip-flop changes after the high-to-low transition of the clock and the propagation delay (t_{CQ}) of the slave latch. Thus the propagation delay of the MS flip-flop is equal to t_{CQ} of the slave latch. Since the data signal must propagate through the master latch before a high-to-low clock transition and still satisfy t_{setup} of the slave latch, $t_{setup}(MSFF) = t_{DQ}(Master\ latch) + t_{setup}(Slave\ latch)$. Note that the master latch (and hence the MSFF) is not sensitive to changes in the input data while the clock is low. For example, the glitch in D while clock is low does not affect the output Q of the MSFF (Figure 3.10(b)). Hence the hold time for the MS flip-flop

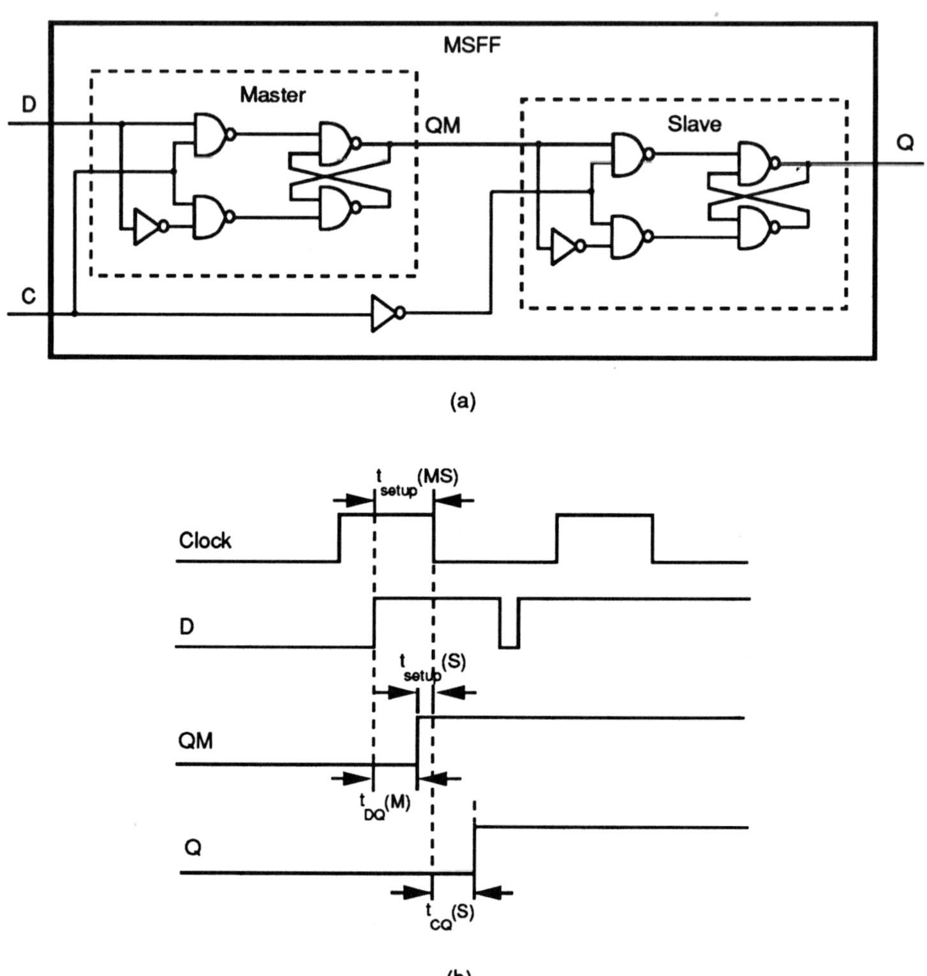

Figure 3.10: Master-slave flip-flop: (a) D-latch implementation, (b) delays and constraints.

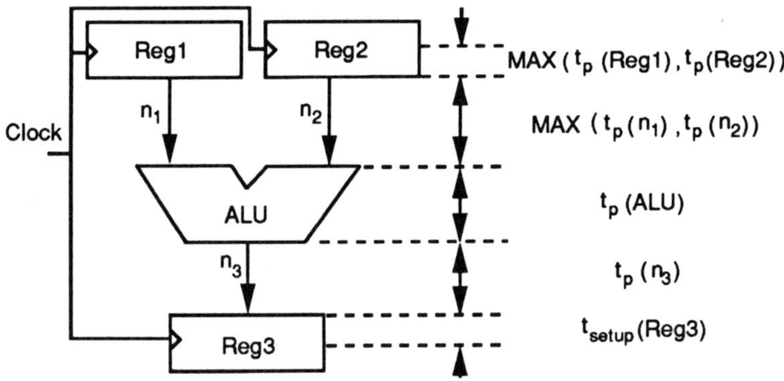

Figure 3.11: The register-transfer path.

is equal to the hold time (t_{hold}) of the master latch. We can formulate similar definitions for an ET flip-flop which propagates data on the low-to-high transition of the clock signal.

3.4.4 System Clock Cycle

Since the execution time of a description is equal to the product of the number of control steps and the clock period, number of control steps is good measure of performance for a fixed clock period. Similarly, clock period is a good quality performance measure for a scheduled behavioral description with a fixed number of states.

The clock cycle is determined by the worst register-to-register delay in the FSMD architecture. For example, a typical register-to-register delay in a datapath is shown in Figure 3.11. It includes the delay through the source registers, ALU and connections, and the setup time of the destination register. In other words, we can specify the clock period by

$$
\begin{aligned}
t_{clock} \;=\; & MAX(t_p(Reg1), t_p(Reg2)) + MAX(t_p(n_1), t_p(n_2)) \\
& + t_p(ALU) + t_p(n_3) + t_{setup}(Reg3).
\end{aligned} \tag{3.15}
$$

If the data transfer has more components Equation 3.15 must be adjusted approximately. For example, if a multiplexer component is inserted at some input port, an extra multiplexer propagation-delay and the corresponding connection delay must be added to the Equation 3.15.

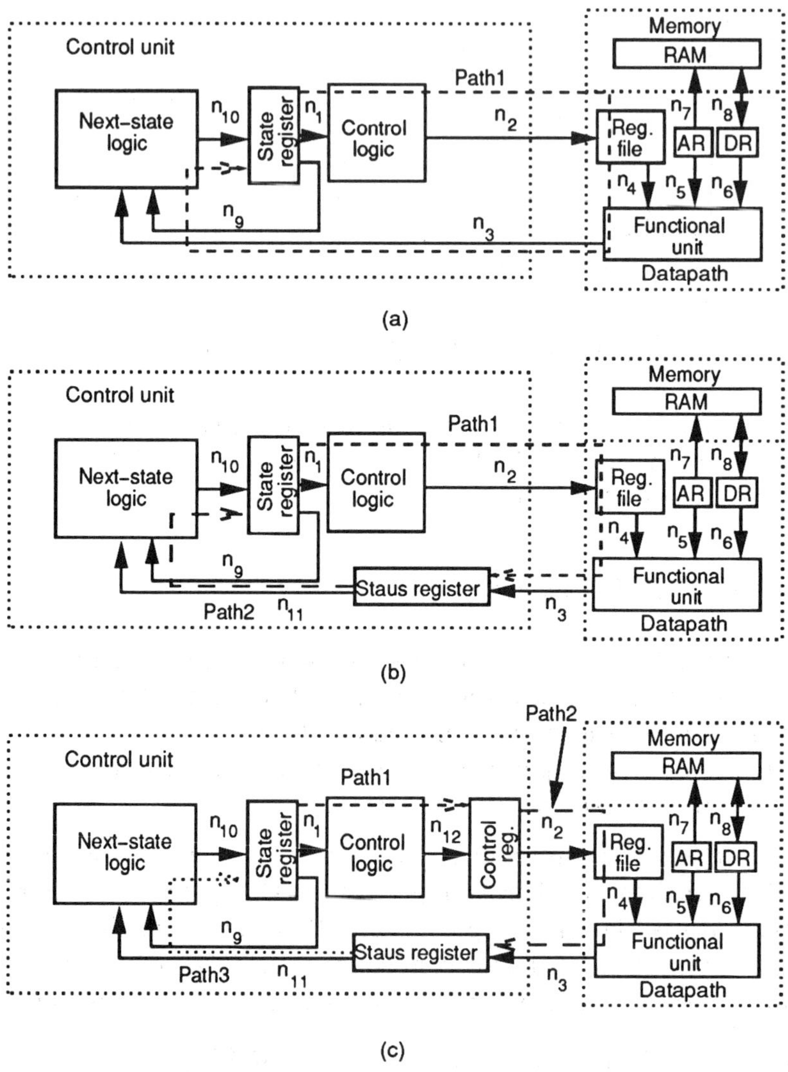

Figure 3.12: System clocking models: (a) nonpipelined control, (b) one-stage pipelined control, (c) two-stage pipelined control.

When the clock period for a FSMD is computed we must include the delay in the control unit in addition to delays in the datapath. A typical FSMD structure is shown in Figure 3.12(a). There is one critical path (*Path1*) from the *State register*, through the *Control logic*, the *Reg. file*, the *Functional unit*, the *Next-state logic* and back to the *State register*. Thus, the clock period is the sum of all the unit delays and all the net delays, that is,

$$
\begin{aligned}
t_{clock} \;=\; & t_p(State\ register) + t_p(Control\ logic) + t_p(Reg.\ file) \\
& + t_p(FU) + t_p(Next\text{-}state\ logic) + t_{setup}(State\ register) \\
& + \sum_{i=1,2,3,4,10} t_p(n_i).
\end{aligned} \tag{3.16}
$$

The clock period calculated by the Equation 3.16 may be too long for many applications. As suggested in Chapter 2, the design must be pipelined to reduce the clock period by insertion of the *Status register* (Figure 3.12(b)). The clock period is now determined by the longer of two paths, *Path1* and *Path2*. *Path1* starts with the *State register* and goes through the *Control logic*, the *Reg. file*, the *Functional unit* and ends with the *Status register*, whereas *Path2* starts with the *Status register*, goes through the *Next-state logic* and ends in the *State register*. Thus, the clock period is

$$
\begin{aligned}
t_{clock} \;=\; MAX \quad & ((t_p(State\ register) + t_p(Control\ logic) + t_p(Reg.\ file) \\
& + t_p(FU) + t_{setup}(Status\ register) + \sum_{i=1,2,3,4} t_p(n_i)), \\
& (t_p(Status\ register) + t_p(Next\text{-}state\ logic) \\
& + t_{setup}(State\ register) + \sum_{i=10,11} t_p(n_i))).
\end{aligned} \tag{3.17}
$$

The design can be further pipelined by introduction of the *Control register* as shown in Figure 3.12(c). The clock period is then determined by the longest of these three paths *Path1*, *Path2* and *Path3*, that is,

$$
\begin{aligned}
t_{clock} = \quad MAX \quad &((t_p(State\ register) + t_p(Control\ logic) \\
&+ t_{setup}(Control\ register) + \sum_{i=1,12} t_p(n_i)), \\
&(t_p(Control\ register) + t_p(Reg.\ file) + t_p(FU) \\
&+ t_{setup}(Status\ register) + \sum_{i=2,3,4} t_p(n_i)), \\
&(t_p(Status\ register) + t_p(Next\text{-}state\ logic) \\
&+ t_{setup}(State\ register) + \sum_{i=10,11} t_p(n_i))). \quad (3.18)
\end{aligned}
$$

As we mentioned in Chapter 2 the functional unit (FU) in the data-path can be further pipelined which makes clock period even shorter. The clock period can be estimated using Equations 3.16, 3.17 and 3.18. The delays of registers, register files, and functional units are known from the given RT component library or predefined component instances. The delay of the control unit depends on the selected layout style. Using the standard-cell implementation, the propagation delay of next state and control logic can be calculated using the Equation 3.14 for the sum-of-product implementation described in Section 3.3. The PLA implementation uses a 2-phase clock as shown in Figure 3.6. The input is latched and the AND array is precharged during the first phase of the clock, whereas the output is latched during the second phase of the clock. Hence, the propagation delay of the PLA is the sum of input latch propagation-delay, AND-OR plane propagation-delay and the setup time and propagation delay of the output latch. Using the NOR-NOR implementation of the AND-OR plane, the propagation delay can be approximated with the propagation delays of two NOR gates.

The propagation delay of particular nets can be computed from wire lengths and input capacitances of the components they drive. An upper bound on the length of the wire is obtained using the layout models described in Section 3.3. We assume that net n_4 that connects the register file and functional unit in Figure 3.12 occupies one vertical track in the datapath layout model shown in Figure 3.3. Thus, the length of net n_4 is less or equal to the height of the datapath H_{dp}. Similarly, control net n_2 and status net n_3 in Figure 3.12 run horizontally in the model of the data path in Figure 3.3. The length of nets n_2 and n_3 is not longer than the datapath width W_{dp}. Note that control and data lines running in metal1 and metal2 layers do not jog. The length of nets n_1, n_9, n_{10}, n_{11} and n_{12} is less than half the perimeter of the control unit, $H_{sc} + W_{sc}$, for the standard-cell implementation or $H_{PLA} + W_{PLA}$ for the PLA implementation. Thus, using Equation 3.16, 3.17 and 3.18 we can compute the clock period for a given FSMD structural description. The clock period may be used to select the best implementation for a given FSMD structure and a set of RT components during register, unit or connection allocation in HLS. The product of the clock period and the number of control steps is used to select the best schedule for a given behavioral description and a set of RT components during scheduling. The number of control steps may not be easy to estimate if the behavioral description depends on data. In this case, we must assign probabilities to different control paths (e.g., different branches in a "if" statement) in the description.

We demonstrated the computation of the clock cycle for arbitrary FSMD architectures. When a chip consists of several communicating FSMDs we must include delays due to communication. If two FSMDs communicate asynchronously as shown in Figure 2.13, then the clock calculation for each FSMD is independent of the other. On the other hand, communication time must be included when computing execution time for a given behavioral description. If several FSMDs use a common clock then we must include register-to-register delays between communicating FSMDs. If such register transfers between two FSMDs are not needed in each control step, the communication between two FSMDs may take several clock cycles and not influence the clock period. In either case we must estimate the wire lengths between FSMDs. Such an estimate requires the floorplan of the entire chip or an estimate of the floorplan. If two FSMDs are on different chips, multi-chip modules (MCM) or printed-circuit boards (PCB), we must estimate pin capacitances and connection delays on the chip, MCM or PCB. The methods for estimation of floorplans described in Section 3.3 may be used for this purpose.

3.5   Other Measures

In the past, high-level synthesis focused on quality measures based on area and performance. Other measures, such as power consumption, testability, verifiability, reliability and manufacturability are not well defined and have not been widely used.

The power consumption of the CMOS digital system consists of two elements: static dissipation and dynamic dissipation. Since the static dissipation of CMOS circuits is in the nanowatt range, it is often assumed to be negligible. The dynamic dissipation of CMOS circuits is caused by charging and discharging of load capacitances during signal transitions. When a CMOS gate changes state from either "0" to "1" or "1" to "0", both n- and p-transistors are turned on for a short period of time. This results in a short current pulse to charge and discharge the output capacitive load. One way to compute the dynamic disspation (P_d) of a CMOS digital system [BrGa90] is

$$P_d = P_{gate} \times N \times f_c, \qquad (3.19)$$

where P_{gate} is the average dynamic dissipation of a gate, N is the maximum number of gates turned on in a clock cycle, and f_c is the clock frequency. P_{gate} is given by the gate manufacturers. f_c is given in the design specification. N is difficult to calculate, since the number of gates changing their output values in each clock cycle is dependent on the function they implement (e.g., addition or subtraction) and on the input data. It is safe to assume that only 50% of all gates switch during the same clock cycle. In reality, much fewer gates are active in each clock cycle.

Power-delay-product (PDP) is another useful and often-quoted metric. PDP of a design is calculated as the product of average power consumption P_d and average gate delay (t_p), that is,

$$PDP = P_d \times t_p. \qquad (3.20)$$

The units of PDP are (watt)(seconds) = joules. It can be viewed as the energy consumed for each logic decision made.

Design for testability is the task that produces a design with minimal test cost by considering the test hardware and the cost of testing during the synthesis process. Design for testability is bounded by three factors: the number of pins available for testing, the test circuitry available on-chip and the performance constraints. The number of pins required for testing is an important factor in the cost of manufacturing a chip because the cost of larger packages and their assembly time is higher. Additional test pins require more chip area and more power dissipation because more test circuits and I/O drivers are needed. Furthermore, increased circuit complexity reduces yield. Thus, increased test circuitry may reduce the cost of testing; however, it may increase the cost of production. In addition, increased test circuitry increases capacitive loads and creates additional delays that may degrade the design performance [KiTH88].

The main objective of design for testability is to increase the "controllability" and "observability" of the internal states of the design and to minimize the hardware overhead. Controllability defines the ability to initialize the internal logic into specific controlled states. Observability defines the ability to externally measure the internal logic states of the design. Several approaches based on controllability and observability measures were proposed to tradeoff test cost and test hardware for datapath and FSM synthesis [ChPa91,EsWu91,PaCH91].

Verification ensures that designs are complete and consistent with respect to the given specification. In recent years, many formal verification methods have been proposed to prove that the design meets the specification at the physical and logic levels [Brya85,HaJa88,MWBS88]. Some methods have been proposed to verify RT designs [Vemu90,FeKu91]. Design for verifiability in HLS is an open problem.

Reliability represents the probability of successful operation of a design over its lifetime. Manufacturability addresses the evaluation of overall manufacturing costs, including the costs of design, fabrication, packaging, testing and labor. Design for reliability and design for manufacturability are not well defined for HLS, and there is much work to be done before these measures can be used to guide HLS.

3.6 Summary and Future Directions

In this chapter, we discussed some design quality measures to be used as cost functions in HLS algorithms. In particular, we discussed the area cost and performance measures for the FSMD architecture. The area cost of a design was approximated with the silicon area of custom layout using handcrafted or standard cells. The FSMD performance measures were approximated with the length of the clock cycle and execution time. We also briefly mentioned some other measures although we did not elaborate their evaluation, since they have not been used frequently or even characterized well.

Since quality measures for HLS must be computed rapidly to be effective, we used a simple design model and some simplifying assumptions. We neglected the impact of logic optimization on the control unit and microarchitectural optimization on the datapath. We derived upper bounds on the area and performance estimates from nonoptimized Boolean expressions. It is reasonable to expect that state minimization, state encoding, logic minimization and technology mapping will reduce control unit area and delays by 50%. Presently, there are no estimation methods that will include the impact of logic and sequential synthesis on area and delay estimates of FSMDs. Future work is needed to develop such methods and extend their scope to complete chips and system.

We modeled chip placement by assuming total clustering of Boolean expressions in the control unit and routing by assuming one signal in each routing track. Better estimates can be obtained by assuming some distribution of pin positions, wire length and number of wires in each track. Those estimates can be further improved by modeling placement algorithms to define the pin positions for each net and routing algorithms to define the position of each wire. The best estimate would obviously use some simple placement and routing algorithm to obtain the exact layout area and wire lengths. Future research will focus on developing much tighter bounds on quality measures for HLS. Those bounds must predict the impact of lower level synthesis tools. Further research is also needed to establish measures of quality for testability, verifiability, reliability, manufacturability and other criteria.

We presented only area and delay measures for custom CMOS technology. Algorithms for estimating quality measures for other architectures or technologies, such as gate arrays or field-programmable gate arrays are open problems in HLS.

3.7   Exercises

1. Explain the difference between the accuracy and fidelity of quality measures. Explain which one is more important for supporting design tradeoffs and why?

2. Explain why the number of multiplexers, the number of equivalent 2-to-1 multiplexers, or the number of multiplexer inputs are not good interconnection area measures.

3. Determine the total number of transistors and routing tracks required to implement the state table shown in Figure 3.4 using:
(a) the two-level AND-OR implementation, and
(b) the two-level NAND-NAND implementation.

4. *Outline an area estimation procedure for standard-cell layouts using a constructive approach. Hint: [Zimm88,KuRa91].

5. *Outline an algorithm for determining area and performance measures by taking into account the impact of floorplanning.

6. *Define FSMD area measures for gate-array and field-programmable gate array (FPGA) layout architectures.

7. **Define the area and performance measures for an FSM specified by a symbolic state table, taking into account the impact of logic optimization.

8. *Outline an area measure for folded PLAs.

9. **Develop an area and delay estimation technique from the sum-of-product expressions that take into account the impact of logic minimization.

10. Describe the data-transfer paths (see Figure 3.11) and define the clock formulas (see Equation 3.15) for the multiple-cycle and the chaining operations described in Chapter 2.

11. *Develop a technique to estimate N in Equation 3.19.

12. *Outline an algorithm for trading-off fault coverage for test hardware. Hint: [ChPa91].

13. *Define quality-measure requirements for reliability and manufacturability.

Chapter 4

Design Description Languages

4.1 Introduction to HDLs

The input to high-level synthesis (HLS) is typically a behavioral description written in a hardware-description language (HDL). HDLs allow the user to specify the design's outputs in terms of the inputs over time using abstract data carriers and language operators. At a glance, such a language might seem similar to a high-level programming language like 'C' or Pascal, in which programs are written using sequences of statements with expressions that assign values to variables. However, the semantics of a behavior specified in an HDL implies a hardware design yet to be implemented (built), rather than a piece of behavior to be executed on an existing machine. Furthermore, these HDLs have additional constructs to accommodate the intrinsic features of hardware, such as the notions of clocking and asynchrony.

One might question the need for such HDLs, since a hardware design can be adequately specified using a purely structural language that specifies hardware components and their connections. In fact, current design methodologies in the hardware design community primarily use schematic capture and simulation, as shown on the left side of Figure 4.1. The design exists initially as a specification, often in plain English annotated with flowcharts, state graphs, timing diagrams and block structures. In the capture-and-simulate methodology, the designer manually refines the specification, adding design information until a complete register-transfer (RT) design is obtained. During this refinement process, the designer does not always maintain the original specification or a consistent design description that correlates with the original specification. This lack of a formalized description makes the task of simulation and verification difficult, since the whole design is not maintained explicitly but instead in the designer's mind. Furthermore, the lack of documentation during the design process makes maintaining and re-targeting the design, as well as incorporating synthesis, simulation and verification tools into the design process, difficult.

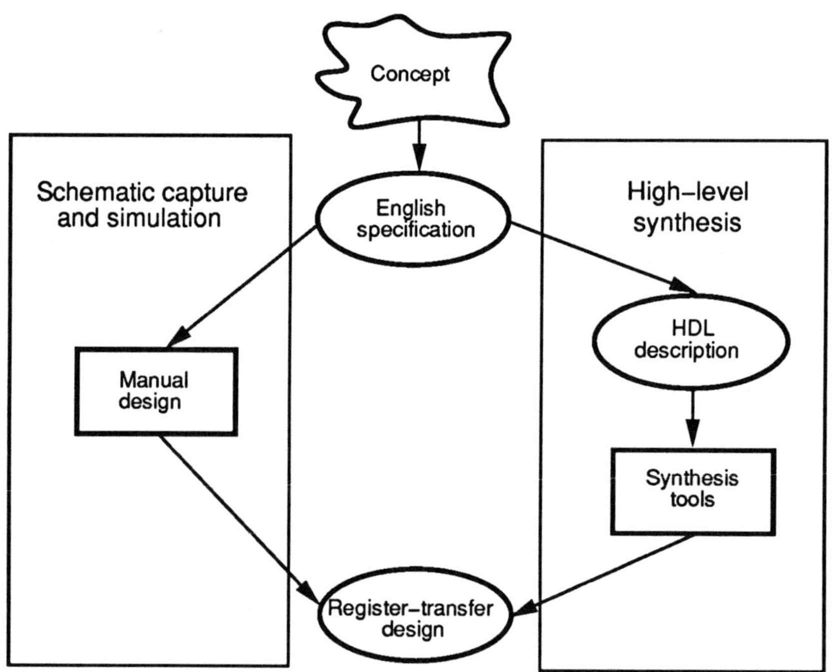

Figure 4.1: Design-specification-to-implementation path.

Many benefits result from writing and maintaining the original specification in an HDL. This formalized input can be used for documentation, simulation, verification and synthesis. The design description also serves as a good exchange medium between different users, as well as between users and design tools. Finally, as designs become increasingly more complex, HDLs provide a convenient mechanism for raising the level of abstraction to cope with the explosion of design complexity.

HDLs have been around for almost as long as hardware has [Comp74, Comp77]. Early HDLs were focused on the gate and logic levels, since designs were specified and implemented at that level (e.g., DDL [DuDi75]). As the abstraction level of design became higher, so did the abstraction level of HDLs, resulting in a proliferation of high-level HDLs [CHDL79, CHDL81, CHDL83]. Unfortunately, this increase in the number of diverse HDLs prevented the design community from exchanging design descriptions, defeating the goals of readability and portability [Lipo77, Hart87]. Such problems led to recent efforts at standardizing HDLs for digital design (Conlan [PiBo85], VHDL [IEEE88], UDL/I [YaIs89]), where the goal was to develop a common language that could be used for modeling, simulation and documentation, thereby achieving portability of designs among different sets of users.

Language standardization, although a worthy goal, is difficult to achieve. Users at different levels of the design process have varying levels of expertise with HDLs as well as varying backgrounds. Furthermore, the HDL has to be used for a variety of design applications, (e.g., digital signal processing (DSP) or real-time control), for different design aspects (e.g. simulation and synthesis), and for a variety of target architectures (e.g., FSM and systems). These differences impose vastly different requirements for the HDL to be used effectively. For instance, a DSP application is naturally expressed in terms of signal flow graphs, while a bus interface is naturally expressed with timing diagrams; a description for simulation needs the notion of simulation time, while a description for synthesis must have synthesizable HDL constructs; a FSM-based design needs the notion of states built into the HDL, while a system description needs constructs for expressing hierarchy and communication protocols. Hence, designers need different types of descriptions such as tables, graphics and text, to fit their backgrounds, the type of design application and different aspects of the target architecture being described. In order to use HDLs effectively, we need to understand some issues in HDL specification and relate these issues to the target architectural models. The rest of the chapter describes the differences between HDL models and target architectural models, and uses examples from a few HDLs to illustrate language features necessary for synthesis.

4.2   Language Models vs. Architectural Styles

In Chapter 2, we discussed some basic architectural design styles and their impact on synthesis. Just as designs are typically targeted towards a particular architectural style, the languages used to describe these designs at higher levels have their own semantic models. Consequently, each construct in an HDL has a particular interpretation based upon the semantics of the language being used.

For instance, VHDL [IEEE88] is a language based on simulation semantics. Signals and variables in VHDL represent data carriers that are assigned values at discrete points of time according to a schedule maintained by the simulator. The simulator goes through an event-driven simulation cycle of evaluating expressions at current simulation time, scheduling events for future time, and advancing the simulation clock to the next discrete point that has activity based on scheduled events. The behavioral VHDL statement

$$A \Leftarrow B + C \text{ after 20 ns};$$

assigns the sum of B and C to the signal A 20 ns after current simulation time. During simulation, the addition operation is performed instantaneously, in zero simulation time, while the result is scheduled to be assigned to A 20 ns into the future. These simulation semantics imply no correspondence to the register-transfer hardware that might implement this statement. For instance, the signals A, B, C and the addition operator do not correspond to registers or adders; the delay of 20 ns is simply a simulator's model of inertial delay and has no direct correspondence to the inertial delay of any physical component, since there is no such component implied. In fact, this delay mechanism is primarily used by the simulator to resolve conflicts resulting from scheduling several values to a signal at the same discrete instance of simulation time.

In high-level synthesis, however, we need to map these language constructs to RT-level hardware. If we use the behavioral VHDL statement as input to high-level synthesis, VHDL signals will have to be mapped to storage elements (e.g., registers) or physical wires, while VHDL operators will have to be implemented by RT units. To perform efficient synthesis, we therefore need to understand how to map the language's semantic model to the architectural models described in Chapter 2.

Obviously, having a close match between the language semantic model and the target architectural model simplifies the task of synthesis and also results in high-quality designs. In general, language models (semantics) have to be tuned towards different applications models (e.g., hardware, event-driven simulators) to make the specific design activities (e.g., synthesis, simulation) more efficient.

An input language for hardware description needs to describe a design's behavior and its relationship to the external world. Thus, an HDL description is typically composed of three parts: definitions, behavior and constraints. Before describing the behavior of the intended design, we must first define the abstract data carriers (variables, signals) used in the specification, the physical ports used by the design to communicate with the rest of the world, any hardware structures used in the behavioral description, and the data types (abstractions) used to simplify the specification of the design behavior. In the behavior, we specify how the outputs are generated as a function of the inputs over time, using language operators and the defined data abstractions. Finally, we must specify constraints on the design behavior necessary for correct, practical and efficient implementation of the synthesized hardware.

The next three sections outline some HDL features and issues necessary to describe designs that span the continuum from simple, combinatorial functional units, through FSMs, FSMDs, and up to communicating systems, i.e., the range of design abstractions described in Chapter 2. Note that not all requirements are complementary; in fact, several may conflict with each other. However, by focusing on a target architecture, some of these inconsistencies may be resolved in favor of one or another requirement.

4.3 Programming Language Features for HDLs

HDLs have traditionally evolved from programming languages [Shiv79, Waxm86, Hart87]; consequently, most HDLs exhibit features common to imperative languages, such as data abstraction mechanisms (using data types and variables), behavioral operators for transforming data, assignment statements to change values of variables, and control and execution ordering constructs to express conditional and repetitive behavior.

4.3.1   Data Types

Data typing in a system is characterized by the formats (e.g., number of bits), types (e.g., Boolean, integer, floating point) and the data representation (e.g., signed/unsigned, 2's complement) of all variables in the behavioral description. Data typing provides a powerful mechanism to provide concise, understandable descriptions. It also allows the compiler to detect semantic errors in the description for instance, assigning a 16-bit value to an 8-bit register. Strict data typing permits consistency checks during the synthesis process, but also burdens the language compiler with more tests. A minimal data type would consist of variables characterized by their bit widths, with the synthesis system assuming a default type (e.g., bit-vector) and data representation scheme (e.g., 2's complement).

4.3.2   Operators and Assignment Statements

The transformation of data in the design description is performed by various types of language operators. Broadly, these may be classified into arithmetic, Boolean, logic, bit manipulation, array access, and assignment operators. Most HDL's have a sufficiently large set of predefined operators. In addition, some HDLs permit overloading of operators whereby the user may redefine the semantics of an existing operator symbol. The most basic language construct for specifying data transformations is the assignment statement, in which the result of an expression on the right-hand side is assigned to a target variable on the left-hand side.

4.3.3   Control Constructs

To specify the behavior of a design over time we need to describe sequences of assignment statements. We use control constructs such as *if-then-else, case* and *loop* to specify conditional and repetitive behavior.

The body of the ISPS [Barb81] description in Figure 4.2 uses the control constructs *repeat* for looping, and *decode* and *if* for conditional control to describe the fetch-decode-execute cycle for a simple processor. The first statement in the *repeat* loop fetches the current instruction into *PI*. The *decode* statement selects the appropriate action for execution based on the value of f (i.e., the first three bits of *PI*). The *if* statement (when f equals six) shows conditional execution.

```
                                    **Instruction Execution**
                                    Icycle\Instruction.Cycle(main) :=
                                    begin
                                        repeat
   Mark1 :=                              PI = M[CR]<0:15> next
   begin                                 decode f =>
                                        begin
   **Memory.Primary.State**
       M[0:8191]<31:0>,                     #0  :=  CR = M[s]
                                            #1  :=  CR = CR + M[s]
   **Central.Processor.State**              #2  :=  Acc = Acc - M[s]
       PI\Present.Instruction<0:15>,        #3  :=  M[s] = Acc
       f\function = PI<0:2>,                 #4, #5  :=  Acc = Acc + M[s]
       s<0:12> := PI<3:15>,                 #6  :=  if Acc < 0 =>
                                                        CR = CR + 1
       Acc\Accumulator<0:31>,               #7  :=  stop()

                                        end next
                                      CR = CR + 1
                                      end
                                    end
                                  end
```

Figure 4.2: ISPS Mark1 processor description.

4.3.4 Execution Ordering

The behavior of a design exhibits sequential and parallel threads of operation. These orderings can be specified implicitly, through the HDLs model of execution ordering, or explicitly, through specific language constructs. For instance, all statements in the ISPS example in Figure 4.2 are assumed to execute in parallel, unless constrained by control constructs or by the keyword *next*, which forces sequential ordering of consecutive statements.

Sequentiality in HDLs is specified in two ways: implicitly, through data dependence of operations, and explicitly, through specific control and order-specific language constructs. For example, Figure 4.3(a) shows a VHDL process description in which assignment statements are executed sequentially. In Figure 4.3(a), the first statement assigns the old value of B to A. The next statement assigns the new value of A to B. Consequently, both the signals A and B receive the old value of B at the end of *PROCESS P1*.

```
P1: PROCESS (clock)

begin

    A <= B;

    B <= A;

end PROCESS P1;
```

```
B1: BLOCK (clock'event AND
                        clock = '1')
begin

    A  <=  guarded  B;

    B  <=  guarded  A;

end BLOCK B1;
```

(a)

(b)

Figure 4.3: Execution ordering in VHDL (a) sequential, (b) concurrent.

Parallelism is an ubiquitous feature of hardware, requiring implicit language semantics and explicit language constructs for its expression. For instance, all assignment statements within a VHDL block construct are performed in parallel. In Figure 4.3(b), the right-hand side of each assignment statement gets evaluated in parallel (using the old values of signals) before assigning the result to the signal on the left-hand side. Hence, the values in signals A and B get swapped after the VHDL block is executed. Note that even in sequential language descriptions, such as in Figure 4.3(a), parallelism is implied between operations that have no control or data dependencies between them.

4.4   Hardware-Specific HDL Features

Standard programming language constructs provide behavioral abstractions, but do not provide ways of expressing hardware-specific properties in a design description. To describe hardware-design behavior at the FSM, FSMD and system levels, we need additional constructs for interface declarations, partial design structure specification, RT and logic operators, asynchrony, hierarchy, interprocess communication, design constraints and user bindings. We briefly describe each of these below.

4.4.1 Interface Declarations

Since the design being described has to obtain inputs from the outside world and supply the results of computation as outputs, we need to define its input and output ports. Port definitions include the size (e.g., num-bits), the mode (e.g., input, output, or input-output), and hardware-specific features (e.g., whether the port is buffered or tristated). Figure 4.4 shows a DSL [CaRo85] specification of an exponentiation circuit, in which the *INTERFACE* declarations specify ports and their attributes.

4.4.2 Structural Declarations

Structural declarations allow the specification of registers, accumulators, counters and other hardware structures that are to be used like variables in the HDL description. Structural declarations are useful when some partial structure is necessary, well understood or predesigned and needs to be used explicitly in specifying the design behavior. For instance, processor designs normally have architectural registers (such as program counters or register files) fixed before the behavior of the processor is described; these architectural registers are declared as physical structures and may be used as variables in describing the processor's behavior.

4.4.3 RT and Logic Operators

In addition to the standard arithmetic and Boolean operators in a programming language, we need to support operational primitives corresponding to RT- and logic-level hardware. Examples of bit-level logical operators include bit shifting, bit rotation, bit-stream extraction and

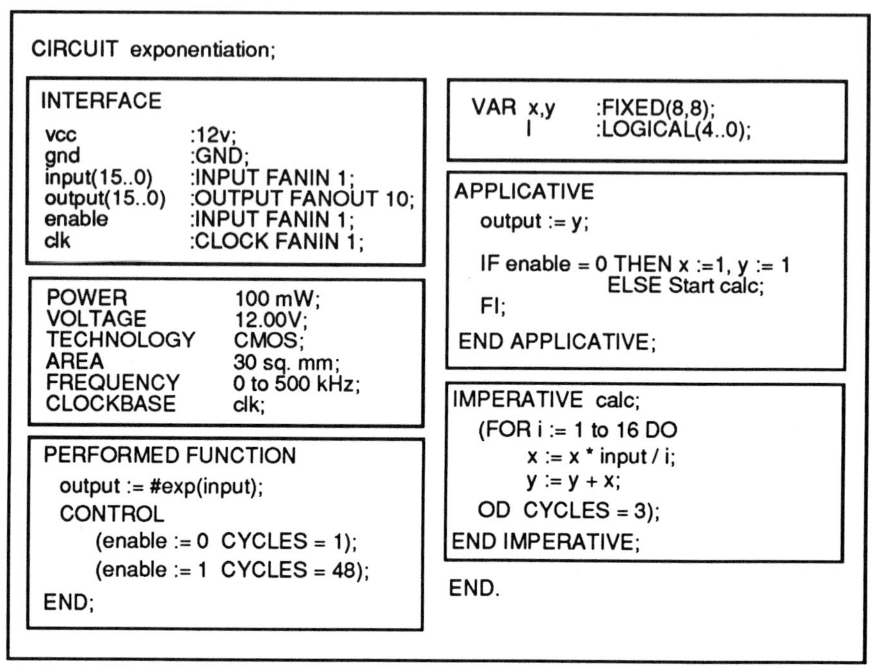

Figure 4.4: DSL specification.

concatenation, and bit-wise logical operations. Sample RT-level operators include the increment and decrement operators for variables.

4.4.4 Asynchrony

RT-level behavior is expressed in terms of register transfers for each clock cycle in the design. However, in addition to synchronous behavior, RT-level hardware typically exhibits asynchronous behavior in the form of sets, resets and interrupts. Such asynchronous behavior can be interleaved with the RT-level behavior using special HDL constructs for asynchrony. Alternatively, all asynchronous behavior for the design can be grouped into a separate description, with semantics indicating that the asynchronous behavior overrides any synchronous behavior in the specification. An example of asynchronous constructs in DSL [CaRo85] takes the second approach: the *APPLICATIVE* part of the design description is applied asynchronously, and the *IMPERATIVE* part of the design operates under the synchronous clock. Figure 4.4 shows the description of an exponentiation circuit in DSL, in which the *enable* signal in the "applicative" part asynchronously determines if the calculation of the exponent in the "imperative" part *calc* should proceed synchronously.

A more general form of asynchrony arises when we consider communicating processes that run on different clocks. In such cases, asynchronous events at the interface of a process can interrupt synchronous behavioral patterns and force the design into an interrupted behavioral sequence. These event transitions can be used to define asynchronous FSMs, in which an event-state is entered on an input-signal change, instead of on a clock-signal change. Languages with such features include Statecharts [Hare87], WAVES [Borr88], SpecCharts [NaVG91] and BIF [DuHG90]. Figure 4.5 shows a simple computer system described in SpecCharts, in which the design asynchronously moves from the *ACTIVE* state to the *RESET* state when the signal *RESET_IN* rises, regardless of which *ACTIVE* substate (i.e., *FETCH, DECODE* or *EXECUTE*) it was in.

4.4.5 Hierarchy

As designs get more complex, we naturally resort to hierarchy as a means of describing, managing and capturing the complex behavior. Hierarchy in an HDL can take several forms: procedural, structural, behavioral and design hierarchy.

The standard programming language constructs of procedural hierarchy use functions and procedures to provide a shorthand notation for describing repetitious behavior. This procedural hierarchy permits decomposition of the behavior in a structured fashion and allows a concise representation. In some HDLs, a particular behavior may be encapsulated into a structure, and may be used later in the structural hierarchy. Figure 4.4 shows how this encapsulation is achieved in DSL using the *PERFORMED FUNCTION* construct.

Structural hierarchy specifies the interconnection of the communicating processes via global signals and ports on the processes. Consider an RT-level datapath design consisting of a register file and an ALU. A hierarchical netlist for this design would describe the interconnections of RT components (e.g., register file and ALU) at the highest level, with each component (e.g., the ALU) being decomposed into a structural netlist of Boolean gates. Structural VHDL descriptions support this form of hierarchy.

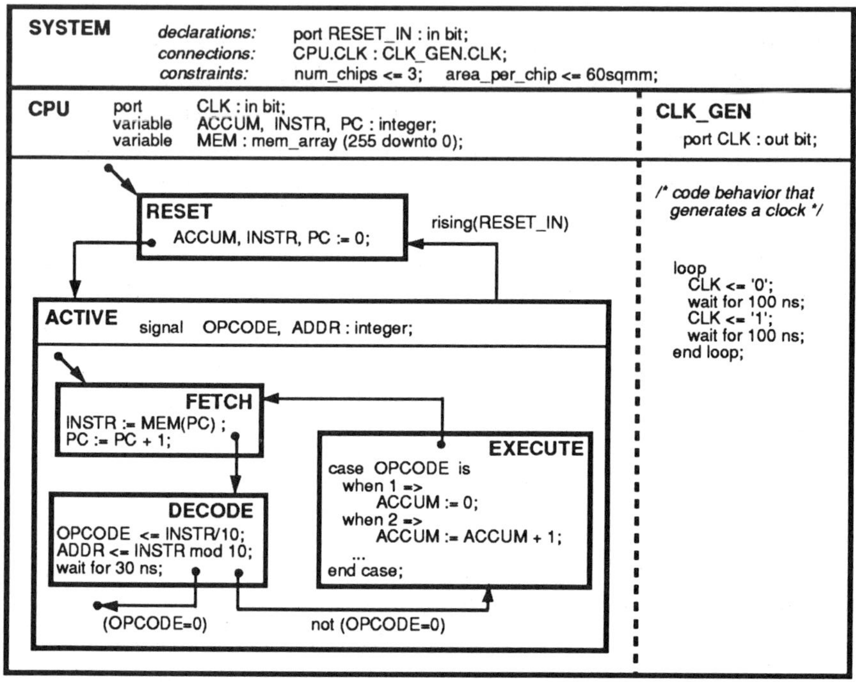

Figure 4.5: Behavioral hierarchy in SpecCharts.

As design behavior becomes more complex, these procedural and structural hierarchical abstractions are not sufficient to cope with the increase in complexity. This is particularly true of state-based systems in which input events cause immediate transitions to different sequences of behavior. In such designs, we use behavioral hierarchy to concisely capture complex behavior. Figure 4.5 shows an example of behavioral hierarchy in a simple computer system described using SpecCharts [NaVG91]. The broken line in the figure indicates that the computer system *SYSTEM* is composed of two concurrently executing processes, the *CPU* processor and the clock generator *CLK_GEN*. The *CPU* consists of two major states: *RESET*, which is the default entry state for the *CPU*, and *ACTIVE*, where the processor is operational. The *ACTIVE* state is further composed of the three substates *FETCH*, *DECODE* and *EX-ECUTE*, which describe the detailed operation of the processor. The behavioral hierarchy shown here allows decomposition of a state into substates and concurrent processes, and also provides a powerful notation for describing global asynchrony (e.g., sets and resets) using events. For instance, the event *rising(RESET_IN)* forces the *CPU* process into the *RESET* state, no matter which *ACTIVE* substate it is currently in.

Finally, a design hierarchy specifies how the whole design is composed in terms of its subcomponents, communicating processes, etc. Figure 4.6 shows the design hierarchy represented by a VHDL configuration. Each design entity in this figure is hierarchically decomposed into smaller blocks, each of which may represent behavior or structure.

4.4.6 Interprocess Communication

Section 2.5 described a target-system architecture composed of interacting hardware processors in which communication is achieved using signals connecting the control units and datapaths of the FSMDs. At a higher level of abstraction, the processors are represented by processes. We need mechanisms in the HDL to enable specification of communication between the interacting processes. For purely synchronous designs, this communication can be embedded within the process' behavioral description, since each process operates in a lock-step manner. In such cases, the user can explicitly describe the communication using standard HDL constructs to force reads and writes on correct clock cycles.

When the communicating processes are not synchronous, we need specific protocols to achieve synchronization between the interacting designs. There are two basic methods for achieving this synchronization: shared medium and message passing. Shared medium synchronization is achieved using shared wires or memories that can be simultaneously accessed by the communicating processes. In an HDL such as HardwareC [KuDe91], shared medium synchronization is described by parameter passing across ports interconnected on the processes. The example in Figure 4.7(a) shows how data written on *port b* of *process P1* is accessible by *process P2* on *port y*. Similarly, data written by *process P2* on *port x* can be read on *port a* of *process P1*.

Message passing in HDLs is achieved using built-in primitives to

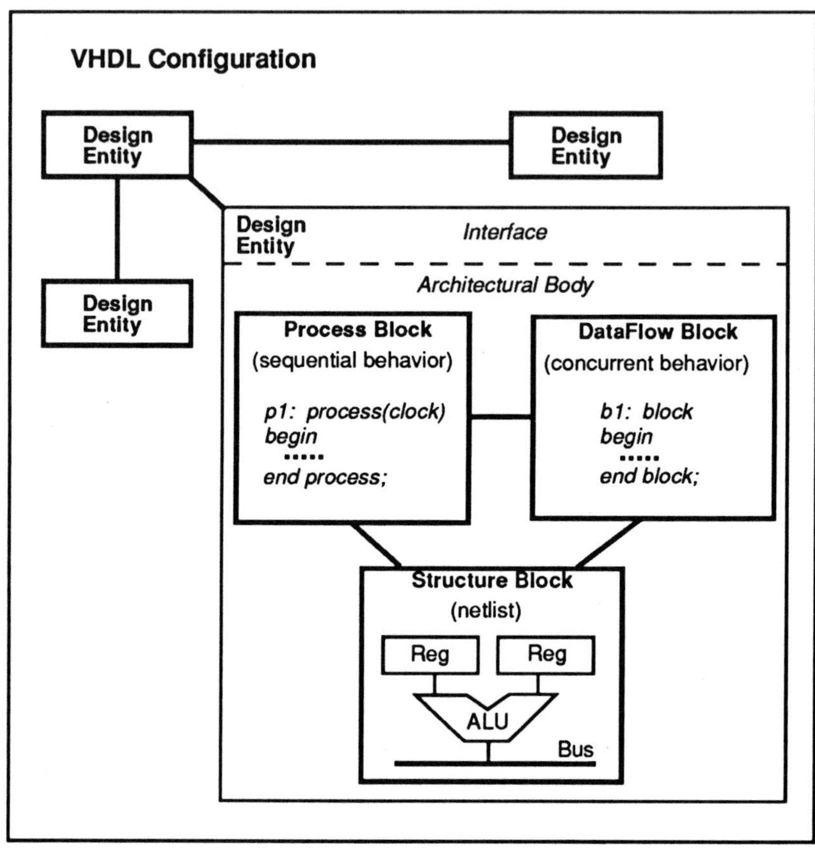

Figure 4.6: Design hierarchy in VHDL.

describe synchronization between the communicating processes. Figure 4.7(b) shows how message-passing synchronization can be described in HardwareC. The channels *a* and *b* specify the communication medium between processes *P1* and *P2*. The HardwareC synchronization primitives *send* and *receive* are used on the channels to specify transfer of data across the processes. The statement *receive(a,buf)* in *process P1* forces *P1* to wait until *process P2* sends data across *channel a* with the statement *send(a,msg)*. The request-acknowledge handshaking protocol described in Section 2.5 can be expressed in an HDL using constructs similar to these primitives.

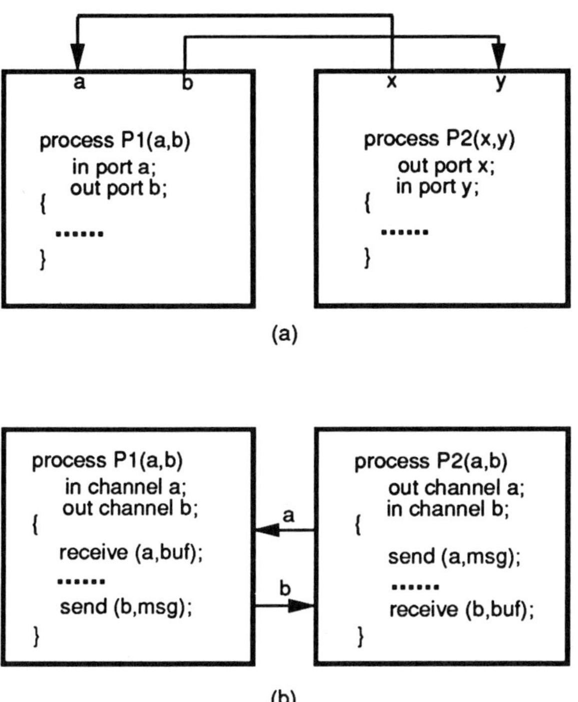

Figure 4.7: Process synchronization in HardwareC: (a) parameter passing, (b) message passing.

4.4.7 Constraints

Constraints on the design behavior guide the synthesis of the design towards feasible realizations in terms of performance, costs, testability, reliability, and other physical restrictions. Design constraints can be specified separately from the design behavior, or can be interleaved with the behavior in an HDL. Physical constraints (e.g., pin-counts, technology and voltage) are typically specified separately from the behavior, either in the declarations or in a separate file. For instance, Figure 4.4 shows how the declarations section of DSL permits the specification of various kinds of physical constraints such as power, technology, voltage and area.

Timing constraints are often critical for achieving correct behavior as well as for specifying performance requirements for the design. Timing constraints can be specified at the interface of a design, or can be interleaved with the operational behavior of the design. In either case, these timing constraints are typically specified as delays. In synchronous designs, timing constraints are normally specified relative to the system clock; the delays are specified as multiples of clock cycles. Figure 4.4 shows how performance constraints are specified in DSL: in the imperative procedure *calc*, the do-loop is constrained to execute within three clock cycles.

When describing communication between processes, we may need to specify delays relative to specific events. These delays represent absolute time values necessary for correct implementation of the communication protocol. For instance, in a simple read-cycle protocol for a memory board, we might first assert the address lines and a request for memory, and then issue a delayed request for the data bus, thereby giving the memory time to react to the memory request and address before getting control of the data bus. Figure 4.9 shows an example of this read-cycle protocol in BIF [DuHG90]. In state 1, *BusReq* is cleared 175 ns after the state entry event *Falling(MemReq)*, that is, after the address *ADDR* and the memory request *MR* are asserted.

Absolute delay values can also be used to specify constraints on design performance, delays on logic paths and timeouts. Simulation languages such as VHDL allow the specification of absolute delays using two delay types: inertial and transport. An inertial delay represents a component's inertia which requires its input value to persist for a specified period of time before the output is activated (e.g., set-up and hold times for a flip-flop). A transport delay represents a wire delay in which an input change is always propagated through the component and can affect the output.

4.4.8 User Allocation and Bindings

The task of high-level synthesis can be viewed as a process of allocation and binding of abstract behavioral entities to RT structures over time. Traditionally, this task has been viewed as a completely automatic procedure over which the human designer has little or no control. However, experienced designers can often provide hints to synthesis tools to improve the quality of the synthesized hardware. Since synthesis tools are based on a finite amount of encoded design-process knowledge, human designers may be able to do better than automatic synthesis tools in identifying problems, detecting critical sections and providing partial design solutions. The input HDL must allow the description of such partial design information in the form of user allocation and bindings. Allocations can be specified as RT structures in the HDL declarations. Four types of user bindings can be specified in the input: state, register, functional-unit and connection. These bindings can be specified both in the declarations and within the design behavior.

State binding assigns an operation in the behavioral description to a state of the synthesized design; the operation can then be annotated with its assigned state. Some HDLs are state-based (e.g., BIF [DuHG90]); state binding is then automatically achieved by describing an operation in its appropriate state. Functional unit, register and connection binding assign operations, variables and assignments to functional units, registers, and connections, respectively. These bindings can be described by annotating language operations, variables and assignments with their allocated components and connections, using special language constructs to indicate the binding. For example, BIF and ISPS [Barb81] use curly braces to indicate component bindings in the behavior. Mimola [Marw85] and HardwareC [KuDe91] also permit varying degrees of user allocation and binding in the input description.

4.5 HDL Formats

Before design synthesis can begin, the designer has to go from a specification to a formalized HDL description as shown on the right half of Figure 4.1. This design conceptualization process plays a major role in high-level synthesis, since the way a design is modeled and described has a direct impact on its final implementation both in terms of cost and performance. A badly modeled design description will often result in an inferior implementation, no matter how good the supporting synthesis tools. Furthermore, conceptual errors that are not detected early in the design cycle can lead to expensive and time consuming efforts to detect these lapses as the design proceeds through the lower levels of logic and layout. Unfortunately, there are few tools and media that facilitate the task of modeling and describing of designs at a higher level; most of this work is currently done in a designer's mind, with no explicit representation or formalism to capture this information in a form amenable to high-level synthesis and simulation.

To support design description and modeling, we need a variety of HDL formats that suit different applications and users. At lower levels of the design process (e.g., RT, logic and layout), graphical and other modes of design description and capture have been prevalent for many years. Behavioral HDLs have taken primarily a textual form, because these HDLs have evolved from textual programming languages. However, even for higher levels of design description, a combination of various input media is more natural for design specification. A designer often supplements a textual design description with flowcharts or other descriptive aids to explain its functionality, since a description written purely in a textual language may not convey much information about the design at a glance. For instance, protocol and interface information is captured naturally through timing waveforms; small state machines are described concisely using graphical or tabular-state diagrams; and pure behavior is expressed naturally in a textual language format. In this section, we review textual, graphical, tabular and waveform-based HDL formats for high-level synthesis.

4.5.1 Textual HDLs

Most higher-level HDLs have borrowed the syntax and some semantics of high-level programming languages such as Pascal, ADA and C. Textual HDLs can succinctly capture behavior that contains assignment statements with complex data transformations (e.g., arithmetic and logical operators). Applications with large amounts of computation are best described using textual assignment statements. Examples of textual HDLs include ISPS [Barb81], Silage [Hilf85], VHDL [IEEE88], HardwareC [KuDe91] and MIMOLA [Marw85]. HDLs based on formal mathematical logic, such as HOL [Gord86], are also best described using text.

4.5.2 Graphical HDLs

Graphical schematic capture of structural designs has been popular with designers for a number of years. Design behavior can also be captured effectively in a graphical format. In particular, execution ordering, parallelism and control flow are easily captured and understood using a graphical notation. A behavioral flowchart is a good example of a graphical capture mechanism; control flow is expressed graphically, while operational behavior is described using textual assignment statements.

When the size of the behavior being described is relatively small, flowchart-based descriptions are a useful mechanism for design capture. The ASM notation [Clar73] and EXEL [DuGa89] are two examples of such languages. When the design behavior becomes more complex, we can capture this behavior hierarchically. StateCharts [Hare87] and Spec-Charts [NaVG91] provide a concise formalism for such design capture. Figure 4.5 shows how SpecCharts can be used to graphically capture the behavior of a simple processor system. Petri net based descriptions have also been used for graphical capture of hardware-design behavior; GDL [DBRI88] is an example of such a specification.

4.5.3 Tabular HDLs

As the sequence of examples in Chapter 2 shows, tabular descriptions provide a concise notation for state-based design descriptions, particularly for FSMDs, where the state-based portion of the design can be clearly expressed in a state table and the datapath operations can be expressed in a textual form appended to the appropriate parts of the state table. The Behavioral Intermediate Form (BIF) [DuHG90] is an example of such an HDL medium. State sequencing in BIF is described in a tabular format, but operational behavior is expressed using textual expressions for data operations in each state.

For instance, Figure 4.8 shows the block diagram for a system composed of a CPU, a bus controller and a memory board. The memory board is internally composed of a memory controller and a ROM. Within the memory board, communication between the memory controller and the ROM is achieved using a signal *Addr* that addresses the ROM array, a signal *MR* that enables the ROM, and an internal *Data* bus that

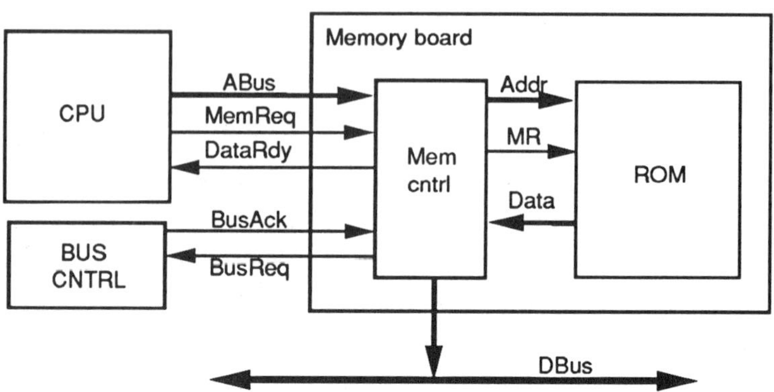

Figure 4.8: CPU-memory board block diagram.

receives the ROM output data. At the system level, several signals and buses are used to achieve a handshaking protocol: *Abus* represents the address lines of the external bus, *DBus* is the data bus, while *DataRdy* indicates that the ROM has valid data on its data bus, and *BusReq* is a handshaking signal for the data bus. All handshaking signals are active low.

Figure 4.9 shows a simple read-cycle handshake protocol for the memory board using a tabular state-based description in BIF. The read cycle is initiated in event-state 1, when the *MemReq* signal falls and the *Board_Id* of the memory board matches the addressing tag on bits 16 through 18 of the address bus *Abus*. At this point, the memory is enabled, the internal address register *Addr* is loaded with the address bus value, and the memory board takes control of the data bus *DBus* by setting *BusReq* to zero. The event *Falling(BusAck)* forces a transition into event-state 2, where the memory board sends the addressed data *Data* on the data bus *DBus*. The handshake is completed in event-state 3, where the memory board is disabled.

4.5.4   Waveform-Based HDLs

The memory-read handshaking protocol described in the previous section could also be succinctly expressed using timing diagrams (Figure 4.10).

Present State	Cond	Val	Actions	Next State	Event
0		T	DBUS = 'X'; DataRdy = 1; BusReq = 1;	1	Falling(MemReq)
1	Abus {18..16} == Board_Id	T	MR = 0; Addr = ABus; BusReq|(delay 175ns) = 0;	2	Falling(BusAck)
		F		1	Falling(MemReq)
2		T	BusReq = 1; DBus = Data; DataRdy = 0;	3	Rising(MemReq)
3		T	MR = 1; Addr = 'X';	0	Rising(BusAck)

Figure 4.9: Memory board read cycle in BIF.

In fact, a cursory glance at any standard component databook shows that timing waveforms commonly are used for expressing interface behavior, protocols and associated timing constraints. Timing diagrams can graphically represent changes on signals (e.g., events such as *rising* and *falling*), can show sequencing of events, and can also effectively show timing relationships between events. Hardware designers frequently use timing diagrams for capturing interface behavior and also for design documentation. Several timing-editor-based languages have been developed to present this type of information for example, Waves [Borr88] and XWAVE [AbBi91].

4.6   A Discussion of Some HDLs

This section describes the features of a few languages that are typical for the language category they represent, or for the target applications they describe.

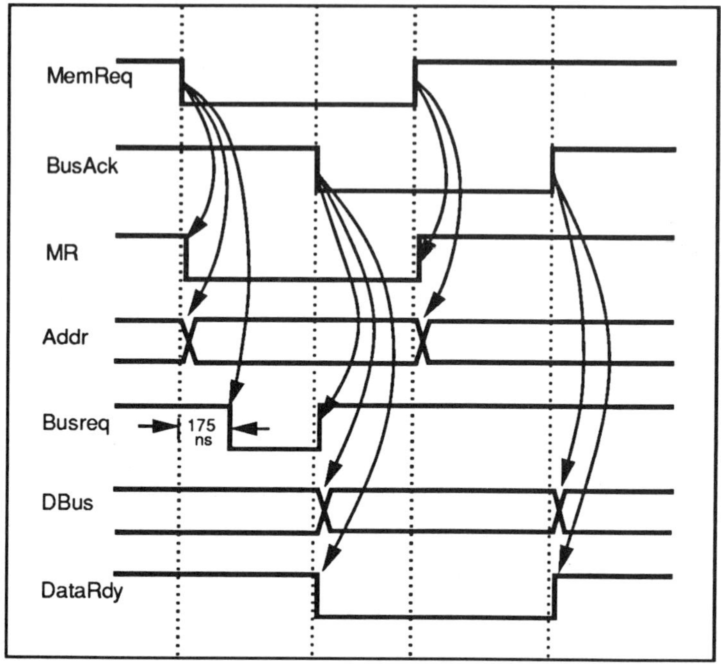

Figure 4.10: Timing diagram for memory board read cycle.

4.6.1 Instruction Set Processor Languages

The Instruction Set Processor Specification (ISPS) language [Barb81] is
an example of an HDL designed for the description of instruction set
processors. It has been used extensively in the set of high-level synthesis
tools being developed at Carnegie Mellon University [TLWN90]. The
architectural design model of the ISPS language is based on a processor
with a fixed instruction set, which repeatedly executes a *fetch-decode-
execute* cycle. The ISPS language was originally designed for architectural simulation, design synthesis and automatic generation of machine
code for specific processors.

An ISPS description includes an interface declaration and the behavioral description of several communicating processes. The interface
declaration specifies the ports to the external world as well as the data
carriers used to communicate between interacting processes in the behav-

ior. The behavior is specified using operators and control constructs in a standard programming language syntax. Figure 4.2 shows an ISPS description for a simple processor, the MARK1. The key control construct is the *decode* statement (similar to a "case" statement), which permits decoding of an instruction register to enable execution of an appropriate instruction based on the decoded value.

ISPS specifies a processor's behavior by describing its instruction set. During execution of an instruction, all operations are typically performed in parallel. Consequently, all ISPS language assignments are assumed to occur in parallel, unless constrained by control constructs. To enforce sequential execution in an ISPS description, we can use the construct *next* between consecutive ISPS statements. This is illustrated in Figure 4.2, where the present instruction register *PI* is loaded from memory before the function bits *f* are decoded; the sequencing is forced by the *next* construct. Although ISPS describes instruction-set processing, clocked machine states are not explicitly specified. Since a processor's instruction may take several states to execute, an ISPS description needs to be scheduled into states of the FSMD model of Chapter 2.

Constraints in ISPS are specified using qualifiers in curly braces. These can be used by application programs (e.g., synthesis or simulation tools) for further interpretation. Bounds on execution times for parts of the ISPS behavior can be specified as constraints, as can binding of physical units to ISPS operators.

ISPS, as originally conceived, was intended for a large range of applications related to instruction set processors. However, surveys [LKVW88] show that a majority of ASIC designs are not instruction set processors, but consist of communicating FSMDs. Experience in using ISPS for high level synthesis of ASICs shows that it does not have a convenient mechanism for specifying interface constraints. Nestor and Thomas [NeTh86] extended the ISPS language for specifying timing constraints between ISPS statements annotated with labels to allow protocol and interface specification.

4.6.2 Programming-Language-Based HDLs

HardwareC [KuDe91] is an example of a behavioral HDL based on the programming language C. Although HardwareC has the basic syntax of C, its semantics are hardware oriented. HardwareC's declarative semantics allow the specification of structural modules and their interconnections, and its procedural semantics permit the specification of abstract behavior over time.

A HardwareC design description is written as a block at the highest level, which is then decomposed into an interconnection of structural blocks and communicating behavioral processes. A *block* corresponds to a structural interconnection of other design entities. A *process* represents a free-running piece of behavior that may be hierarchically described in terms of procedures and functions. All processes run in parallel with each other and may communicate explicitly through shared variables or through message passing using special HardwareC communication primitives such as *send* and *receive*. *Procedures* and *functions* are logical groups of operations that may call other procedures and functions.

The user can specify timing and resource constraints within a HardwareC description. Tags (i.e., labels) can be associated with HardwareC statements. Timing constraints (absolute values or clock cycles) can then be specified on a tag or between tags, representing the delay of a statement or the timing constraint between two statements, respectively. Resource constraints specify limits on the number of component types and instances that may be used during synthesis. The user can also bind certain pieces of behavior to structural components using HardwareC model calls.

Several other HDLs have their syntax borrowed from standard programming languages. We mention a few here for completeness: Pascal-based HDLs include MIMOLA [Marw85], DSL [CaRo85] , and FLAMEL [Tric87]; HARP [TaKK89] uses a Fortran-based HDL, while ELF [GiBK85] is ADA based.

4.6.3   HDLs for Digital Signal Processing

High-level specification of digital signal processing (DSP) applications is typically done at the signal flow-graph level, where streams of data are transformed by data operators at fixed intervals of time, called the sampling rate. These DSP applications have little control flow and exhibit a relatively large number of data computations. Therefore, behavioral HDLs designed for DSP applications tend to focus on high-level primitives supporting DSP-specific features. Silage [Hilf85], FIRST [Berg83] and ALGIC [ScGH84] are examples of high-level HDLs designed for DSP applications. We use Silage to illustrate some features of a DSP-based HDL.

Silage is an applicative language in which operations do not produce side-effects. This is due to the fact that a Silage description consists of equations that are similar to algorithmic definitions (e.g., the Fibonacci sequence) rather than assignment statements. A typical Silage description attempts to describe data transformations on signals for a single sample interval. Besides providing some standard data transformation operators, Silage has primitive constructs for sample delays (e.g., @) and sample rate conversions (e.g., *Decimate* and *Interpolate*). The term *a@1* in a Silage definition refers to the value of the operand *a* during the previous sample interval. The *Decimate* function allows the selection of a few elements of an input data stream (e.g., every fifth element of a data stream), thereby reducing the sample rate. Conversely, the *Inter-polate* function increases the sample rate of a data stream by inserting more values into an existing sample, i.e. increasing a sample's size by repeating data.

Since Silage is an applicative language with statically bound loops, no recursion, and limited function calls, it is possible to treat such constructs as language macros. The Silage description is then flattened into a single, large description in which data-flow analysis, optimizations and synthesis can be easily performed. Another interesting feature in Silage is the *pragma* construct, which permits the user to specify hints or bindings to the compiler or synthesis system. Silage pragmas can be used to bind specific operators to hardware, or to indicate the degree of parallelism allowed for data streams by binding Silage functions to processors.

4.6.4 Simulation-Based HDLs

Simulation-based HDLs have language semantics for driving an event-driven simulator, which simulates physical hardware behavior over simulation time. Section 3.2 introduces an example using VHDL [IEEE88], an IEEE standard hardware description language that has strong simulation semantics.

VHDL is a broadband language designed to capture hardware behavior at several levels in the design spectrum, spanning logic gates to complex architectures consisting of communicating hardware processes. At the highest level, a VHDL description consists of an interconnected network of entities (Figure 4.6). Each entity is composed of an interface description, called the entity declaration, and an architectural body. An architectural body can be hierarchically decomposed into subentities. At the bottom of the hierarchy, an architectural body can be described as a process block, a data-flow block, or a structural block; these are the three basic description facilities in VHDL.

In the behavioral domain, VHDL can be used to model a system architecture as a set of concurrently executing process and data-flow blocks. Statements within a process block execute sequentially. The execution of each process can be controlled (that is, started and suspended) by making the process sensitive to changes on certain signals (e.g., a VHDL sensitivity list). Statements within a data-flow block execute concurrently. A guard condition can be associated with a dataflow block, enabling conditional assignment of selected statements in the block. Within each block, standard programming language control constructs such as *IF, CASE* and *LOOP* can be used for conditional and repetitive behavior.

In the structural domain, a VHDL structural block models components, component instances, interconnection nets and the design interconnectivity as a mapping of component instance ports to the interconnection nets. The design structure can be maintained hierarchically.

In the temporal domain, the notion of VHDL time corresponds to that of discrete simulation time. During simulation, a behavioral VHDL model responds to external stimuli by executing the behavior instantaneously but scheduling the outputs as transactions in future simulation time as dictated by delays specified in the behavior. Since VHDL processes run concurrently, it is possible that a signal (or data carrier) can be scheduled to receive many values at a particular instance of simulation time. A VHDL driver refers to such a collection of time/value pairs for the signal. When the simulation clock is advanced to a time instance where a driver has multiple candidate values for a signal, a VHDL resolution function is used to assign a unique value to the signal for that instance of simulation time.

From a language viewpoint, VHDL is a strongly typed language, with an extensive user-definable typing capability. A VHDL package is a facility for defining types and other attributes that can be reused in different design descriptions. A predefined set of VHDL types is available in the *STANDARD* package. VHDL attributes can be used to define language extensions and to specify constraints on the design behavior. Detailed descriptions of the use of VHDL for modeling and simulation can be found in [Arms89, Coel89, LiSU89, Perr90, HaSt91].

4.7   Matching Languages to Target Architectures

From the previous sections, we see that behavioral HDLs can have a range of design models and associated semantics. However, if these languages are to be used efficiently for high-level synthesis, we need a good match between the language's semantic model and the target-hardware architectural model. When languages are specifically developed for a target architecture (e.g., Silage), these models are closely matched. However, when we look at broadband languages designed to cover a large spectrum of designs and applications, the language semantic model may be quite different from the target architecture generated by the high-level synthesis tools.

This discrepancy between language semantic models and target architectural models is particularly true with VHDL. It was intended for a broad range of design descriptions, from logic to communicating systems, and was also based on simulation semantics. Consequently, it has a rich set of language constructs that allow the description of the same behavior in many dissimilar ways. All of these descriptions may produce the same simulation results because the simulation semantics are primarily concerned with projecting future behavior in simulation time based on the current behavior being executed. Unfortunately, these correctly simulated descriptions may have no relationship to the actual hardware being designed by high-level synthesis tools because VHDL descriptions are simply *models* of real hardware.

For example, consider the specification of design behavior for a full adder in VHDL for which an interface specification is shown in Figure 4.11(a). Depending upon the background and design experience of users, they can develop different behavioral descriptions for this simple

```
entity FULL_ADDER is
   port (X,Y:  in BIT;
         CIN:  in BIT;
         SUM:  out BIT;
         COUT:  out BIT );
end FULL_ADDER;
```

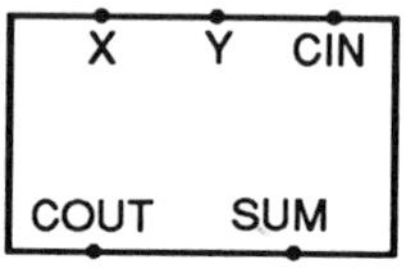

(a)

```
architecture FA_BOOLEAN of
              FULL_ADDER is
signal S1, S2, S3: BIT;
begin
   S1  <=  X xor Y;
   SUM  <=  S1 xor CIN  after 3 ns;
   S2  <=  X and Y;
   S3  <=  S1 and CIN;
   COUT  <=  S2 or S3  after 5 ns;
end;
```

(b)

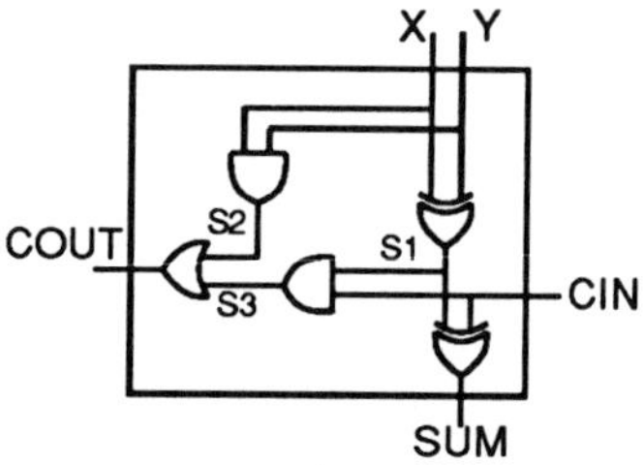

(c)

Figure 4.11: VHDL full adder data-flow description (a) VHDL full adder entity interface description, (b) full adder data flow, (c) synthesized structure from data-flow description.

full-adder design. If users have some hardware or logic design background, they probably will produce a description similar to Figure 4.11(b), since the full-adder is naturally described in terms of the lower-level Boolean operators *and, or,* and *xor.* A user with extensive software background might prefer a description style similar to Figure 4.12(a), which uses arithmetic operators on variables with higher-level data types (e.g., integers, arrays) to describe the ones-counting property from the full adder truth table.

Both of these descriptions produce the same behavior for simulation. However, the first description style is easier to understand and to synthesize automatically for a hardware engineer because it uses Boolean operations. This Boolean-level description of the full-adder has some hints on how to synthesize the structural design from the Boolean operators *and, or,* and *xor.* Figure 4.11(c) shows the synthesized hardware structure for a data-flow description in which each boolean logic operator is mapped to the corresponding Boolean gate structure.

On the other hand, if the behavioral process description in Figure 4.12(a) is used as input to high-level synthesis tools, it may initially result in the structure shown in Figure 4.12(b). This design has unnecessary hardware for type conversion between Boolean and integer signals, and uses three incrementers for implementing a simple Boolean full adder! These inefficiencies have to be removed by the high-level synthesis tools to generate a reasonable design. This requires an expensive phase of recognition and optimization – something that could easily have been avoided by the use of an appropriate language description style for this design (Figure 4.11(b)). Furthermore, there is no guarantee that we can always recognize and optimize such inefficiencies in general. Users have the freedom to describe the behavior of the design in any way they choose, thus making it difficult to account for every possible description style.

A second problem arises from the timing model of the language versus the target architecture. A simulation language models the values of data carriers (i.e., wires) over simulation time. Delays specified in such languages refer to the scheduling of events on these carriers in future simulation time, with the actual execution of statements taking zero simulation time. Consequently, the input-to-output delay for a set of assignment statements can be distributed in any fashion over the individual

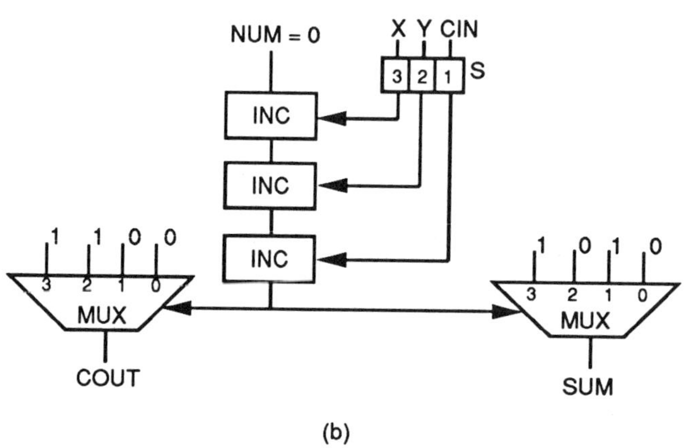

Figure 4.12: VHDL full adder behavioral description (a) VHDL full adder behavioral process description, (b) synthesized full adder from VHDL behavior.

statements. They can even be lumped together in one statement. All of these descriptions will yield the same (i.e., correct) simulation result. However, these timing delays have ambiguous semantics for synthesis because of the different ways in which the delays can be specified.

We illustrate this difference in timing semantics with VHDL. Consider the full adder descriptions in Figure 4.11(b) and Figure 4.12(a) that specify a delay value of 5 ns for the carry output. The 5 ns delay for the carry in Figure 4.11(b) is lumped with the *or* operator in the assignment statement. Hence the simulation semantics of Figure 4.11(b) suggest that the generation of $S2$ and $S3$ (i.e., the *and* operations) takes zero time, while the *or* operator is constrained to execute in 5 ns. However, in the context of synthesis, these delay values are ambiguous, since they represent the delay from intermediate signals to the output $COUT$, instead of from the inputs X, Y and CIN to the output $COUT$. Similarly, the corresponding 5 ns delay for the carry in Figure 4.12(a) is lumped with the last signal assignment which has no operators at all, while the generation of *Ctemp* is assumed to take zero time. Although both descriptions simulate correctly, they clearly have ambiguous semantics for synthesis because of the way in which delays are lumped together for specific assignment statements.

We can also develop a VHDL structural model for the full adder in which no behavior is directly implied. Figure 4.13 shows a VHDL structural description in which the full adder is constructed using two half adders and an OR gate. Simulation of this structural design is accomplished by developing VHDL simulation models for each of the structural components. In this context, synthesis involves technology mapping of the half adder and OR-gate structures into lower-level primitives, followed by placement and routing for the whole design.

Note that the structural description in Figure 4.13 explicitly refers to components and their connections, while the data-flow description in Figure 4.11 describes abstract concurrent behavior using Boolean expressions. Although the data-flow description in Figure 4.11 seems to describe structure, in fact it does not. The abstract behavioral Boolean operators such as *or* and *xor* have to be bound to physical components that perform these operations.

In addition to the problems of description style affecting the quality of synthesized designs and the ambiguity of delay constructs for synthesis, a

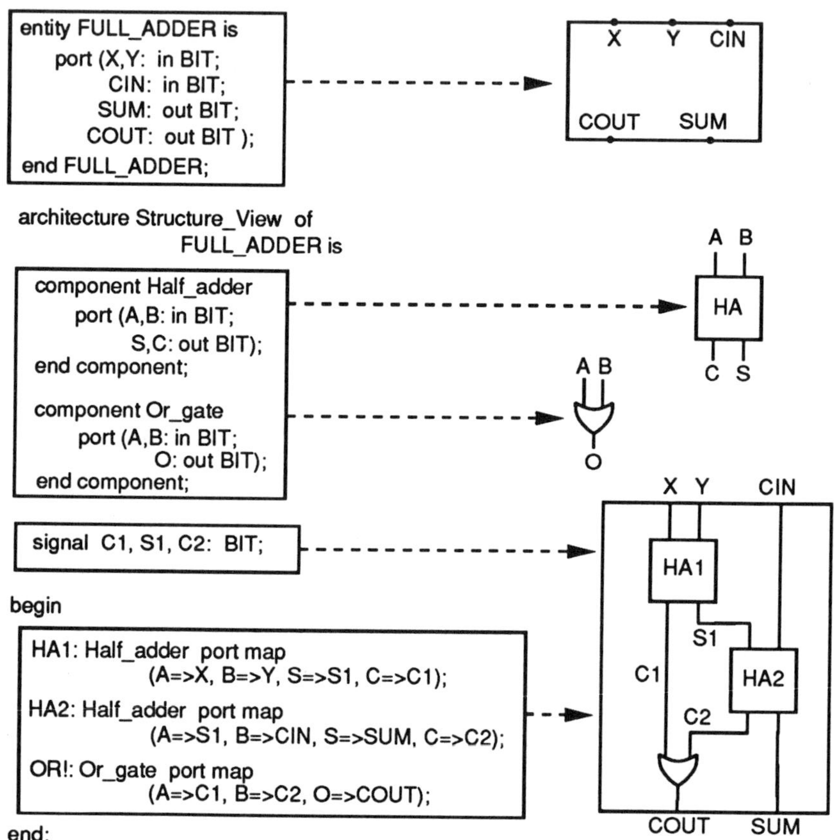

Figure 4.13: VHDL full adder structural description.

simulation-based language such as VHDL may have language constructs that do not have equivalent or feasible hardware realizations.

Consider the segment of VHDL code in Figure 4.14(a), which describes the *clear* and *count_up* functions for a simple counter. *CNT_CLR* is a VHDL block that describes the asynchronous clearing of the counter when the signal *CLR* is high. *CNT_UP* is a VHDL block that describes the synchronous *count_up* operation when the counter is enabled and the *INC* signal is high. Finally, the VHDL block *SEL* assigns the output *CNT* of the counter, the cleared value *CNT1*, or the incremented value *CNT2*, in that order, ensuring that the asynchronous *clear* overrides the synchronous *count_up*. The third block is necessary only because of VHDL's simulation semantics: we cannot assign *CNT* values in different blocks without writing a resolution function. Designers may find this description a fairly natural way to model the behavior of the counter for simulation, since the asynchronous reset and the synchronous counting operations can be specified separately. Different delay values can also be associated with individual operations to ensure that the simulation of the behavior faithfully follows the required timing constraints for the counter.

However, a straightforward attempt to synthesize hardware from this description initially results in the structure shown in Figure 4.14(b). This structure has four unnecessary multiplexors, *mux1*, *mux2*, *mux3* and *mux4*, as well as two hardware structures that have to detect changes on the signals *CNT1* and *CNT2*. The generated structure does not resemble a standard counter. In fact, considerable optimizations would be necessary to transform this structure into a reasonable counter-like component shown in Figure 4.14(c).

4.8   Modeling Guidelines for HDLs

As the examples in the previous section show, the flexibility of an HDL allows the user to describe design behavior in several different ways. In general, it may not be possible to optimally synthesize hardware for every HDL description, but certain description styles clearly match the target architectural model better than others. Furthermore, both the quality of synthesized designs, and the complexity of synthesis tools used are influenced by the styles of HDLs used to describe the input behavior.

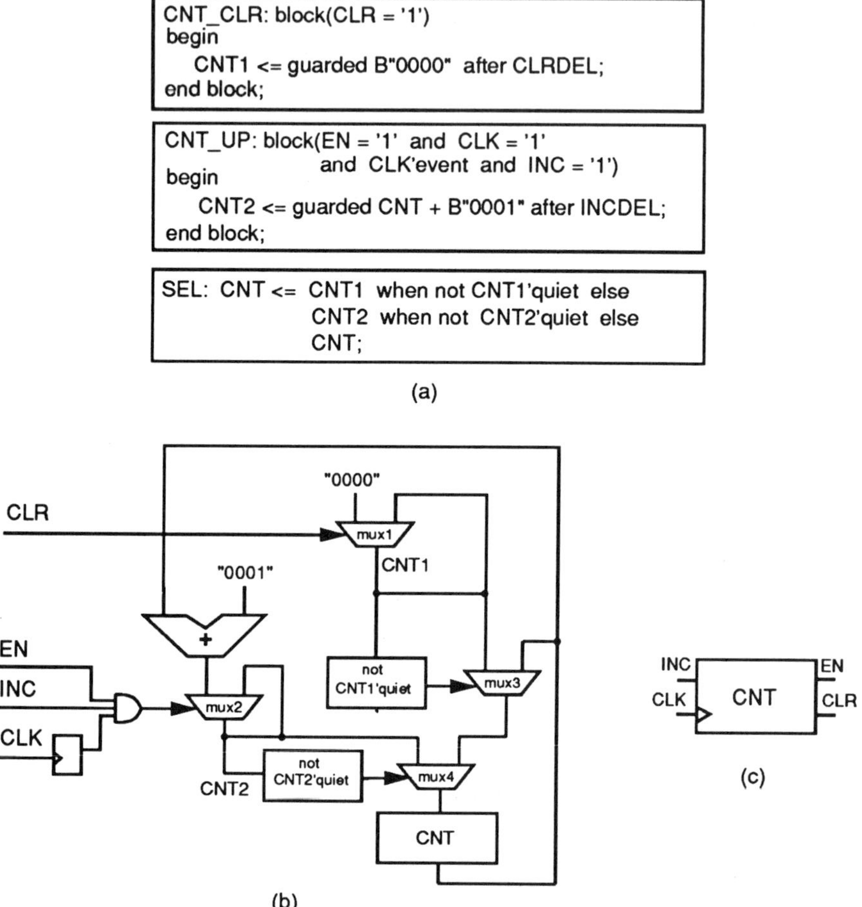

Figure 4.14: Up counter: (a) VHDL counter description, (b) initial hardware synthesized , (c) RT counter.

In order to achieve efficient hardware synthesis, we need to match the model of the language to that of the underlying target architecture. When the HDL model differs greatly from the hardware model, or when the HDL is designed for a wide variety of design applications, the user is more likely to describe the behavior of a design in ways that produce inefficient hardware implementations.

There are two solutions to the problem of matching HDL semantic models to architectural models. First, we could build specialized languages whose semantic models closely match the target architecture or application on top of the HDL. These specialized languages can be customized towards different user backgrounds, and so allow for a *look* and *feel* appropriate for the HDL user [Pilo91, Bout91, Agne91]. SpecCharts [NaVG91], StateMate [ArWC90] and BIF [DuCH91] exemplify specialized languages and formalisms built on top of VHDL.

Second, we could use the HDL directly, but restrict ourselves to following modeling styles and guidelines in the language to guarantee high quality synthesis. In addition to making synthesis more efficient, modeling styles and guidelines improve documentation and communication between designers, by producing a small set of well understood, documented description styles. In fact, HDL modeling styles may be compared with structured programming guidelines in software for improving readability and portability.

Since VHDL is a standard broadband language with semantics that are different from the hardware model, we will use it to illustrate guidelines for efficient synthesis. There are several research efforts attempting to understand modeling guidelines and practices for synthesis from VHDL [LiGa88, HaKl89, CaST91]. In what follows, we illustrate some basic modeling philosophies in VHDL for a few target architectures described in Chapter 2. We deliberately avoid going into details of the VHDL language constructs; instead, we focus on types of descriptions suitable for synthesizing specific levels of target architectures.

4.8.1 Combinatorial Designs

In the previous section, we showed three ways of modeling a full adder in VHDL. From this example, we saw that the description used Boolean language operators such as *and, or,* and *not.* (Figure 4.11(a)) was closer to the underlying hardware model. For combinatorial logic, this is obvious: the hardware design is composed of an interconnection of logic gates. Therefore, a language description that uses behavioral abstractions of these logic gates, that is Boolean VHDL operators, is closely matched to the underlying hardware model consisting of a combinatorial logic network.

Furthermore, combinatorial designs exhibit concurrency; since there is no clock to sequence pieces of behavior, the design's outputs are sensitive to changes in the input at any time. Consequently, combinatorial designs are often best described using the data-flow constructs in VHDL because these have concurrent execution semantics.

4.8.2   Functional Designs

Functional designs are characterized by a mixture of synchronous and asynchronous behavior, in which asynchronous events may override synchronous operations. Typical functional designs include sequential logic components such as registers, shift registers, counters and memories. Each functional design is a single-state FSMD, which may have different modes of operation. We can illustrate a functional design using an up/down counter with asynchronous set and clear. The counter is typically in one of a few synchronous modes of operation: *count_up, count_down,* or *parallel_load.* The operations *set* and *clear* affect the behavior of the counter asynchronously.

To describe functional designs in VHDL, it is necessary to identify which parts of the behavior operate synchronously, and which parts operate asynchronously. Therefore, we type each signal at the interface of the design based on its purpose (e.g., *clock, reset, control, data*). This link to purpose makes the design description clearer and easier to synthesize because individual signals have hardware semantics associated with them. Because of signal typing, synchronous and asynchronous behavior can be described in a single data-flow block. Furthermore, since functional designs represent single-state FSMDs, VHDL data-flow block statements are best used to describe this architectural model.

Let us now describe the behavior of the counter shown in Figure 4.14(a) using a functional description style in VHDL. First, we need to assign types to some of the signals: *CLR* gets type reset, *CLK*

```
CNT_UP_CLR: block( CLR = '1'
                      or (EN = '1'  and  CLK'event  and  CLK = '1'))
    begin
      CNT  <=  guarded
                  B"0000" after CLRDEL   when CLR='1'  else
                  CNT + B"0001" after INCDEL  when INC='1' else
                  CNT;
    end block;
```

Figure 4.15: Sample VHDL functional description style.

gets type clock, and *EN* gets type control. Next, we describe the behavior of the up-counter using a data-flow block (Figure 4.15). This block is activated only when *CLR* is high, or when *EN* is high and *CLK* has risen. Within the block, *CNT* is assigned a value with a guarded signal assignment statement. This assignment ensures that if *CLR='1'*, *CNT* is cleared and not incremented, although *EN='1'* and *CLK* has risen. This functional description will make synthesis of a standard counter structure much simpler.

4.8.3 Register-Transfer Designs

RT designs correspond to the FSMD model defined in Chapter 2. Hence, the behavioral description of an RT design is naturally expressed on a state-by-state basis so that each state specifies the conditions to be tested in the control unit, the operations to be performed in the datapath, and the next state for the design. RT designs have an implicit notion of states and state transitions. If an HDL has built-in semantics for describing states, then RT behavior can be described appropriately using that language abstraction.

Unfortunately, VHDL does not have the concept of states built into the language. Consequently, we can model state-based RT behavior in various ways. Not all of these description styles will be appropriate ways of describing the state-based behavior, nor will all result in efficient synthesis results. One approach to modeling RT states in VHDL is to associate every state of the design behavior with a separate block. Since blocks execute concurrently when unrestricted, we need to selectively activate only that block which corresponds to the current state. We do so by assigning a unique state name to each state block and maintaining a state signal for the current state. A state block is then sensitized if there is a change on the clock signal and if the current state signal is equal to the state name. On entering the current state block, the datapath operations can be described using concurrent constructs. This description is appropriate because register transfers in the datapath occur in parallel during a single state. At the end of the state block, the current state signal can be assigned the next-state name.

Figure 4.16 shows a piece of VHDL code for the *Fetch* and *Decode* states of a simple processor, in which each block corresponds to a single state. A state block is activated when the clock rises and the *state* signal equals the current state code. For instance, the *Fetch* state block is entered when *state* = *S0*. Sequencing to the next state is achieved by assigning a new value to the *state* signal within the current state block.

```
State_Fetch:  block ( (CLK'event and CLK='1')  and  (state=S0))
begin
        IR <= M(PC);
        state <= S1;
end block;


State_Decode:  block ( (CLK'event and CLK='1')  and  (state=S1))
begin case IR is
          when "0000"  =>   ACC <= ACC + 1;
                            state <= S2;
          when "0001"  =>   ACC <= 0;
          .....             state <= S3;

        end case;
end block;
```

Figure 4.16: Sample VHDL RT description style.

4.8.4 Behavioral Designs

Behavioral descriptions do not have states assigned, and need to be scheduled before they can be mapped to the FSMD architectural model. Design behavior is typically expressed in a sequential language style using sequential assignment statements. Control constructs such as *if, case* and *loop* are used to force conditional and iterative sequencing of groups of assignment statements. This type of behavior comes closest to programs written in a standard imperative programming language. However, we need additional constructs in HDLs to specify timing and other constraints.

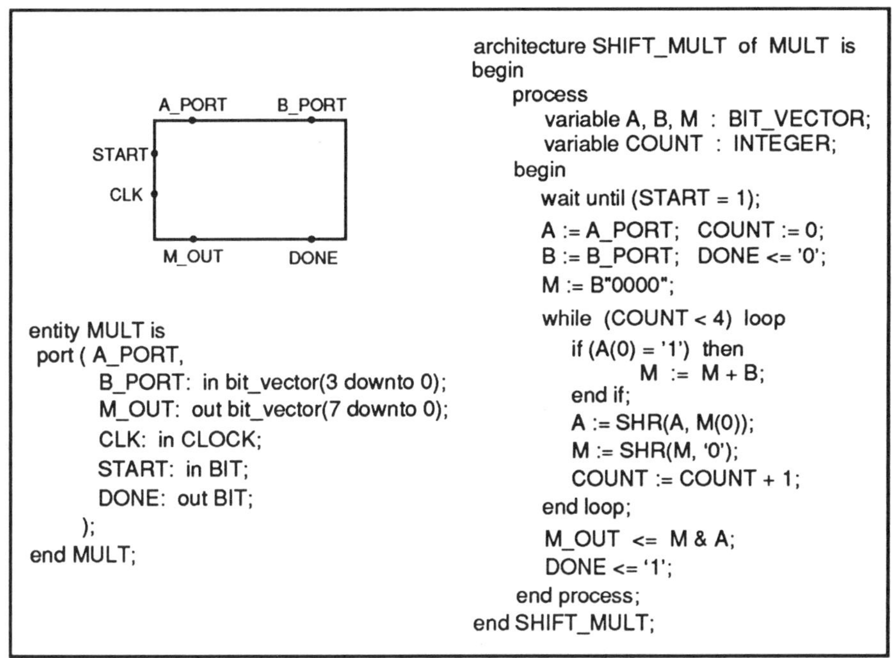

Figure 4.17: Behavioral description of a shift-multiplier.

In VHDL, the process statement fits most of these requirements: statements within a process are executed sequentially, and conditional and iterative constructs can be used for control of behavioral flow. In using the process statements, the user may need to specify which signals correspond to the clock and which to asynchronous events. Furthermore, specific guidelines have to be developed for the use of timing constructs, such as "after" and "wait", since these constructs have simulation semantics for scheduling future events or for suspending simulation execution. Without specific guidelines, these timing constructs have ambiguous semantics for high-level hardware synthesis.

Figure 4.17 is an example of a VHDL behavioral description for a simple multiplier design. The entity description shows input data ports *A_PORT* and *B_PORT*, a *START* signal and the synchronous *CLK*. The design multiplies the values on *A_PORT* and *B_PORT*, places the result on *M_OUT*, and signals *DONE* on completion. The behavioral description, using a simple shift-and-add multiply algorithm, is naturally specified in a *process* statement.

4.9 Summary and Future Directions

In this chapter, we described the need for high-level HDLs, and explained the difference between language semantic models and target architectures. We then described desirable features of HDLs to enhance design description for high-level synthesis. Since existing HDLs have their own semantic models, we described how language models need to be matched to the target architectural models for achieving high quality synthesis results. Finally, we showed some modeling guidelines and practices for using a broadband language such as VHDL in high-level synthesis.

As we move to higher levels of design description and synthesis, users will need specialized languages and formalisms that best suit their level of expertise and the target architecture being synthesized. Future research will investigate the development of such user-friendly languages and front-ends for design capture and synthesis. Since one language, formalism or interface cannot fit all possible end-users and design scenarios, research is necessary for developing synthesis frameworks that will enable users to capture design intent in a variety of languages and formats, each suitable for a particular aspect of the design specification. Finally, as HDL standards evolve, further research is required to understand efficient use of these languages and establish guidelines for specific applications and architectures.

4.10   Exercises

1. Manually synthesize RT hardware for the DSL specification of the exponentiation circuit shown in Figure 4.4.

2. The applicative behavior of DSL can asynchronously override any imperative behavior specified in DSL. How would you resolve synthesis of the applicative and imperative behaviors in general? Hint: see [CaRo89].

3. If an HDL does not support behavioral hierarchy, how would you specify a global, asynchronous event that forces the design to move from any working state to a special interrupt state? For instance, how would you describe the transition *rising(RESET)* from any *ACTIVE* state to a *RESET* state (as shown in Figure 4.5)?

4. Recall that VHDL process descriptions assume sequential execution, but that operations within a VHDL process which do not have control or data dependencies can be executed concurrently. Determine which operations in the body of a process description shown below can be executed concurrently.

$$
\begin{aligned}
x1 &:= x + dx; \\
u1 &:= u - (3 * x * u * dx) - (3 * y * dx); \\
y1 &:= y + (u * dx); \\
x &:= x1; \\
u &:= u1; \\
y &:= y1;
\end{aligned}
$$

5. Two straightforward ways of synthesizing the behavioral hierarchy shown in Figure 4.5 are given below:

(a) implement each behavioral hierarchy level as a process and make these processes communicate to implement the overall behavior, or

(b) flatten the whole behavioral hierarchy and implement the complete design as a single process with no hierarchy.

Discuss and compare the effects of these synthesis schemes on the communication overhead, the hardware generated and the complexity of the synthesis tasks.

6. Handshaking protocols between communicating behavioral processes are easily specified using the message-passing paradigm shown in Figure 4.7(b). Use pseudo-code to describe a simple handshaking protocol using the parameter-passing paradigm shown in Figure 4.7(a).

7. Using pseudo-code, describe the communication between the CPU, the bus controller, and the memory board as shown in Figure 4.8, Figure 4.9 and Figure 4.10. Use the message passing paradigm for describing protocols.

8. Figure 4.11 shows a data-flow behavioral description of a full-adder design. How would you reassign delays for the sum and carry bits in the VHDL description so that they semantically represent the total delay from inputs to outputs? Is it possible to do this for any behavior described in VHDL?

9. (a) Draw a truth table for a Boolean full adder.

   (b) Describe this truth table directly using the VHDL behavioral process description style (do not use the one's counting property shown in Figure 4.12(b)).

   (c) Synthesize hardware from this description and compare this with the hardware shown in Figure 4.12(c).

10. Develop language constructs to express pipelining features in VHDL. In particular, you should be able to specify the number of stages, the latency, and control for each pipelined stage.

11. *Develop a language for specifying different types of timing constraints for the behavior of a design. The language should allow the specification of event types (e.g., rising, falling), event relationships (e.g. before, after) and duration (e.g., minimum, maximum, nominal). Show how you can represent timeouts, set-up and hold times, pin-to-pin delays and throughput constraints in this language.

12. Describe the state machine in Figure 4.18 with VHDL in three different ways.

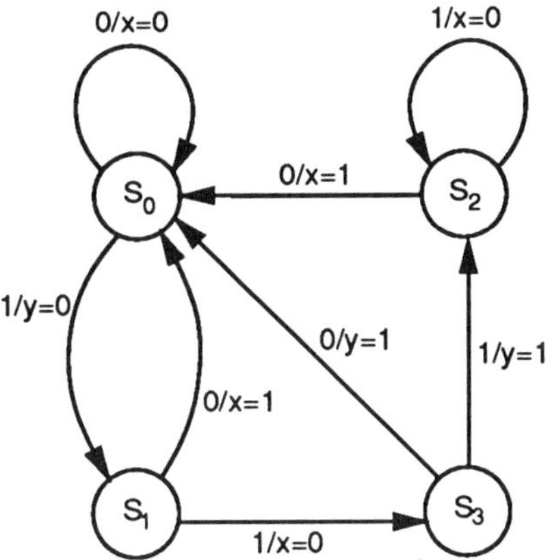

Figure 4.18: Sample FSM state diagram.

13. **Develop a set of modeling guidelines for high-level synthesis of VHDL descriptions.

14. **Develop a set of modeling guidelines to ensure testability during high-level synthesis from VHDL descriptions.

Chapter 5

Design Representation and Transformations

5.1   Introduction

Since there can be substantial variation between the semantic models of input hardware description languages (HDLs) and the synthesized target architectures, we need a canonical intermediate representation that facilitates efficient mapping of the input HDL descriptions into different target architectures using a variety of synthesis tools (Figure 5.1). The *Canonical intermediate representation* needs to preserve the original behavior of the input HDL specification, while allowing the addition of synthesis results through various refinements, bindings, optimizations and mappings. Since the outputs of high-level synthesis consist of a set of register-transfer (RT) components and a symbolic control table, the canonical representation must also capture these outputs of synthesis. The input behavior, the synthesized structure and the synthesized control represent objects in different design domains as well as different design levels, as illustrated by the behavioral and structural axes of the Y-chart in Figure 1.1. We therefore need to correlate those objects so that we can perform multi-level simulation and debugging.

An ideal intermediate representation that supports the tasks involved in high-level synthesis has the following requirements:

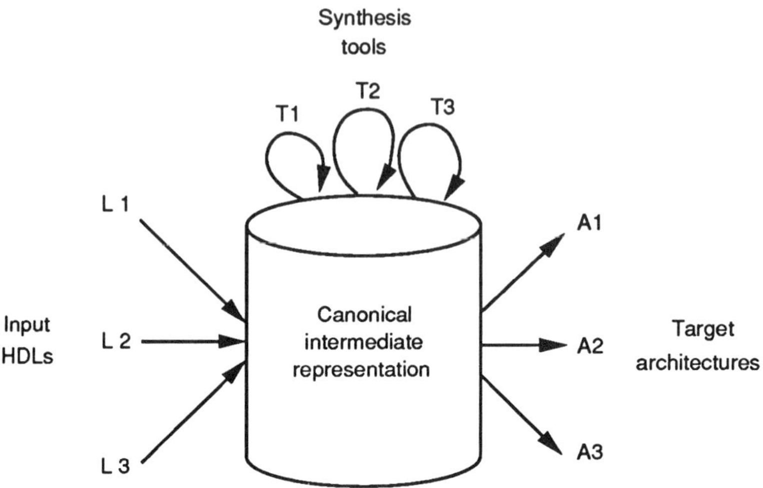

Figure 5.1: Matching HDL models to target architectures.

(a) It should be a repository of the complete design information during synthesis, including the original behavior, design constraints, synthesized state information, synthesized structures and the bindings between different design objects.

(b) It must have a canonical form that provides a uniform view of the representation across tools and users.

(c) It should be HDL independent to support a variety of input languages.

(d) It should be powerful enough to support numerous target architectural styles.

Some of these requirements for an intermediate representation may seem to contradict each other. For instance, we saw in Chapter 4 that there is no standard input behavioral languages since HDLs use different types of operational primitives and different levels of data abstractions, resulting in different semantics for representation. However, in practice, we can maintain an intermediate representation that is useful for a certain class of HDLs and that also supports a certain class of target architectures. This chapter primarily focuses on the representation of the input design behavior for the general class of HDLs described in Chapter 4, and the intermediate and final results of high-level synthesis for the FSMD target architecture described in Chapter 2. We begin by taking a simple example through various phases of high-level synthesis to illustrate what kinds of information have to be represented. We then focus on issues in the compilation of the input HDL description into an intermediate flow-graph representation, and describe the additional information generated on completion of high-level synthesis. Next, we describe design views for maintaining the complete representation, including behavior, structure and bindings. We conclude with a description of some transformations used to optimize the intermediate representation.

5.2 Design Flow in High-Level Synthesis: An Example

The first step in high-level synthesis is compilation of the input behavior into an intermediate graph representation. We then perform several high-level synthesis tasks such as scheduling, unit selection, functional, storage and interconnection binding, and control generation. In this section, we use the design of a shift multiplier to illustrate some of these synthesis tasks.

Consider the behavioral VHDL description of a shift-multiplier shown in Figure 4.17. The VHDL process description specifies an algorithm for computing the 8-bit product M_OUT of two 4-bit inputs A_PORT and B_PORT. Recall that statements in a VHDL process description are executed sequentially, but operations that do not have data dependencies between each other may execute concurrently. Hence, the sequential VHDL description of the shift-multiplier design has implicit parallelism hidden within the description.

To make this parallelism more explicit, we can compile this description into a control/data flow graph (CDFG) representation as shown in Figure 5.2. The control-flow graph of the CDFG representation captures sequencing, conditional branching and looping constructs in the behavioral description, while the data-flow graph captures operational activity described by the VHDL assignment statements (Figure 5.2). Each node

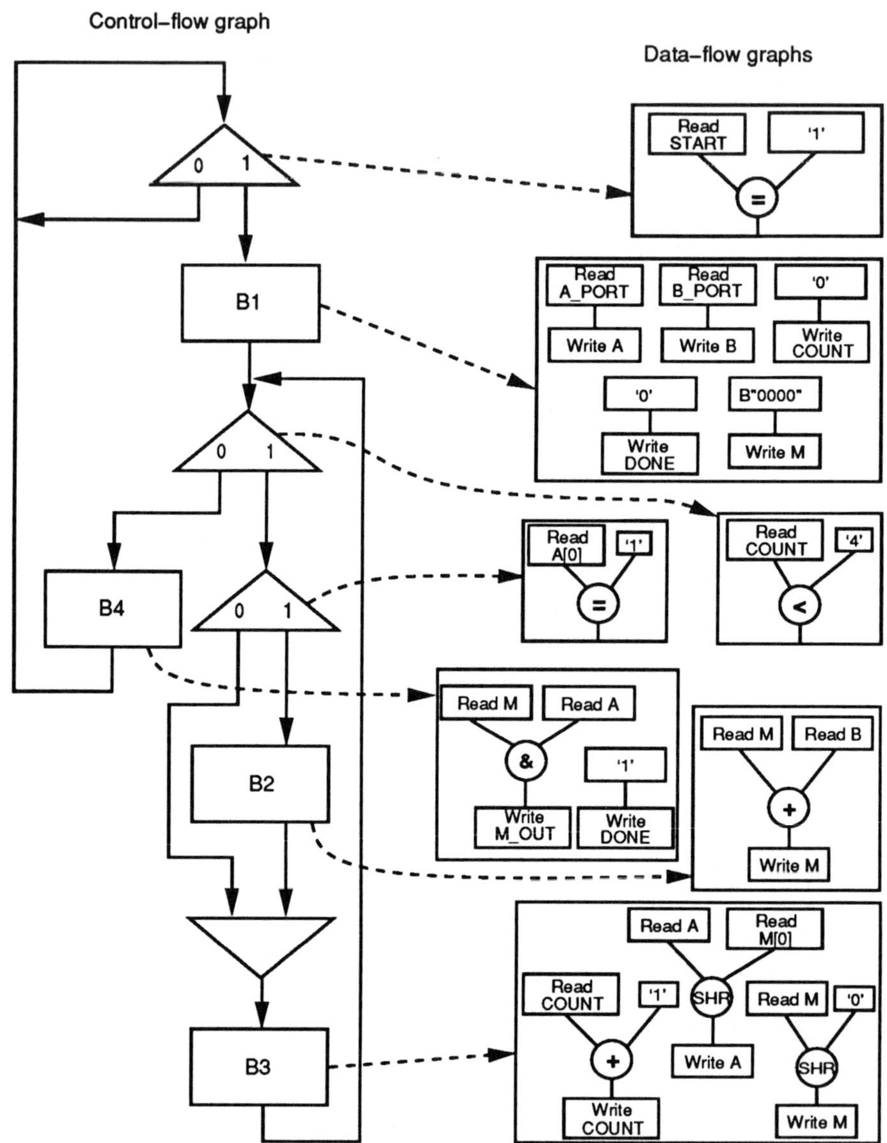

Figure 5.2: CDFG for the shift-multiplier example.

of the control-flow graph can have a data-flow block associated with it that represents the operations performed in that control-flow node. The CDFG data-flow blocks are similar to basic blocks in structured programming languages. Note that the CDFG scheme is only one example of many possible intermediate graph representations; we use the CDFG for illustration since it clearly reflects the structure of the input description. In this shift-multiplier description, the assignment statements within a basic block have no data dependencies between each other. For instance, block *B1* has five assignments in the data-flow graph, all of which can be performed in parallel. The data flow representation shown in Figure 5.2 makes explicit the parallelism inherent within the sequential VHDL process description.

Before we synthesize hardware from this input behavior, we note that neither the original VHDL description nor the corresponding CDFG representation indicates how the design is implemented in hardware. Variables, such as A, B and M in Figure 5.2, are not bound to storage elements. Similarly, operations, such as the addition in block *B2*, are not bound to functional units. Furthermore, the VHDL description and the CDFG do not specify the sequencing of states or the control signals for activating datapath components in each state.

Let us implement this shift-multiplier design using the FSMD target architectural model from Chapter 2. First, we schedule the CDFG into four synchronous time steps labeled S_0, S_1, S_2 and S_3 (Figure 5.3). In state S_0, the shift multiplier busy-waits for signal $START$ to rise. When $START$ is asserted, the shift multiplier is initialized with the two input operands from A_PORT and B_PORT, and all internal variables are cleared. We saw that the assignment statements in the data-flow graph *B1* of Figure 5.2 have no data dependencies between each other. Hence, we schedule all statements in *B1* into a single state S_0. State S_1 is the head of the shift-multiply loop. If the least significant bit of A is one, variable M accumulates the current partial product in state S_2. In state S_3, the counter is incremented and the partial product is shifted right. When the shift-multiply cycle is over, the result is sent out on port M_OUT. When $COUNT$ equals four in state S_1, the multiplication is completed by signaling $DONE \Leftarrow$ '*1*' and sending out the result on M_OUT. After scheduling, we can annotate each node in the CDFG with the state in which it is scheduled, thus maintaining a link between the

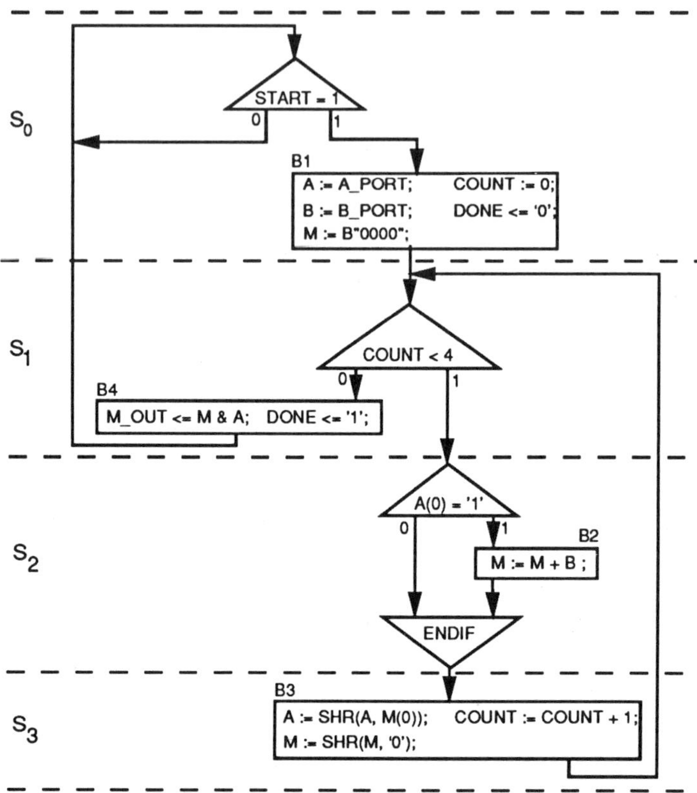

Figure 5.3: Scheduled CDFG for the shift-multiplier example.

abstract behavior and the states of the design.

The next task in high-level synthesis is unit selection, which determines the types and numbers of RT units to be used in the design. We select storage units (i.e., registers) for variables that are used in more than one state. Since four variables (i.e., *COUNT, A, B* and *M*) are used in more than one state, we select four registers named *Count_Reg*, *A_Reg*, *B_Reg* and *Mult*. Furthermore, we need to implement the operations specified in each state using functional units. We select one adder (labeled *Adder*), two shifters (labeled *Shift1* and *Shift2*) and a comparator (labeled *Compar*) as shown in Figure 5.4.

Unit binding follows unit selection: we need to bind behavioral vari-

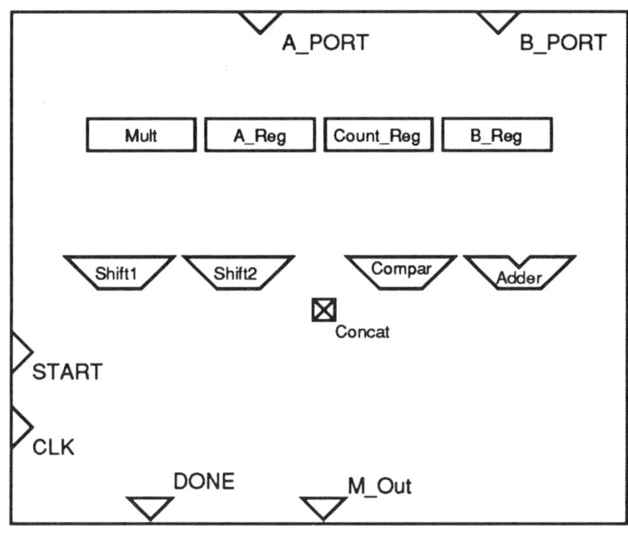

Figure 5.4: Initial unit selection for the shift-multiplier design.

ables and operators in the CDFG representation to the selected storage
and functional units. We bind the variables *COUNT, A, B* and *M* to the
registers *Count_Reg, A_Reg, B_Reg* and *Mult*, respectively. We bind the
shift operations for *M* and *A* in state S_3 to the shifters *Shift1* and *Shift2*,
respectively. We share the adder *Adder* between the addition in state S_2
and the incrementing of *COUNT* in state S_3. The comparator *Compar* is
allocated to the loop test for *COUNT* in state S_1. We then connect these
RT functional and storage units properly to ensure correct data transfers
within the datapath. In doing so, we find that we have multiple sources
at the inputs of the units *Mult, A_Reg* and *Adder*. We disambiguate
these multiple sources at inputs by allocating the multiplexors *Mux1,
Mux2, Mux3* and *Mux4* (Figure 5.5).

Finally, we derive the controller that sequences the design and controls functional and storage units in the datapath. The symbolic control
table (*Control unit* in Figure 5.5) shows the next state information and
the unencoded control lines for controlling the datapath units in each
state. Each column of the symbolic control table corresponds to a state
and condition combination, while each row (except the last row) represents a control signal for a datapath unit. The last row of the control

table shows the next state for the current state and condition value. To understand how this table is generated, consider the first column, which represents the state S_0 and the condition $START = 1$. From the scheduled behavior in Figure 5.3, we see that variable A, which is bound to A_Reg, should be loaded with the value read from A_PORT. This data transfer is achieved in Figure 5.5 by transferring data from A_PORT through the right input of $Mux2$ to A_Reg and loading A_Reg. Hence, the control-signal entry for $Mux2$ is set to 1 (i.e., the right input of the multiplexor is selected) and the $Load\ A\_Reg$ control signal is set to 1 (i.e., A_Reg is loaded). The last entry in the first column of the control table indicates that the next state is S_1. The rest of the symbolic control table is derived in a similar fashion.

Starting with the input behavior, we added several types of information during high-level synthesis: states, RT units and connections, datapath-control information, and state-sequencing information. Furthermore, we need to correlate these additional pieces of information to the input behavioral description using annotations and/or binding links. The intermediate design representation must thus explicitly represent the state bindings, unit selection and bindings, interconnected RT structure, and symbolic control. Such a representation will support all phases of high-level synthesis in the trajectory from abstract behavior to the final design implementation by representing the intermediate results of synthesis as well as links between intermediate stages of synthesis.

5.3   HDL Compilation

In this section, we illustrate the compilation task using a VHDL description as input and the CDFG representation as output. In the CDFG representation, HDL control-flow constructs, such as loops and conditionals (e.g., "if" and "case"), are mapped to control-flow nodes, and blocks of assignment statements between these control-flow constructs are represented by data-flow graphs.

We show the generation of data-flow graphs from the sequential VHDL description in Figure 5.6(a). First, the HDL parser generates parse trees by a one-to-one mapping of individual language statements to the parse trees (Figure 5.6(b)). Since these parse trees describe the

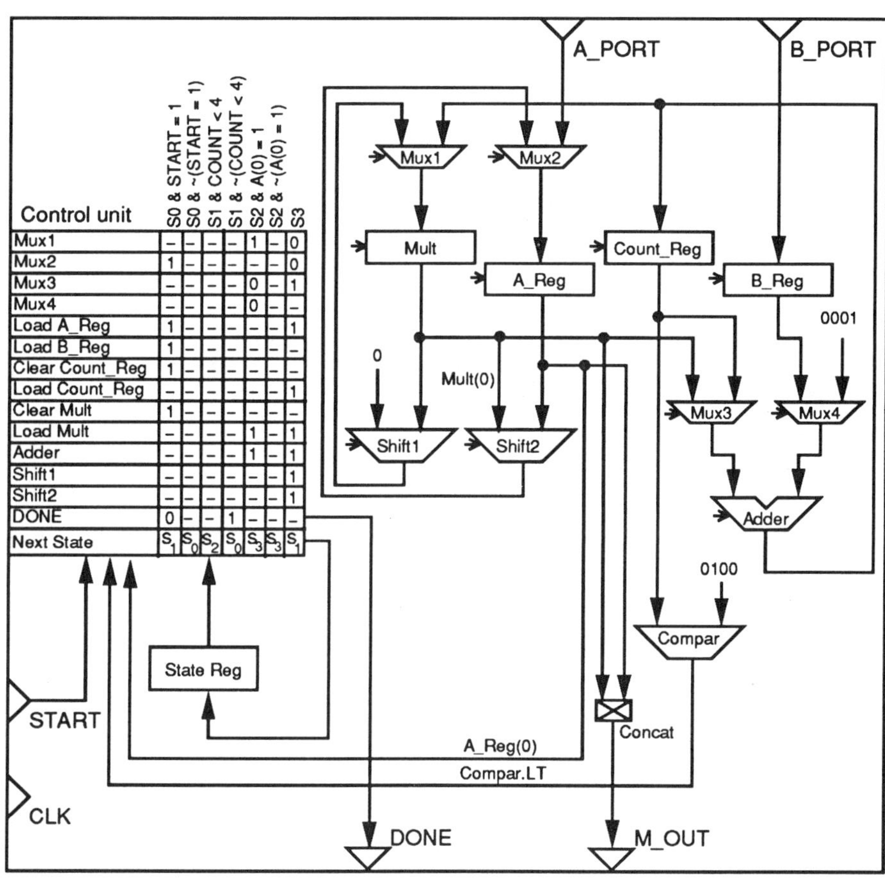

Control unit table:

	S0 & START = 1	S0 & ~(START = 1)	S1 & COUNT < 4	S1 & ~(COUNT < 4)	S2 & A(0) = 1	S2 & ~(A(0) = 1)	S3
Mux1	–	–	–	–	1	–	0
Mux2	1	–	–	–	–	–	0
Mux3	–	–	–	–	0	–	1
Mux4	–	–	–	–	0	–	–
Load A_Reg	1	–	–	–	–	–	1
Load B_Reg	1	–	–	–	–	–	–
Clear Count_Reg	1	–	–	–	–	–	–
Load Count_Reg	–	–	–	–	–	–	1
Clear Mult	1	–	–	–	–	–	–
Load Mult	–	–	–	–	1	–	1
Adder	–	–	–	–	1	–	1
Shift1	–	–	–	–	–	–	1
Shift2	–	–	–	–	–	–	1
DONE	0	–	–	1	–	–	–
Next State	S_1	S_0	S_2	S_0	S_3	S_3	S_1

Figure 5.5: Synthesized design for the shift-multiplier example.

behavior of sequential assignment statements in VHDL, we use the statement list *(Stmt1,Stmt2,Stmt3)* to show the execution ordering of the parse trees.

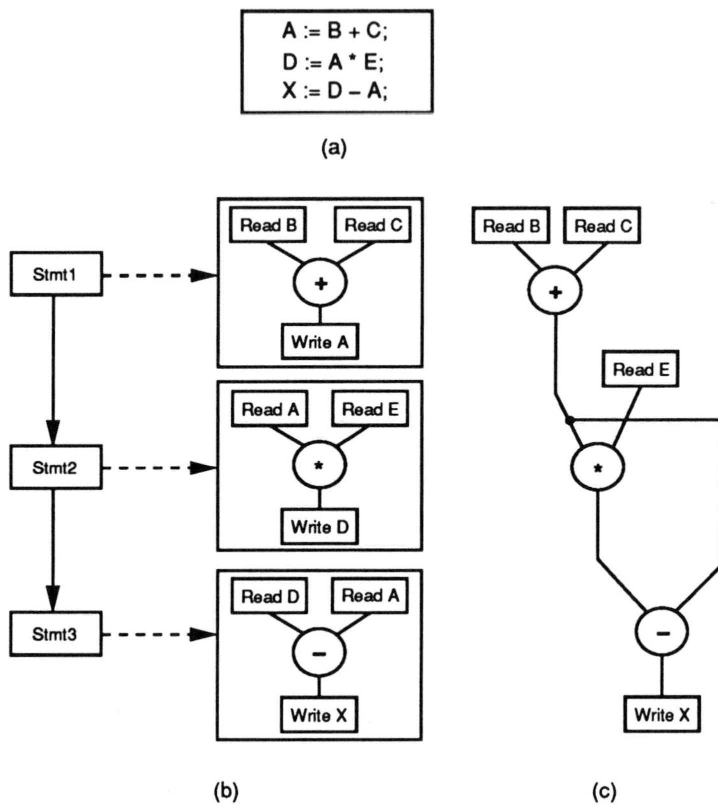

Figure 5.6: Data-flow graph generation: (a) HDL description, (b) parse trees, (c) DFG.

Next we interpret the execution ordering of the parse trees based on the semantics of the input HDL description style. When the HDL semantics demand that statements be executed concurrently (e.g., within a VHDL data-flow description), the parse trees are analyzed to ensure that all expressions on the right-hand side (RHS) of assignment statements are evaluated concurrently, before assigning values to the left-hand side (LHS) variables. The parse trees are kept intact, except for accesses to common RHS source variables that are merged.

If the HDL constructs have sequential execution semantics (e.g., within a VHDL process description), we perform data-flow analysis between the parse trees to reveal the concurrency between the sequential HDL statements. This step is similar to data-flow analysis in a traditional programming language compiler [AhSU86]. Since the description in Figure 5.6(a) has sequential execution semantics, we apply data-flow analysis to the parse trees shown in Figure 5.6(b). We see that variable A is defined in statement *Stmt1* and is used within statements *Stmt2* and *Stmt3*. Therefore, we have forward data dependences for variable A between statement *Stmt1* and both statements *Stmt2* and *Stmt3*. Similarly, variable D is defined in statement *Stmt2* and is used in statement *Stmt3*, resulting in a forward data dependence for variable D between statements *Stmt2* and *Stmt3*.

We complete the data-flow analysis by fusing together all the parse trees into a single data-flow graph, while maintaining any data dependencies that exist between different statements. Figure 5.6(c) shows the resulting data-flow graph generated from the set of sequential VHDL assignment statements in Figure 5.6(a).

5.4 Representation of HDL Behavior

HDL descriptions are typically captured with flow graphs, such as the CDFG scheme described previously. These flow-graph schemes differ in their representation of control constructs and in the representation of data transfers within a data-flow graph. In this section, we first use some simple schemes to intuitively illustrate the capture of control-flow constructs. Next we discuss some options for the representation of sequencing and timing information. Finally, we use examples to illustrate three specific classes of flow-graph schemes used by high-level synthesis systems: disjoint control and data, hybrid control and data, and parse-tree representations.

5.4.1 Control-Flow Representation

Let us examine two simple representations of control constructs to intuitively show that we can capture control flow in different ways. The segment of VHDL code in Figure 5.7(a) shows a *case* statement that selects a set of assignments based on the value of the signal C. In the first scheme (Figure 5.7(b)), control constructs are mapped to control-flow nodes that maintain the explicit sequencing and flow of control specified in the input description. The control-flow graph consists of conditional branch nodes, conditional join nodes and statement assignment blocks, symbolically represented by triangles, inverted triangles and rectangles, respectively. Each control-flow node can have a data-flow block associated with it that describes its operational activity (i.e., assignments or tests). For example, Figure 5.7(b) shows the *case* statement mapped to a conditional branch construct (i.e., the triangles) and the assignments within each conditional evaluation mapped to a data-flow block (i.e., the rectangle). In this control-flow representation, each path of the *case* statement explicitly shows mutual exclusion, making it easy for synthesis algorithms to share resources across different branches of the conditional. This representation also closely matches the flow of control in the input description, and so is easy to debug and link to the original description. This representation method is similar to the control/data flow graphs with basic blocks used by software compilers.

In the second representation scheme, we map control constructs into a data-flow graph by evaluating all branches of a conditional test in parallel and choosing the correct values for assignments after all branches have been executed. To do so, we compute all possible values for a target LHS variable and select the appropriate one based on the value of the condition variable. In this data-flow scheme, we use circles to denote operations, arcs to denote data-flow, rectangles to denote the reading and writing of data, and inverted triangles to denote selection of data based on the value of a control line. Figure 5.7(c) shows the data-flow representation of the *case* statement from Figure 5.7(a). The data-flow selection nodes (represented by reverse triangles in Figure 5.7(c)) select the appropriate values for the target variables X and A, which appear on the LHS of at least one assignment statement.

Like the control-flow representation, the data-flow representation also explicitly shows the concurrency due to mutual exclusion of different con-

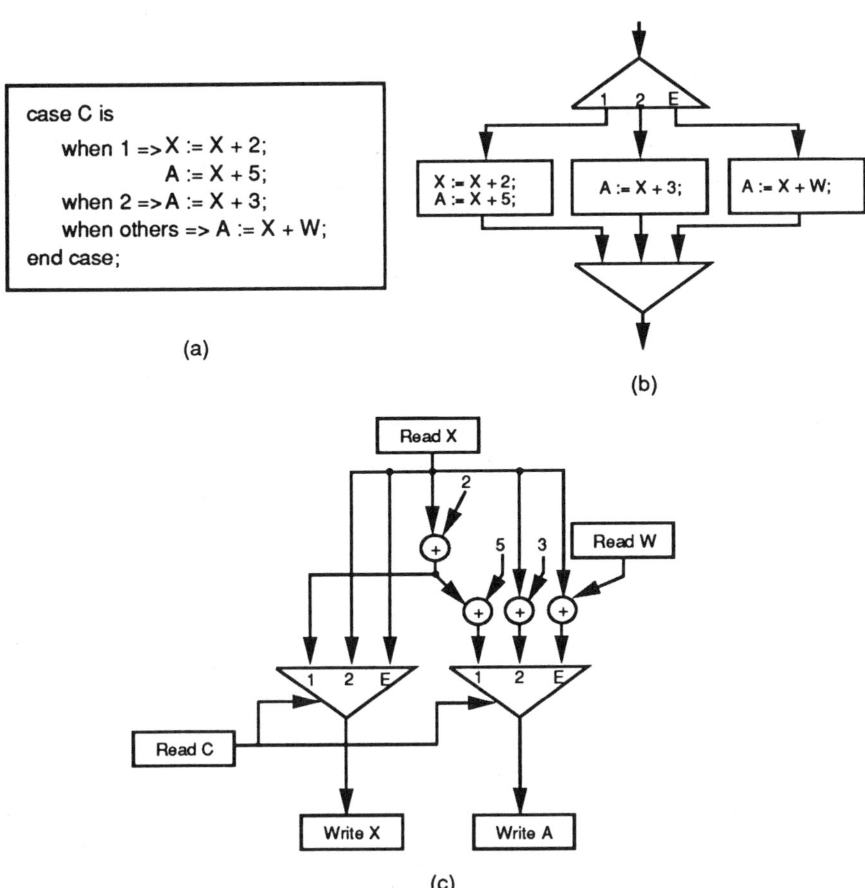

Figure 5.7: Case statement example: (a) VHDL description, (b) control-flow representation, (c) data-flow representation.

ditional paths. However, since the data-flow representation evaluates all branches of a conditional test in parallel, it makes a larger segment of the graph available for refinement and optimization than the control-flow representation. Consequently, the data-flow representation is better suited for the task of scheduling straight-line code than the control-flow representation. On the other hand, reconstruction of the original description from the data-flow representation becomes more difficult because all the control-flow information is embedded within the data-flow graph. Also, because all nested conditions and loops are flattened into a single data-flow graph, deeply nested structures generate several levels of selectors that choose between conditional paths. This generation can result in data-flow graphs that are too large and cumbersome to handle.

The first scheme described above assume a partitioning of the behavior into data flow and control flow, based primarily on the basic block structure of the input description. We can also use a hybrid representation scheme, in which control flow and data flow are intermixed; the control flow then dictates explicit sequencing of the data-flow nodes. We give examples of several different representation schemes after discussing the representation of sequencing and timing constraints.

5.4.2   Representation of Sequencing and Timing

In addition to control and data-flow dependencies, the intermediate representation also has to maintain ordering relationships that are specified implicitly or explicitly within the input description. We can use precedence arcs to show explicitly the ordering between nodes of a flow graph. Precedence arcs are needed to enforce the ordering of external read and write operations (e.g., two successive port read operations) in the input description. Precedence arcs can also be used to enforce ordering of array accesses when the index values are not constant. Figure 5.8(a) shows a segment of a concurrent VHDL description in which signals a and b are incremented and swapped. Since these statements execute concurrently, we must ensure that the read operations precede the write operations for signals a and b. The bold arcs in the data-flow representations in Figure 5.8(b) show these precedence relationships.

Flow-graph representation schemes also differ in their data-flow representation of variable accesses (i.e., reads and writes). Some schemes

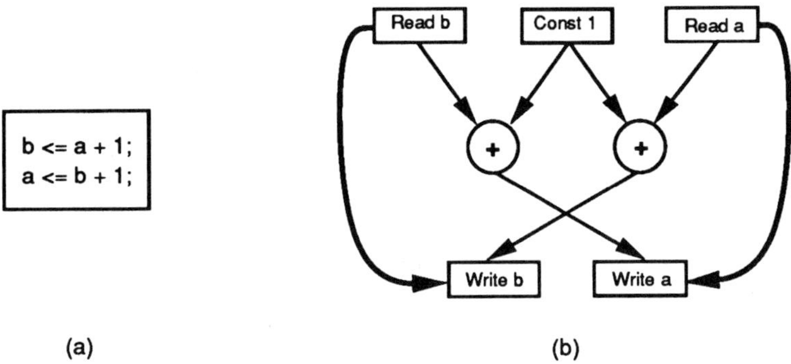

Figure 5.8: Data-flow with precedence arcs: (a) concurrent description, (b) DFG representation.

explicitly represent every read and write of a variable by a variable access node. In a block of sequential behavior, we must ensure that variable values are read before a new value is defined (i.e., written). Conversely, a read access of a variable should not be executed before its value is defined (i.e., written). Precedence arcs can be used to enforce such ordering of read and write nodes to the same variable (Figure 5.9(b)). Since variable accesses may get bound to storage elements that require control, this explicit representation scheme allows variable accesses and language operators to be treated uniformly, making the tasks of unit selection and binding much easier. Other flow-graph based schemes represent variable accesses implicitly as data traces by using arcs in the data-flow representation(Figure 5.9(c)). In contrast to the explicit scheme, data-flow optimizations are much easier to perform on traces of the values than if these accesses were explicitly represented as operation nodes.

We often specify timing constraints on the design behavior to ensure correct operation or to meet performance goals. The timing information represented in the control-flow and data-flow graphs is used as constraints for scheduling, unit selection and unit binding. It also places constraints on the performance of a design implementation, (e.g., the clock cycle). Depending upon the representation scheme used, we can capture such timing information in several ways. We illustrate some sample schemes for the representation of timing constraints in the data-flow

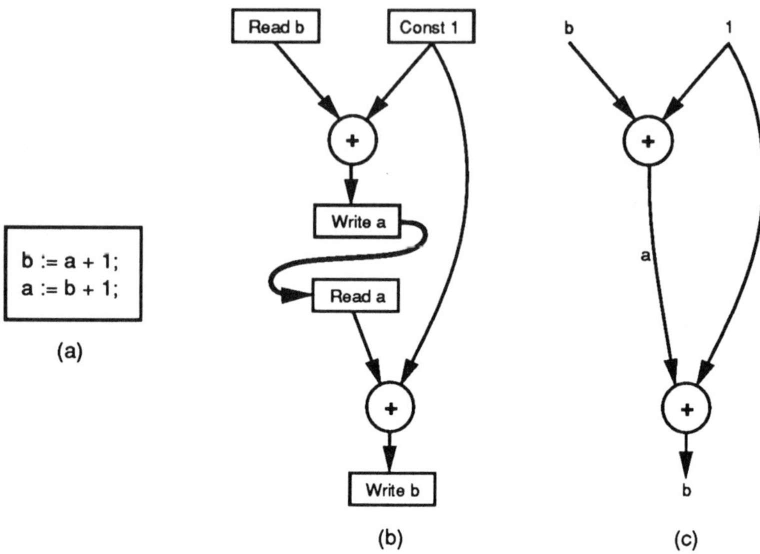

Figure 5.9: Variable access representations: (a) sequential VHDL description, (b) DFG with variable access nodes, (c) DFG with variable traces.

and control-flow graphs.

At the data-flow graph level, we can use two simple schemes: by annotating timing information on precedence arcs between two nodes of the graph, or by creating a timing node between two arcs of the graph. Annotating timing information on the precedence arcs between nodes of the data-flow graph is useful for representing minimum, maximum and nominal timing constraints between the occurrence of two operations (i.e., the nodes). Such a timing constraint often occurs when read and write operations to the design's external ports have to be ordered and spaced apart in time for correct operation (e.g., in communication protocols). For example, Figure 5.10(a) shows a timing constraint *min=500,max=1000* annotated on the precedence arc between the reading of input port *req* and the writing of output port *ack*. The second scheme, uses explicit timing nodes between arcs of the data-flow graph and describes point-to-point path delays in the data-flow graph. Such delays provide constraints on minimum and maximum execution times for individual operators, as well as for groups of operators along the data-flow path from the source to the sink of the timing node. For example, Figure 5.10(b) shows a timing node between the left input of the addition operation and the output of the shift operation *shr*, which constrains the execution of the addition and shift operations to a minimum of 50 ns and maximum of 90 ns. When timing nodes are used from the inputs to the output of a single node, they model the pin-to-pin delays for the component implementing the operation.

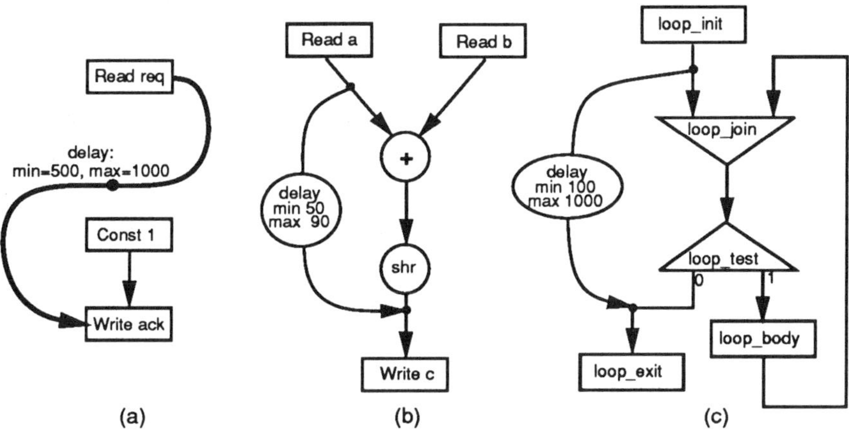

Figure 5.10: Representation of timing: (a) annotating DF precedence arcs, (b) using a DF timing node, (c) using a CF timing node.

Similarly, at the control-flow graph level, we can insert a timing node between two control-flow graph arcs. The semantics of such timing nodes vary with the representation scheme used. In the CDFG representation, a timing node between two control-flow arcs provides a constraint on the execution of all data-flow blocks along the control path between the source and sink of the timing node. Hence, this timing node constrains performance requirements over several data-flow blocks or even over the complete design. For instance, Figure 5.10(c) shows the minimum and maximum delay bounds for the execution of a loop specified by inserting a delay node from the loop entry arc to the loop exit arc of the control-flow graph.

We have shown that different aspects of the HDL input behavior can be represented in several ways using a flow-graph based scheme. We now look at three different flow-graph schemes used in high-level synthesis.

5.4.3   Disjoint Control and Data-Flow Representations

The disjoint control and data-flow representation (CDFG) scheme is similar to the intermediate flow-graph representations used in standard software compilers, where control constructs are mapped to control-flow nodes and assignments within basic blocks are mapped to data-flow nodes. The control-flow nodes are maintained separately from the data-flow nodes in this representation, which makes the overall sequencing of the design easier to follow. Descart [OrGa86] and VSS [LiGa88] use a separate control-flow node to represent a basic block of data-flow nodes. The control-flow graph is directly generated from the control constructs in the input language. Several types of control-flow nodes are used. Language constructs such as "if", "case" and "loop" are mapped to control-flow diverge and merge nodes. Procedure and function calls are represented by control-flow call nodes. Additional control demarcation nodes are used to define the boundaries of behavioral programs, functions and procedures. The data-flow graph also consists of several types of nodes. Data-flow operation nodes represent language operators, variable reference nodes represent reading and writing of variables, and subscript nodes represent array accesses. Since a single control statement block node represents a data-flow graph with several data-flow nodes, we have a one-to-many relationship of control to data-flow nodes. Consider the *case* statement shown in Figure 5.11(a). The CDFG representation for this input is shown in Figure 5.11(b). This scheme explicitly represents variable accesses using data-flow nodes.

The DDS scheme [KnPa85] is another representation that uses separate graphs for data flow and control/timing flow, with links between them to indicate dependencies and sequencing.

5.4.4   Hybrid Control and Data-Flow Representations

The hybrid control and data-flow representation embeds all control and sequencing information explicitly within the data-flow graph. An example of this scheme is the concurrent data-flow representation shown in

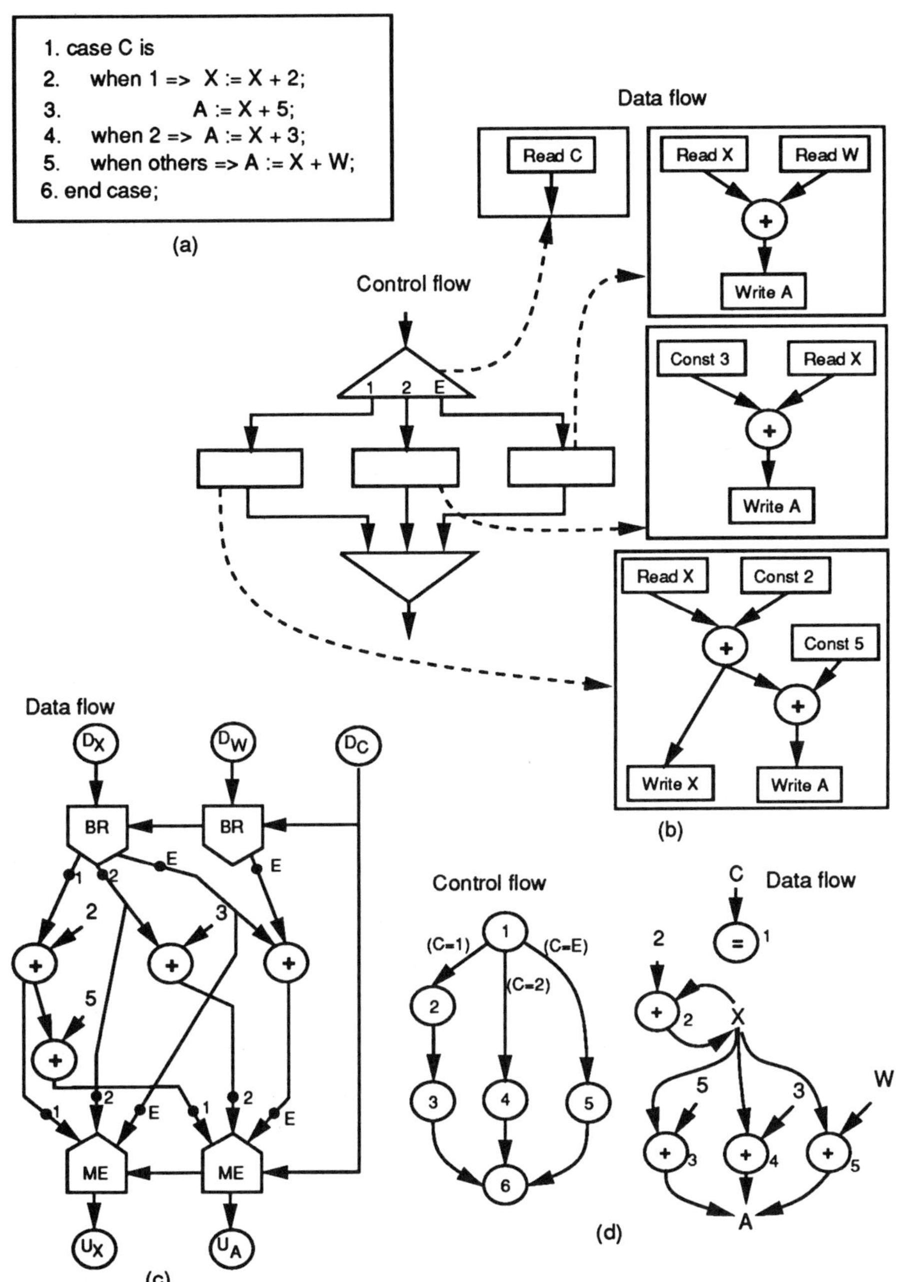

Figure 5.11: Flow-graph representations: (a) VHDL code, (b) partitioned CDFG, (c) DeJong's hybrid flow graph, (d) SSIM flow graph.

Figure 5.7(c), in which all mutually exclusive paths are executed concurrently and the final result is selected based on the value of the conditional. The Value Trace (VT) [McFa78] follows this method.

DeJong's representation [DeJo91] is another data-flow based scheme, in which only active conditional paths are executed, rather than all branches of a conditional test. The representation uses branch and merge nodes to determine which path of the data-flow graph to execute. An example of DeJong's representation for the VHDL *case* statement in Figure 5.11(a) is shown in Figure 5.11(c). In this example, D_X, D_W, and D_C represent the RHS values of variables X, W and C, respectively, while U_X and U_A represent the assignments to the target LHS variables X and A, respectively. The nodes labeled BR represent branching, and the nodes labeled ME represent conditional assignments. Arcs in this representation are annotated with the condition values to indicate the appropriate path for execution.

The DSL system [CaRo89] uses a variation of the hybrid control and data-flow scheme, where the underlying graph nodes represent language operations and variables. Three distinct graphs are built on top of these nodes using three types of arcs: sequencing arcs capture the control and sequencing information in the input description, data-flow arcs show data dependencies between operations, and timing arcs show timing constraints between various operations. SSIM [CaTa89] is a extension of DSL that keeps the control and data-flow parts physically disjoint by duplicating the operator nodes and by displaying the control-flow sequencing arcs in the first graph and the data-flow dependency arcs in the second graph. This one-to-one relationship between control and data-flow nodes for the VHDL behavior in Figure 5.11(a) is shown in Figure 5.11(d).

The DSFG representation [LGPV91] is used to capture signal-flow graph behavior derived from Silage descriptions of digital signal processing designs (see Section 4.6.3). This is another hybrid representation in which the signal-flow graph (i.e., the data-flow graph) is annotated with control information and design constraints. Similarly, the SIF representation used in the Olympus synthesis system [DKMT90] uses a hierarchical sequencing graph to show data-flow and control-flow dependencies.

5.4.5 Parse-Tree Representations

In Figure 5.6(b), we showed that parse trees are first generated in the process of compiling an input description into a flow-graph representation.
These parse trees can themselves be used as the intermediate behavioral
representation, and can be annotated with bindings and structure as
synthesis proceeds. Since the parse trees are maintained without data-
flow analysis between statements, an advantage of this scheme is that it
closely matches the input behavioral description. This makes it easier for
human designers to understand. However, the parse-tree representation
does not explicitly show the potential parallelism between different statements. Hence, synthesis tools have to extract such information from the
tree representation when necessary. An example of the parse-tree intermediate representation is TREEMOLA [BSPJ90], which maintains lists
of parse trees derived from the input MIMOLA [Marw85] behavioral
description.

5.5   Representation of HLS Outputs

The output of high-level synthesis is a structural datapath of interconnected RT components and a symbolic control table. The RT components and their connections represent objects in the structural domain
of synthesis, and the symbolic control table represents the synthesized
FSM behavior at a lower level than the original input description. Since
we are dealing with objects in different design domains and at different
design levels, the synthesized structure and control are generally maintained separately from the input description, and are correlated with the
input behavior by means of binding links.

The synthesized structure is typically stored as a netlist or nodelist
composed of three parts: a list of components, a list of signals, and
the interconnections between the components and signals. For each net
(i.e., signal), a netlist specifies the source and sink ports on components.
Conversely, for each node (i.e., component), a nodelist specifies the signals that are connected to the input and output ports of the component.
Netlists and nodelists can vary greatly in syntax, but they all contain
the same basic interconnection information. An example of a nodelist is
the VHDL description of the full-adder structure in Figure 4.13. We can

maintain correlation between entities in the input behavioral description (e.g., operations and variables) and the synthesized RT structure by annotating each flow graph node with a binding link to its corresponding RT structural component. For example, the addition node in statement block *B2* of Figure 5.2 can be annotated with the binding of the RT component *Adder*. Similarly, every read or write access node to variable *M* in the data-flow graphs of Figure 5.2 can be annotated with the binding of the register *M_Reg*.

A symbolic control table describes states, transitions between states, and the control signals activated to enable appropriate datapath actions in each state. It corresponds to the FSMD state table described in Chapter 2. A symbolic control table for the shift-multiplier design in Section 5.2 is shown in the control unit of Figure 5.5. The control table uses a symbolic representation of its states and control outputs, without using any particular encoding. This illustrated in Figure 5.5, where both the states (e.g., S_0) and the control outputs (e.g., *Load A_Reg*) are represented symbolically. We can maintain links between the original behavior and the symbolic control table by annotating each flow-graph node with the state and condition under which it is executed. For example, since statement block *B2* is scheduled into state S_2 when $A(0)='1'$ (Figure 5.3), we can annotate every operation node in the data-flow block *B2* of Figure 5.2 with the scheduled state S_2 and the condition $A(0)='1'$.

5.6 Design Views and Complete Representation Schemes for High-Level Synthesis

We saw that several types of information have to be maintained during the process of high-level synthesis: the initial design behavior specified in the input HDL, the design constraints and the results of various synthesis tasks in the form of states, state sequencing, datapath structures, and their bindings to behavioral variables and operators.

Vertically integrated high-level synthesis systems (described in Chapter 9) typically maintain the complete design information during high-level synthesis using a variety of intermediate representation schemes. Since several synthesis tools are used during different stages of synthesis, such as scheduling and allocation, we need a mechanism to allow the interchange of design information between tools. Furthermore, a designer might want to manually override or replace certain phases of automatic synthesis. Therefore we need to support different design views that closely match the information required for each step of synthesis. We illustrate the need for such a set of design views by taking the shift-multiplier example (Figure 4.17) through various steps in high-level synthesis. There are many representation schemes that carry the complete design information necessary for high-level synthesis and that provide design views for specific stages of synthesis. At the end of this section, we describe a few such schemes.

Present State	Condition	Value	Actions	Next State
S0	START = 1	T	A := A_PORT; B := B_PORT; COUNT := 0; DONE := '0'; M := 0000";	S1
		F		S0
S1	COUNT < 4	T		S2
		F	M_OUT := M @ A; DONE := '1';	S0
S2	A(0) = 1	T	M := M + B;	S3
		F		S3
S3			A := SHR(A, M(0)); M := SHR(M, '0'); COUNT := COUNT + 1;	S1

(a)

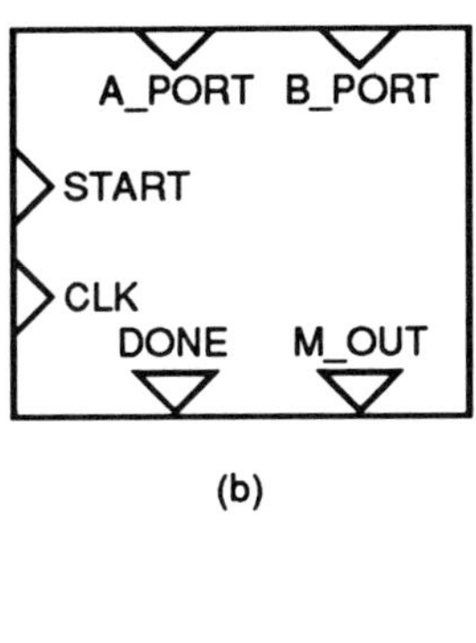

(b)

Figure 5.12: Shift-multiplier example: (a) FSMD state table, (b) I/O ports.

Consider the shift-multiplier design, which was first specified using VHDL in Chapter 4 (Figure 4.17). This description was compiled into the CDFG shown in Figure 5.2. In Section 5.2 of this chapter, we saw this abstract behavioral design scheduled into four states, S_0, S_1, S_2 and

S_3, as shown in Figure 5.3. At this point, the design is at the level of an FSMD, where we know the actions to be performed in each state as well as the transitions between states. However, there is no correspondence between the abstract operations performed on variables in each state and the RT units that implement the behavior. We can represent the results of this schedule in an FSMD state table (Figure 5.12(a)). As discussed in Chapter 2, the FSMD state table describes, for each state, the conditions as test expressions, the outputs as behavioral assignment statements, and the next state for the design. Consequently, this state table is a natural way for designers to capture RT design behavior. Since we have not added any structural information to the design, Figure 5.12(b) shows the initial structure of the shift-multiplier design containing only the ports defined in the original VHDL behavior.

After the shift-multiplier design is scheduled, we select functional units and registers as shown in Figure 5.4. We then perform the tasks of unit and connection binding to complete the RT datapath shown in Figure 5.13(a). In this design, we bind the shift operation for M to *Shift1* and the shift operation for A to *Shift2*. We also share the functional unit *Adder* between incrementing M in state *S2* and incrementing *COUNT* in state *S3*.

The results of datapath allocation can also be captured using an FSMD state table, in which the behavioral assignment statements are replaced by a trace of the data transfers through components in the datapath. For each state and condition combination, we can describe each datapath unit activated, along with details of the operation performed and the inputs used. An example of this component-based state table for the shift-multiplier design is partially shown in Figure 5.13(b) for states S_1 and S_2.

For an explanation of the operation field of this component-based state table, consider the statement $M := M + B$ in state S_2 of Figure 5.3. In the datapath of Figure 5.13(a), this addition is performed by selecting *Mult* and *B_Reg* using *Mux3* and *Mux4*, respectively, performing an *add* operation on *Adder*, and storing the result back into *Mult* via the right input of *Mux1*. This data transfer takes place only when $A(0)={}'1'$ in state S_2. The component-based state table concisely captures this data transfer in its *actions* field, as shown in state S_2, condition $A\_Reg(0) = 1$, of Figure 5.13(b). The first action indicates that the functional unit

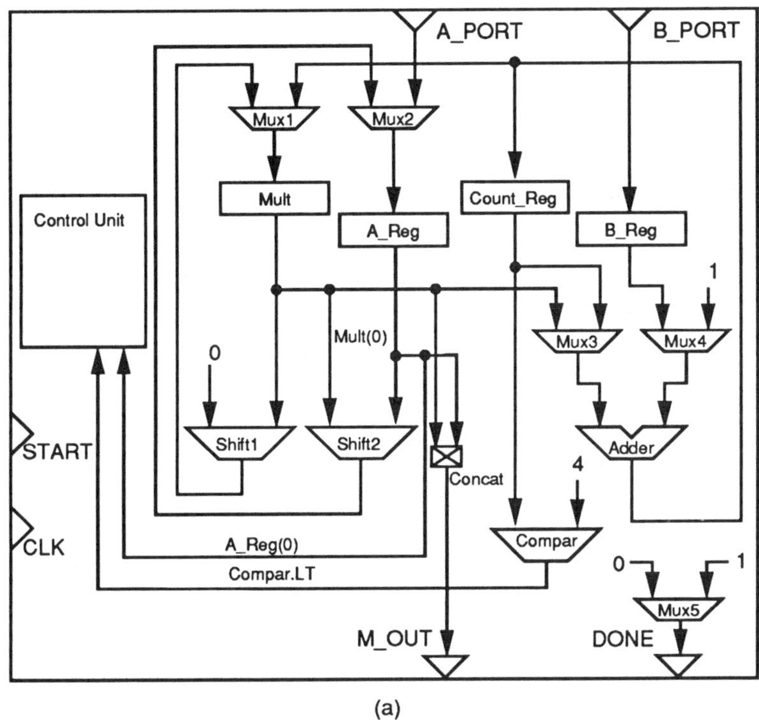

(a)

Present State	Condition	Value	Actions	Next State
		1		S2
S1	Compar.LT	0	Concat(OP: concat, INPS: Mult, A_Reg); M_OUT(OP: load, INPS: Concat); Mux5(OP: c1, INPS: '0', '1'); DONE(OP: load, INPS: Mux5);	S0
S2	A_Reg(0)	1	Mux3(OP: c0, INPS: Mult, Count_Reg); Mux4(OP: c0, INPS: B_Reg, "0001"); Adder(OP: add, INPS: Mux3, Mux4); Mux1(OP: c1, INPS: Shift1, Adder); Mult(OP: load, INPS: Mux1);	S3
		0		S3

(b)

Figure 5.13: Shift-multiplier design after datapath allocation: (a) partially synthesized structure, (b) FSMD component-based state table.

Mux3 selects its left input *Mult*. Similarly, *Mux4* selects *B_Reg* as its source, and *Adder* performs the *add* operation using *Mux3* and *Mux4* as inputs. Finally, *Mux1* switches the output of *Adder* through to *Mult*, which loads the result of the addition.

Present State	Condition	Value	Actions	Next State
S1	Compar.LT	1		S2
		0	DONE := 1;	S0
S2	A_Reg(0)	1	Mux3.sel := 0; Mux4.sel := 0; Adder.add := 1; Mux1.sel := 1; Mult.load := 1;	S3
		0		S3

Figure 5.14: FSMD symbolic control table for the shift multiplier design.

The task of allocation results in a complete datapath, but does not explain how this datapath is controlled over time as represented by the empty *Control unit* in Figure 5.13(a). The last phase of high-level synthesis is control generation where we derive the symbolic control table for the *Control unit* of the synthesized design. For each state and condition, the symbolic control table indicates the control signals needed to activate functional and storage units in the datapath, as well as the next state into which to sequence. The FSM state table model from Chapter 2 naturally describes this symbolic control table, if we use unencoded states and control outputs. Figure 5.14 shows a partial symbolic control table of the shift-multiplier design that indicates the control signals activated in each state of the design and the next-state information. The complete synthesized design with an embedded symbolic control table is shown in Figure 5.5.

In order to better understand the specification of the symbolic control table, let us trace the execution of the behavior $M := M + B$ in state S_2 using the complete design and the symbolic control table shown in Figures 5.5 and 5.14, respectively. We first note that M and B are bound to *Mult* and *B_Reg*, respectively. Hence, we have to steer the outputs of *Mult* and *B_Reg* through *Mux3* and *Mux4*, respectively, set the control line for *Adder*, steer the output of *Adder* through *Mux1*, and finally, load the result in *Mult*. State S_2 of the symbolic control table in Figure 5.14 shows the control lines that have to be active for this data transfer to be effected.

Using the simple shift-multiplier example, we have shown that the input description is gradually refined into a design implementation by adding more structure to the datapath and modifying the type of design behavior to indicate the operation and control of datapath components. Therefore, the intermediate representation has to capture state-based sequencing information together with a mixed behavioral and structural description to support high-level synthesis. At each step of the high-level synthesis process, we have shown a sample design view based on the FSMD model of Chapter 2, which described the complete behavior and structure of the design. In the rest of this section, we briefly look at some representation schemes that maintain this complete design information as well as provide design views that are user-friendly for humans.

The Behavioral Intermediate Format (BIF) [DuHG90] is an example of a canonical intermediate representation for synthesis that captures design behavior of the FSMD target architectural model introduced in Chapter 2. BIF uses state tables annotated with behavioral assignments, functional units and connections to comprehensively represent the behavior, structure and timing constraints for the design at each stage of the high-level synthesis process. BIF provides design views based on an annotated CDFG representation that is used by synthesis tools. The initial, unscheduled behavior is captured using a partitioned CDFG representation. After scheduling the design behavior into states, the CDFG is annotated with the state information. The BIF operations-based state table (OBST) captures the scheduled FSMD behavior using a format similar to the FSMD-based state table illustrated in Figure 5.12. After datapath allocation, BIF's unit-based state table (UBST) describes the register-to-register datapath transfers in each state using a design view closely modeled on the FSMD-based datapath allocation table shown in Figure 5.13. Finally, BIF's control-based state table specifies the symbolic control table for a design in a format closely modeled on Figure 5.14. The BIF representation allows a user to interact with the synthesis tasks by permitting a focused view of the complete design during each phase of synthesis, while the annotated CDFG allows automatic synthesis tools to access the complete design information.

The CMUDA system [TLWN90] uses ISPS as the input language (see Section 4.6.1) and the Value Trace (VT) as the intermediate representation. The synthesized datapath structures, along with the control information, are stored in a separate representation called the Structure and Control Specification (SCS). Binding links are maintained between these three levels of design (i.e., the ISPS description, the VT representation, and the SCS design) by using a tagging scheme [ThBR87]. The CMUDA system also provides a graphical view of bindings between RT structures, the VT carriers, and the input ISPS description, with a separate window for each representation. Users can visualize different types of bindings by clicking on specific icons (e.g., a register) to find out its corresponding binding links (e.g., to ISPS variables or to control states).

The SSIM design representation model [CaTa89] maintains similar links between the following four graphs to keep track of the synthesis bindings: the control-flow graph, the data-flow graph, the control-automaton graph, and the datapath graph. The control-automaton graph can be viewed as a state transition graph, while the datapath graph represents the structural interconnections of datapath connections. Stok [Stok91] also uses a similar scheme to maintain bindings of behavioral variables and operators to control steps and datapath structures.

5.7 Transformations

The initial flow-graph representation closely matches the input description and consequently may represent syntactic language constructs that are not useful or even relevant for synthesis. We can apply transformations to the original flow graph before the high-level synthesis tasks of scheduling and allocation, creating another flow graph more amenable to these high-level synthesis tasks. Each transformation has different effects. A clean-up transformation removes redundancies based on the input-language syntax. An optimizing transformation improves the representation by appropriately deleting or by replacing segments of the representation with a more efficient one. A restructuring transformation modifies the representation to suit a specific style (e.g., pipelining) or to achieve different degrees of parallelism.

Transformations can be applied to the design representation at various stages of high-level synthesis. During compilation of the input description into a flow graph, several compiler optimizations can be performed to remove the extraneous syntactic constructs and redundancies in the HDL specification. Flow-graph transformations are used to convert parts of the representation from one style (e.g., control flow) to another (e.g., data flow) and to change the degree of parallelism. Hardware-specific transformations use properties of register-transfer (RT) and logic components to perform optimizations (e.g., replacing a data-flow graph segment that increments a variable with an increment operation). We will briefly review each type of transformation and illustrate a few transformations using simple examples.

5.7.1 Compiler Transformations

Since the flow-graph representation typically comes from an imperative language description, several standard compiler optimization techniques can be performed on the representation [AhSU86]. Figure 5.15 shows two sample compiler optimizations. Constant folding (Figure 5.15(a)) replaces a flow-graph segment containing an arithmetic operator having constant operands with the result of the operation (i.e., a single constant). We can propagate the resulting constant down a chain of data-flow arcs to perform further folding of constants; this process is called constant propagation.

Redundant operator elimination (Figure 5.15(b)) is another compiler optimization that deletes an operator in a flow graph if there is another flow-graph operator with identical inputs. We can generalize this transformation to search for and eliminate common subexpressions that occur at the outputs of all operators in the flow graph. Global data-flow analysis on the flow graph can uncover many more optimizations, such as dead code elimination, loop invariant detection and code hoisting.

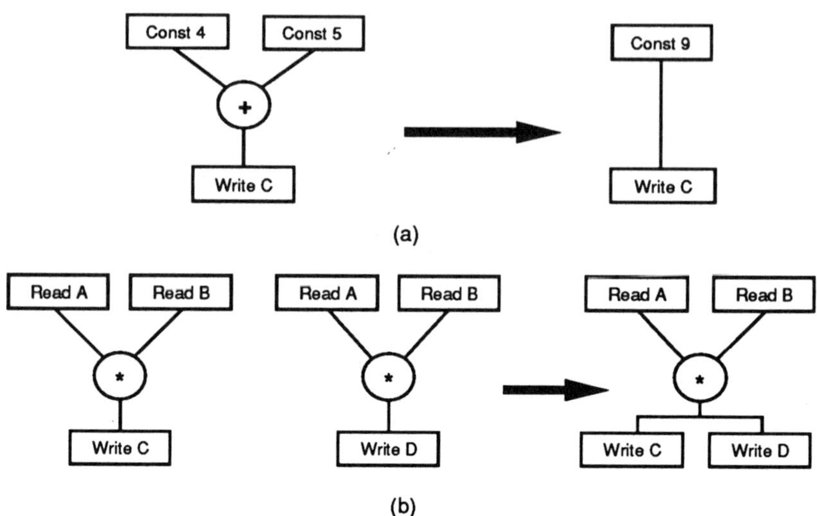

Figure 5.15: Sample compiler optimizations: (a) constant folding, (b) redundant operator elimination.

Arrayed variables are another rich source of compiler-level optimizations for high-level synthesis [OrGa86]. Since arrays in the behavioral representation get mapped to memories, a reduction in the number of array accesses decreases the overhead resulting from accessing memoried structures. Array references with variable indices result in long chains of dependencies between consecutive array accesses since the variable value may result in identical index values along the chain. We can use optimizing transformations to disambiguate and reduce these chains of dependencies. If a chain is on a critical path, such an array optimization reduces the length of the critical path and improves the performance of the design.

Certain compiler transformations are specific to the HDL used for describing the design. When VHDL is used for design description, we can identify specific syntactic constructs and replace them with attributes on signals and nets to indicate their functions [BhLe90, LiGa91]. For instance, Figure 5.16(a) shows a sample segment of VHDL code that tests for the rising edge of signal X. The compiler initially generates a flowgraph structure that includes nodes for reading constants *1* and *stable*, as well as logic operator nodes such as *AND* and *NOT* for evaluating

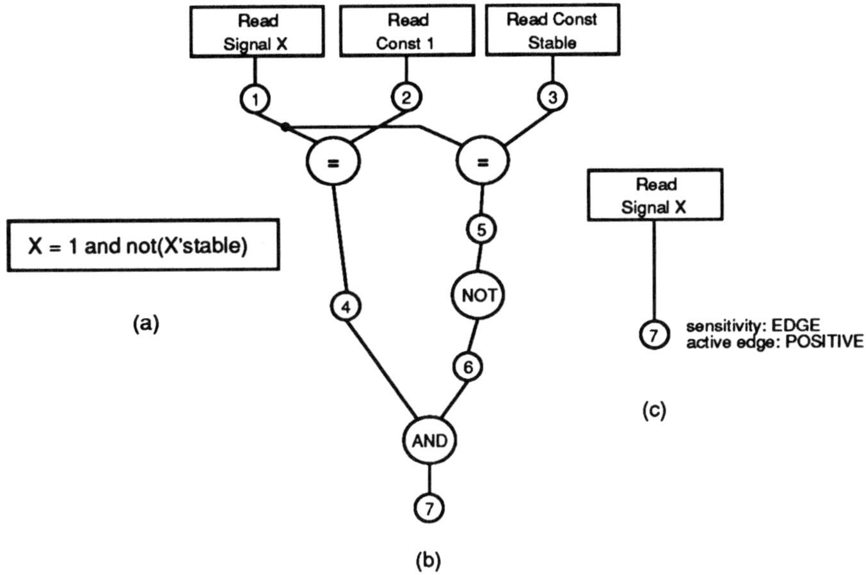

Figure 5.16: Signal attribute transformation: (a) code for rising edge of signal X, (b) initial FG, (c) transformed FG.

the rising-edge condition (Figure 5.16(b)). The small numbered circles in the data-flow graphs of Figure 5.16 indicate signals. Since we simply want to indicate a test for the rising edge of signal X, the flow-graph transformation collects the attributes for the signal change (e.g., *sensitivity* and *signal change*) and attaches these attributes to the output arc of the *READ* operation (signal 7 in Figure 5.16(c)).

5.7.2 Flow-Graph Transformations

We can perform several transformations at the flow-graph level to improve the parallelism of the design. Some flow-graph transformations are independent of the representation schemes, while others are applicable to specific flow-graph schemes. We discuss three types of flow-graph transformations: tree height reduction, control-flow to data-flow transformation, and flattening of control/data flow graphs.

Tree height reduction uses the commutativity and distributivity properties of language operators to decrease the height of a long expression chain, and exposes the potential parallelism within a complicated data-flow graph [HaCa89, PoNi91].   Figure 5.17 shows an example of tree

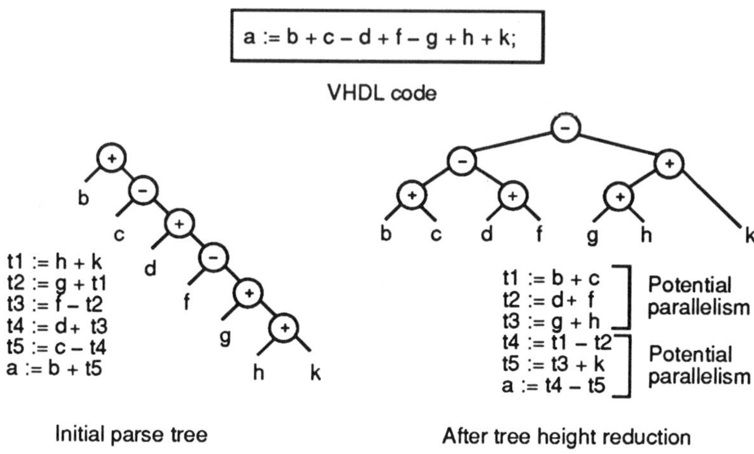

Figure 5.17: Tree height reduction of expressions.

height reduction as applied to an assignment statement with six operators. The initial parse tree generated by the compiler has a height of six and does not show the potential parallelism within this assignment statement. After tree height reduction, we get a tree of height three on which several operations can be performed in parallel.

Another flow-graph transformation (specific to the CDFG representation scheme) converts control-flow constructs to an equivalent data-flow representation. This transformation is useful when we test boolean variables in control flow. Since every control-flow conditional test forces the creation of a new state during synthesis (see Chapter 2),we get several dummy states corresponding to the testing of boolean variables in control flow. We can eliminate these dummy states by transforming boolean control-flow tests to a data-flow representation. Figure 5.18 illustrates this transformation on a sample *if* statement. Figure 5.18(b) shows the control-flow representation derived from the input description in Figure 5.18(a). An added benefit of this transformation is increased explicit parallelism in the data-flow graphs. Since the data-flow blocks in Figure 5.18(b) are disjoint, synthesis tools may not fully exploit the parallelism between the mutually exclusive data-flow blocks *Stmt_Blk1* and *Stmt_Blk2*. Converting this control-flow representation to the equivalent data-flow representation shown in Figure 5.18(c) makes explicit the parallelism within a single data-flow graph.

If the input description is hierarchical, the CDFG can be recursively flattened into a data-flow graph. This hierarchical transformation is particularly useful when all the control-flow conditions test boolean variables; we can eliminate several dummy states in the final schedule. Figure 5.19(a) shows a sample hierarchical CDFG. We can apply the control-flow to data-flow transformation recursively on this flow graph to flatten the representation into a single data-flow graph, as shown in Figures 5.19(b) and (c). Recall that the CDFG scheme captures loop constructs in the control-flow graph. Loops with fixed bounds can thus be unrolled to generate larger data-flow graphs. However, if a loop's bound is not fixed at compile-time, we cannot completely unroll the loop. In such cases, we cannot transform the whole flow graph to a single data-flow block as shown in Figure 5.19. However, loop unwinding techniques [GoVD89] can be used to partially unroll the loop to expose more parallelism.

5.7.3 Hardware-Specific Transformations

Hardware-specific transformations at the logic, RT and system levels can be applied to the intermediate representation. In general, these are local transformations that use properties of hardware at different design levels to optimize the intermediate representation.

At the logic level, we can apply local Boolean optimization techniques [DaJo80] to the intermediate representation. These transformations use properties of Boolean algebra to locally optimize parts of a flow graph by pattern matching and replacement. Figure 5.20 shows a sample piece of flow graph that gets reduced to a simple data transfer because the equivalent Boolean function reduces to a single Boolean variable. Operations that compare values to zero or to constants can also be optimized in this fashion.

At the RT level, we can use pattern matching to detect and replace portions of the flow graph with simpler flow-graph segments. The pattern matching transformations are based on RT semantics of the hard-

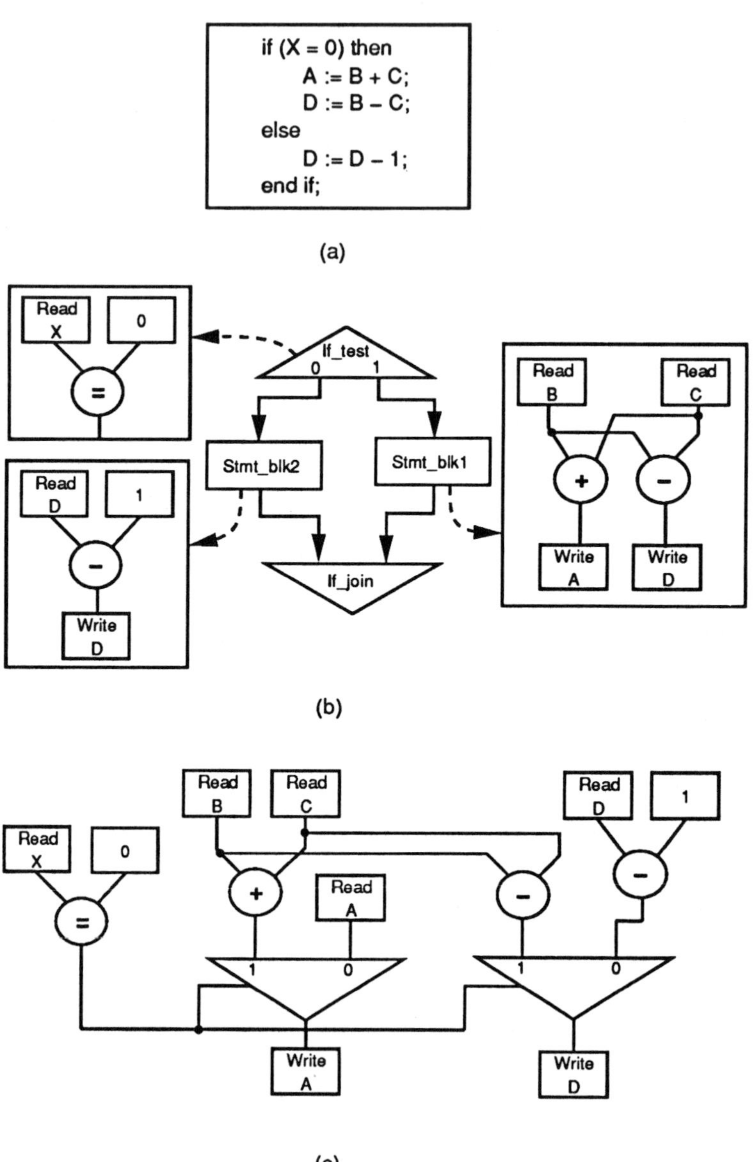

Figure 5.18: CF to DF transformation: (a) "if" statement, (b) initial CF representation, (c) transformed DF representation.

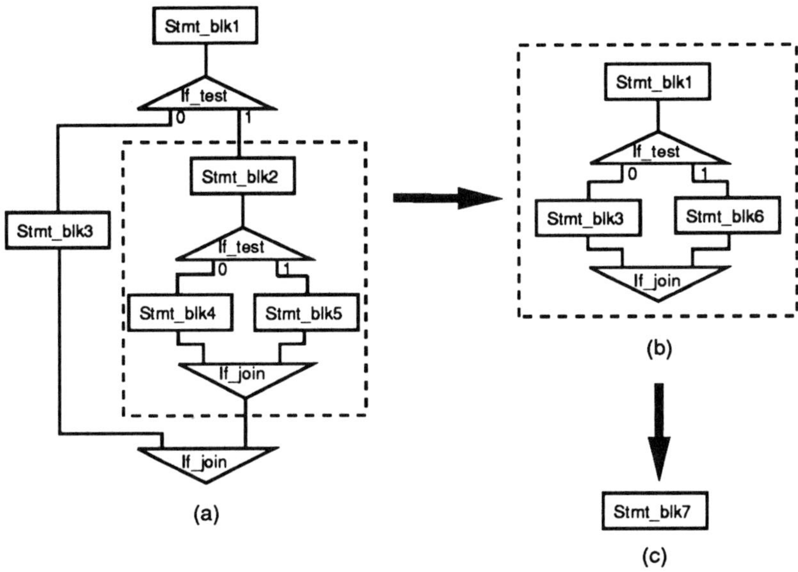

Figure 5.19: Flattening of control flow: (a) initial flow graph, (b) innermost "if" flattened, (c) final DF block generated.

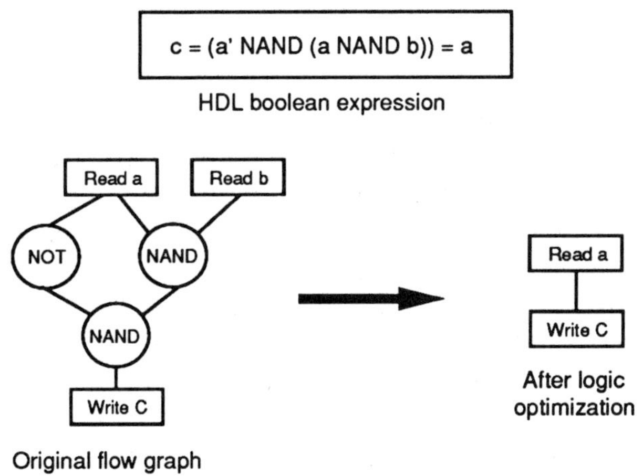

Figure 5.20: Logic-level flow-graph transformations.

ware components corresponding to the flow-graph operators [Rose86]. The transformations can be extended to recognize whole segments of the flow graph that correspond to an RT functional unit. Figure 5.21(a) shows a simple example in which signals *A* and *B* are added together and the result is incremented. If the RT library has a functional unit that combines an addition with an increment, we can replace the flow-graph segment with the *add_inc* operation shown in Figure 5.21(b). Similarly, transformations can be used to convert multiplication or division by powers of 2 into shift operations (i.e., strength reduction).

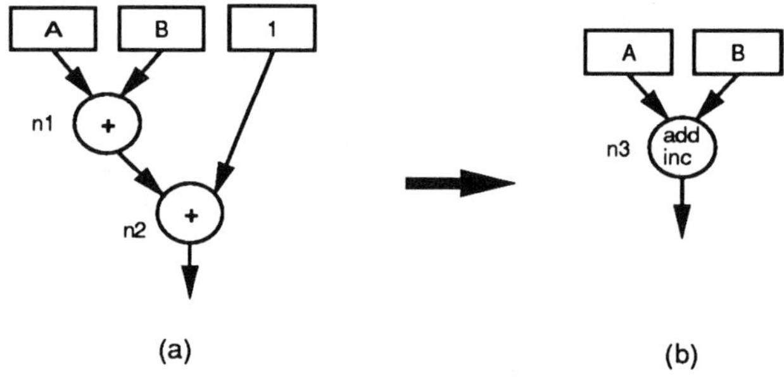

Figure 5.21: Simple RT-level flow-graph transformation: (a) flow-graph pattern, (b) library-specific function node.

A series of such transformations can be used to maximally merge flow graph nodes into commonly used RT-level functional units such as ALUs [RuGB90]. Figure 5.22(a) shows a segment of VHDL code in which signal *OUT* is assigned a different value based on *F*'s value. Figures 5.22(b), (c) and (d) show how the corresponding data-flow graph representation of the VHDL code gets progressively merged into one multifunctional node that corresponds to an adder-subtracter functional unit. These optimizations are particularly useful when the RT functional unit library contains complex multifunctional units such as ALUs. In such cases, we can view this series of transformations as a local preprocessing phase for allocation, where groups of operations corresponding to a complex RT component are transformed into a complex function node; the allocator can then directly map the complex function node to the RT component.

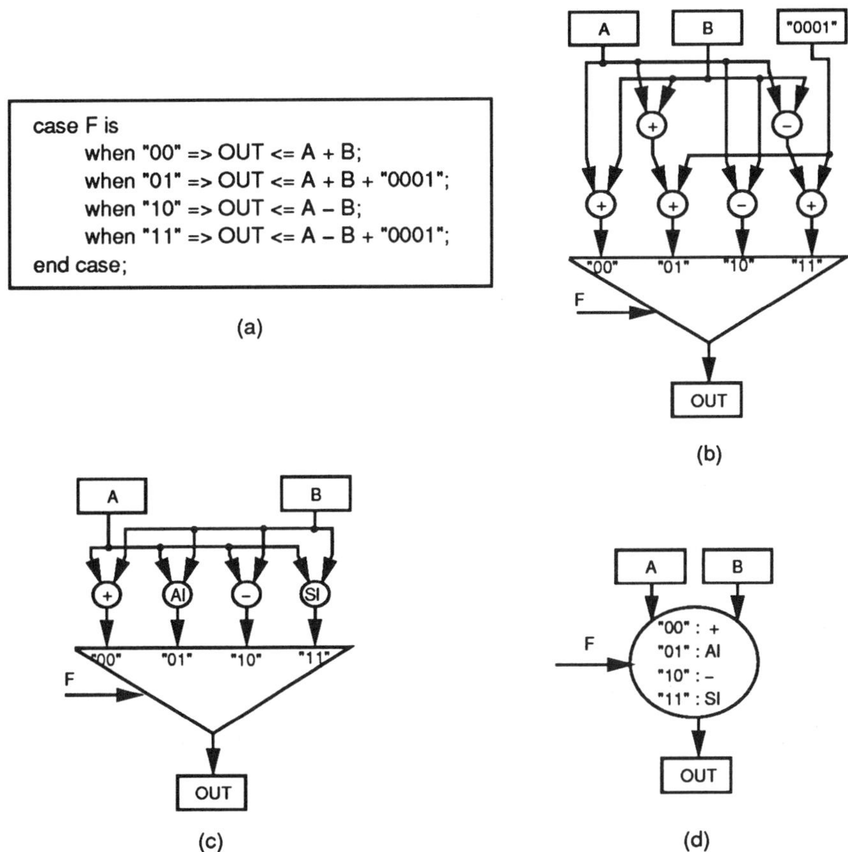

Figure 5.22: Complex function recognition: (a) VHDL description, (b) initial DF representation, (c) simplified DF representation with "add_inc" and "sub_inc" operators, (d) final DF representation with ALU operator.

System-level transformations can be used to divide parts of the flow graph into separate processes that run concurrently or in a pipelined fashion [WaTh89]. Transformations at this level can also be used to partition parts of the flow graph into functional blocks that get mapped into chips or physical partitions in the final layout. Chapter 6 further discusses different types of partitioning techniques and their use in high-level synthesis.

5.8   Summary and Future Directions

This chapter described the need for a canonical intermediate representation that allows efficient mapping of input HDLs to different target architectures using a variety of synthesis tools. We reviewed the design trajectory for high-level synthesis from abstract behavior to a synthesized structural datapath with a symbolic controller. We used a simple shift-multiplier example to highlight the information needed to maintain the complete design during the high-level synthesis process. Next we described the compilation of input HDL specifications into a flow-graph representation and outlined some specific representation schemes for the HDL behavior, RT structure and symbolic controller. We then described a canonical intermediate representation that captures both design behavior and the partial results of synthesis, and provides design views for high-level synthesis. We concluded the chapter with a description of various transformations that can be applied to optimize the design before the scheduling and allocation tasks in high-level synthesis.

Although there have been several attempts at developing a common intermediate representation for high-level synthesis, future research is needed to accommodate different HDLs as well as different synthesis tools targeted towards specific architectures. With the growing acceptance of HDL standardization efforts (e.g., VHDL), we need representation schemes that support the tasks of synthesis using these standardized HDLs. Further research is also necessary on uniform representation of both synchronous and asynchronous behavior, as well as on representing design constraints consistently through all phases of synthesis.

An important area of research, crucial to the commercial success of high-level synthesis, is the development of a mixed-mode interactive synthesis environment. This research will require the development of intermediate representations that support manual design, as well as formats that can be used to exchange designs between human designers and synthesis tools.

In the area of design transformations, more work is needed to understand the real effects of individual transformations on the final design. Furthermore, we need to understand the effects of coupling and ordering individual transformations on the synthesized design. This will allow us to develop transformation scripts that combine individual transformations to suit a particular design style or target architecture.

5.9   Exercises

1. Consider the VHDL description of the shift-multiplier given in Figure 4.17. Set $A\_PORT =$ "1111" and $B\_PORT =$ "1100".
   (a) Execute the VHDL description by hand and compute the result M_OUT.
   (b) Is the result correct? If not, identify the error in the VHDL description.
   (c) Rewrite the VHDL description to correctly capture the behavior of the unsigned shift-multiplier design.
   (d) Schedule this new description using the unit selection given in Figure 5.4.
   (e) Complete the design of the shift-multiplier by generating a detailed RT structure and the symbolic microcode table shown in Figures 5.5 and 5.14, respectively.

2. Using the new shift-multiplier description from the previous problem, change the selection of Figure 5.4 to include four registers, one shifter and one ALU. Schedule the shift-multiplier description using this unit selection, and manually synthesize the RT structure and control.

3. Develop algorithms for merging parse trees generated from statements in
   (a) concurrent code, and
   (b) sequential code (see Section 5.3).

4. Represent the shift-multiplier behavior described in Figure 4.17 using
   (a) DeJong's data-flow scheme (e.g., Figure 5.11(b)), and
   (b) SSIM flow-graph scheme (e.g., Figure 5.11(c)).

5. Consider the segment of sequential VHDL code given below:

$$
\begin{aligned}
Y &:= X(2) + Z; \\
Y &:= X(2) + Z; \\
X(2) &:= 5; \\
X(7) &:= Y + Z; \\
X(I) &:= W + 4;
\end{aligned}
$$

(a) Use the disjoint CDFG scheme to represent this description, assuming that control dependencies exist between successive array accesses.

(b) Since several array accesses in the description have constant indices, optimize the representation in (a) by deleting unnecessary precedence arcs.

(c) Develop a general scheme to optimize the precedence ordering of array accesses in the CDFG representation. Hint: [OrGa86].

(d) Use DeJong's method to represent the description.

(e) Optimize this representation, if possible.

(f) Develop a general scheme to optimize the precedence ordering of array accesses in DeJong's hybrid data-flow representation. Hint: [Stok91].

6. *Develop an algorithm to optimize array references with linear recurrences in the variable indices. Hint: [Bann79].

7. Represent the loop timing constraint of 3 clock cycles for the IMPERATIVE procedure *calc* shown in Figure 4.4. Using the SSIM scheme. How would you partition the timing constraint within statements (or operations) within the loop? Hint: [CaKu86].

8. **Develop a general scheme for partitioning design performance (or throughput) timing constraints into individual timing constraints on flow graph operators.

9. Consider the timing diagram for the memory board read cycle given in Figure 4.10. Note that this diagram describes the behavior of an asynchronous bus interface as shown in Figure 4.9.

    (a) How would you represent the asynchronous behavior in a flow-graph scheme?

    (b) *Develop a general representation that combines both internal synchronous behavior and peripheral asynchronous behavior using an extension of the disjoint CDFG scheme. Hint: [Borr88].

10. **Develop a uniform representation scheme for synthesis that captures different types of timing constraints such as timeouts, event ordering, throughput delays, set-up and hold delays, individual operator delays, etc. Hint: [AmBS91].

11. *Develop an algorithm for tree-height reduction.

12. *Develop an algorithm for optimal loop unwinding. Hint: [AiNi88].

13. Develop an algorithm for flattening nested "if" statements in a structured HDL description, using the CDFG representation.

14. Establish a set of criteria to determine when the CF-to-DF transformations are beneficial.

15. *Develop an algorithm for complex function recognition in a CDFG for a VHDL concurrent block description:

    (a) Using a clique-partitioning approach (hint: [RuGB90]), and

    (b) Without using a clique-partitioning approach.

Chapter 6

Partitioning

6.1 Introduction

In the context of computer-aided design, partitioning is the task of clustering objects into groups so that a given objective function is optimized with respect to a set of design constraints. Partitioning has been used frequently in physical design. For example, it is often used at the layout level to find strongly connected components that can be placed together in order to minimize the layout area and propagation delay. It can also be used to divide a large design into several chips to satisfy packaging constraints.

Partitioning is used in high-level synthesis (HLS) for scheduling, allocation, unit selection, and chip and system partitioning. First, partitioning can be used to cluster variables and operations into groups so that each group is mapped into a storage element, a functional unit or an interconnection unit of the real design. The result of this partitioning can be used for unit selection before scheduling and binding, or it can be used for allocation. It is particularly useful in unit selection since the sum of all unit areas gives a rough estimate of the chip area. Similarly, partitioning can be used to cluster operations into groups so that each group is executed in the same control state or control step. This type of partitioning is used for scheduling.

Second, partitioning is used to decompose a large behavioral description into several smaller ones. One goal is to make the synthesis problem more tractable by providing smaller subproblems that can be solved efficiently. Another goal is to create descriptions that can be synthesized into a structure that meets the packaging constraints.

In this chapter, we will first describe several basic partitioning methods. Then we will describe the application of these partitioning methods to high-level synthesis. Finally, we will suggest some future directions for high-level partitioning.

6.2   Basic Partitioning Methods

Generic partitioning techniques are based on a graph model of the design. Application of these basic partitioning methods to HLS requires that we map behavioral descriptions or design structures into graph models. In this section, we will demonstrate how to create graph models of behavioral descriptions and highlight several common graph partitioning algorithms.

6.2.1   Problem Formulation

To illustrate how specific partitioning problems can be formulated as graph partitioning problems, consider the partitioning of a logic circuit into two blocks. Figures 6.1(a) and (b) show a design structure and its corresponding graph representation. Each node in the graph represents a physical component, such as a gate, flip-flop, register or adder, and each edge represents a physical connection between two components. Multi-terminal nets between several components are decomposed into several two-terminal nets. For example, net n_i in Figure 6.1(a) is represented by two edges e_{24} and e_{25} in the graph of Figure 6.1(b). After partitioning, each subgraph represents a gate cluster, a module, a chip or a subsystem. The edges across the cut line between the two subgraphs represent the external connections between two physical packages. The main objective of partitioning is to decompose a graph into a set of subgraphs to satisfy the given constraints, such as the size of the subgraph, while minimizing an objective function, such as the number of edges connecting two subgraphs. Figure 6.1(b) shows graph G divided into two subgraphs G_1 and G_2 corresponding to $Chip_1$ and $Chip_2$ respectively. Edges e_{24} and e_{36}

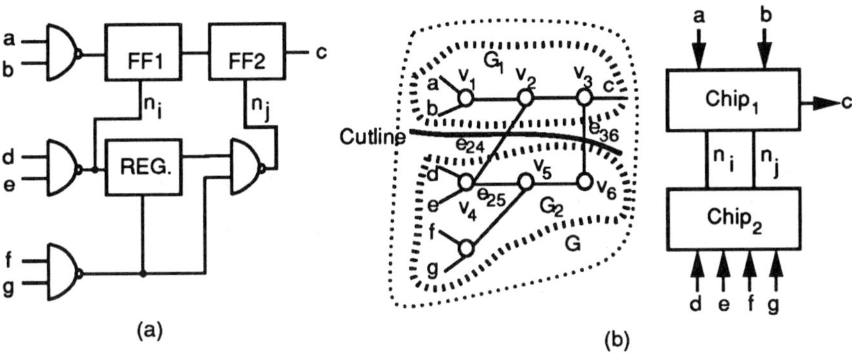

Figure 6.1: Circuit partitioning: (a) a simple example, (b) graph and physical representations.

across the cut-line represent the two nets, n_i and n_j, connecting $Chip_1$ and $Chip_2$.

Similarly, behavioral partitioning can be formulated as a graph partitioning problem. In general, design behavior can be described using two different forms: a textual description or a control/data flow graph (CDFG). A typical system design implementation consists of a set of interconnected components such as processors, memories, controllers, bus arbiters and interface circuits. A behavioral description describes each component using one or more processes. The communication between processes can be specified through a set of global variables. Figure 6.2(a) shows a VHDL behavioral description that contains three concurrent processes. Each process contains sequentially ordered statements, including "if" and "case" statements, assignment statements, procedures and functions. The global *port* and *signal* variables, $(I1, I2, I3, O1, B, F, H)$, are used for communication between processes. Each process can be viewed as a hardware module that executes the process description. Modules can be represented as nodes in a graph where edges represent global variables used for communication between processes (Figure 6.2(b)).

The CDFG description explicitly shows control and data dependencies as described in Chapter 5. Figure 6.2(c) shows the control/data flow graph derived from the VHDL behavioral specification. Each circle represents an operation, each triangle represents a control decision,

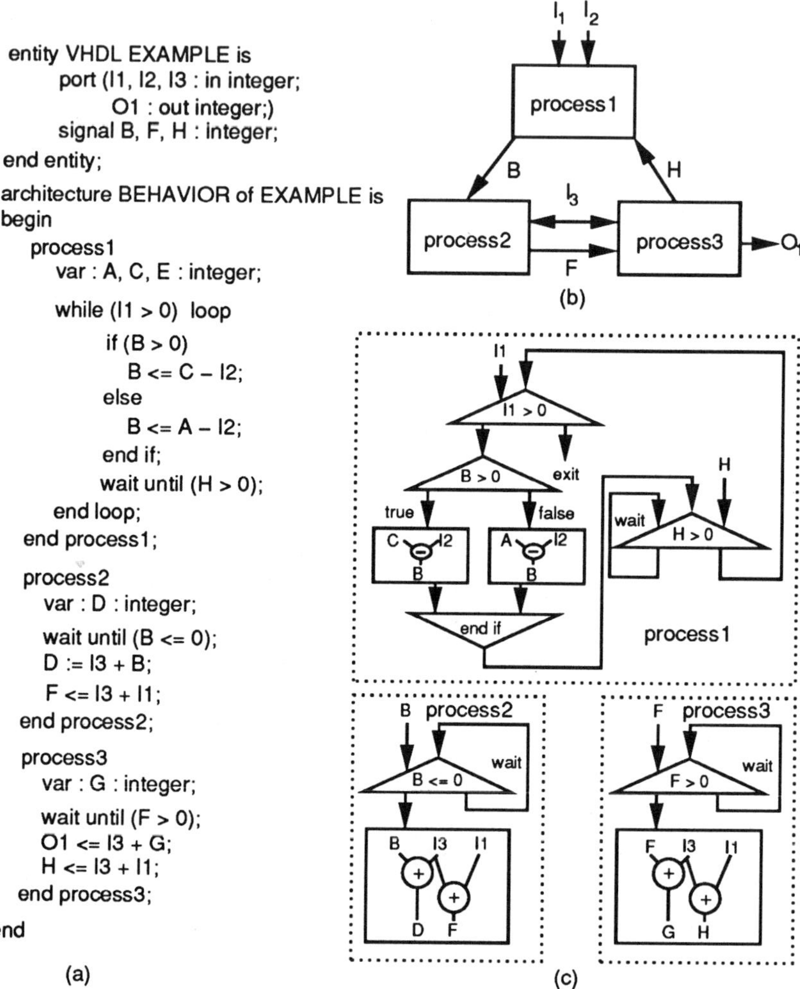

Figure 6.2: A VHDL example: (a) behavioral description, (b) process communication, (c) control/data flow graphs.

and each edge represents a control or data dependence. Control decisions and blocks of straight-line code can be represented as nodes in control-flow graph with edges representing control dependencies. Similarly, operations can be viewed as nodes in a data-flow graph with edges representing data flow.

Behavior graphs and control/data flow graphs differ in the granularity of their nodes. Each node in a behavior graph represents a program or description block that may include several assignment statements, functions or procedures, while a node in a control/data flow graph represents an operation or decision in the program or description.

Both graphs can be partitioned along the time or physical dimension, that is, for performance or for physical size. When partitioning for performance we cluster graph nodes on critical paths while minimizing communication defined by the number of times control or data is passed between clusters. This partitioning may result in several control paths being executed concurrently under one global controller or several local controllers (corresponding roughly to SIMD and MIMD styles of parallel processing).

On the other hand, when partitioning for physical cost we cluster graph nodes by the type of operations they perform while minimizing connectivity (i.e., the number of wires) between different clusters. Note that communication and connection minimization are two different and independent goals. We can have high communication with low connectivity and low communication with high connectivity, as illustrated by a serial communication over a single wire and parallel communication over a bus for the same data rate of n bits/second.

The difference between partitioning for performance and for physical cost can be explained by its efficiency in time and space. Partitioning for performance optimizes time utilization while partitioning for physical cost optimizes component utilization.

In general, there are two basic partitioning techniques: constructive methods and iterative-improvement methods. The constructive method partitions the graph by starting with one or more seed nodes and adding nodes to the seeds one at a time. The iterative-improvement method starts with some initial partition, and then successively improves the results by moving objects between partitions.

In the following sections, we first describe three constructive methods: random selection, cluster growth and hierarchical clustering. Then we describe two iterative improvement methods: min-cut partitioning and simulated annealing.

6.2.2   Random Selection

Random selection is the simplest constructive method and is frequently used to generate an initial partition for an iterative-improvement algorithm. The algorithm randomly selects nodes one at a time and places them into clusters of fixed size until the proper size is reached. This procedure is repeated until all nodes have been placed. Such an algorithm is fast and easy to implement; however, it often produces poor results.

6.2.3   Cluster Growth

Cluster-growth algorithms start with a non-partitioned set of objects and place objects into a given number of clusters according to some closeness measure. A cluster growth algorithm consists of three tasks: seed selection, unplaced node selection and node placement. The first step of cluster growth selects the seed nodes for each cluster in order to guide the clustering process. The seed nodes may be specified by the designer, chosen randomly or selected based on attributes such as size or number of connecting edges. Second, the algorithm determines the clustering order of the unplaced nodes. The order is determined by some closeness measures, such as the number of connections between unplaced and placed nodes. Finally, the algorithm adds the unplaced node with the highest closeness score to the appropriate cluster. This process is repeated until all nodes are placed into clusters.

Let $G = (V, E)$ be a graph with a set of nodes V and a set of edges E. For each pair of nodes $v_i, v_j \in V$, let c_{ij} be the cost of the edge e_{ij} between v_i and v_j. We will partition such a graph into clusters of m nodes each using the cost of edges as the closeness measure. Variable *num_cluster* represents the number of clusters. The procedure $CONNECTED(V_1, V_2)$ returns the node $v_i \in V_1 \subset V$ for which $\sum_{v_j \in V_2 \subset V} c_{ij}$ is maximal, or in other words, it returns the node which is maximally connected to V_2. Variables *seed* and *temp* are temporary variables for nodes in V.

Algorithm 6.1: Clustering Growth.

> num_cluster = $\lceil | V | /m \rceil$;
> **for** $i = 1$ to num_cluster **do**
>   /*Select a seed node*/
>   seed = CONNECTED(V, V);
> $V_i = \{\text{seed}\}$;
> $V = V - \{\text{seed}\}$;
>   /*Construct a cluster of m nodes*/
>   **for** $j = 1$ to $m - 1$ **do**
>   temp = CONNECTED(V, V_i);
> $V_i = V_i \cup \{\text{temp}\}$;
> $V = V - \{\text{temp}\}$;
>   **endfor**
> **endfor**

The cluster-growth algorithm is easy to implement but produces poor results because the closeness measure is computed only for nodes already assigned to clusters and not for all nodes. Like the random-selection method, this method is frequently used to generate an initial partition for an iterative-improvement algorithm.

6.2.4 Hierarchical Clustering

The hierarchical-clustering algorithm considers a set of objects and groups them according to some measure of closeness [Snea57, John67]. The two closest objects are clustered first and considered to be a single object for future clustering. The clustering process continues by grouping two individual objects, or an object or cluster with another cluster on each iteration. The algorithm stops when a single cluster is generated and a hierarchical cluster tree has been formed. Thus, any cut-line through the tree indicates set of subtrees or clusters to be cut from the tree.

As an example, consider the set of five vertices $\{v_1, v_2, v_3, v_4, v_5\}$ shown in Figure 6.3(a). The set of five edges $\{e_{12}, e_{13}, e_{23}, e_{24}, e_{25}\}$ indicates the non-zero closeness measure between vertices. The closeness values are shown as edge weights and as a matrix in Figure 6.3(a). For instance, the matrix shows a maximum closeness of 6 between vertices v_2 and v_4. Thus, v_2 and v_4 are merged into vertex $v_{(24)}$ first.

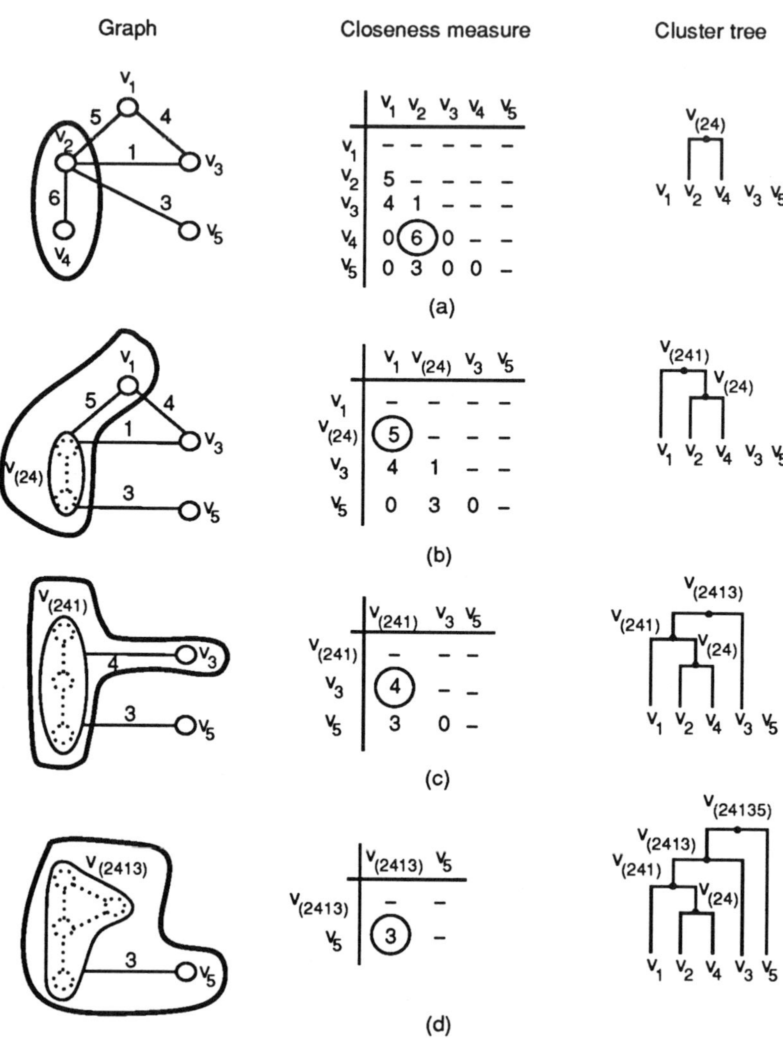

Figure 6.3: Cluster tree formation.

The closeness measures are recomputed for the new set of vertices as shown in Figure 6.3(b). Vertices $v_{(24)}$ and v_1 are merged next as shown in Figure 6.3(b). Note that, the closeness score is based on the maximum weight between two nodes or two clusters. Thus, the closeness measure between the vertex v_3 and the cluster $v_{(241)}$ is $MAX(4,1) = 4$ as shown in Figure 6.3(c). Finally, a single cluster is formed by merging nodes v_5 and $v_{(2413)}$ (Figure 6.3(d)). The final cluster represents the clustering tree. Any cut-line through the clustering tree defines a partition of the original graph. A cut-line close to the leaves of the tree generates many clusters with few "close" nodes in each cluster, whereas a cut-line close to the root generates fewer clusters with more distant nodes.

Let $G = (V, E)$ be a graph where V is a set of n vertices and E is a set of edges representing non-zero closeness of two vertices. For each $e_{ij} \in E$, c_{ij} represents the closeness measure of vertices v_i and v_j. Let $N = \{n_k \mid 1 \leq k < n\}$ be the set of nonterminal nodes in the clustering tree. We denote by $LSucc(n_k)$ and $RSucc(n_k)$ the left and the right successors of each nonterminal node n_k and by $L(n_k)$ the level of the nonterminal node n_k. Let $C = \{c_{ij} \mid 1 < i, j <\mid V \mid , i \neq j\}$ be a closeness matrix. $L(v_i)=0$ for all vertices $v_i \in V$. The function $CLOSENESS(v_i, v_j)$ computes the closeness measure for any two vertices in V. The procedure $MAXCLOSE(C)$ returns the maximal closeness value in the closeness matrix C. If two pairs of vertices have the same closeness measure the one with fewer edges is returned. v_{max1} and v_{max2} are temporary variables storing vertices with maximal closeness on each iteration. Similarly, $c_{max1,max2}$ stores the maximal closeness value. The hierarchical clustering algorithm is outlined in Algorithm 6.2.

A variation of the hierarchical clustering method is the multi-stage clustering method used in architectural partitioning [LaTh91]. This algorithm performs clustering in several stages. Each clustering stage uses a different closeness criteria, such as operator similarity, connection commonality or operation concurrency. The clustering stages are successively applied so that each stage uses clusters generated by the previous stage.

For example, Figures 6.4(a) and (b) show a graph and its corresponding clustering tree based on *Criterion A* and *Criterion B* respectively. To take into account various criteria, the multi-stage clustering algorithm first implements the clustering procedure based on the highest priority

Algorithm 6.2: Hierarchical Clustering.

for all $v_i \in V$ **do** $L(v_i) = 0$; **endfor**
/*Compute closeness measure*/
for all $v_i, v_j \in V$ **do** $c_{ij} = \text{CLOSENESS}(v_i, v_j)$; **endfor**
/*Construct the cluster tree*/
for $k = 1$ to $n - 1$ **do**
    $c_{max1,max2} = \text{MAXCLOSE}(C)$;
    $V = V \cup \{n_k\}$;
    $V = V - \{v_{max1}, v_{max2}\}$;
    $LSucc(n_k) = v_{max1}$;
    $RSucc(n_k) = v_{max2}$;
    $L(n_k) = MAX(L(v_{max1}), L(v_{max2})) + 1$;
    /*Compute closeness measure for n_k*/
    for all $v_m \in V$ **do**
        $c_{km} = \text{MAX}(c_{max1,m}, c_{max2,m})$;
        $c_{max1,m} = c_{max2,m} = 0$;
    endfor
endfor

criterion, then the second highest priority criterion, and so on.

For instance, assuming that *Criterion A* has higher priority than *Criterion B*, the algorithm first produces the cluster tree based on *Criterion A* (Figure 6.4(a)). The algorithm then cuts the resulting cluster tree at the first level which produces four new clusters, $\{f,c\}$, $\{a,e\}$, $\{d\}$, and $\{b\}$. Next, the algorithm groups these clusters further based on *Criterion B* (Figure 6.4(c)). But, if *Criterion B* has higher priority than *Criterion A*, the algorithm can first produce the cluster tree based on *Criterion B* (Figure 6.4(b)). The algorithm then cuts the resulting cluster tree at the first level which produces four new clusters $\{c,e\}$, $\{f\}$, $\{b\}$, and $\{a,d\}$. Then, the algorithm groups these clusters further based on *Criterion A* (Figure 6.4(d)).

We can also implement the clustering procedure on groups at different levels of the cluster tree. For example, Figure 6.4(e) shows result of cutting the cluster tree based on *Criterion A* at the first and second levels and continuing with *Criterion B* afterward.

The main advantage of the multi-stage clustering algorithm is that

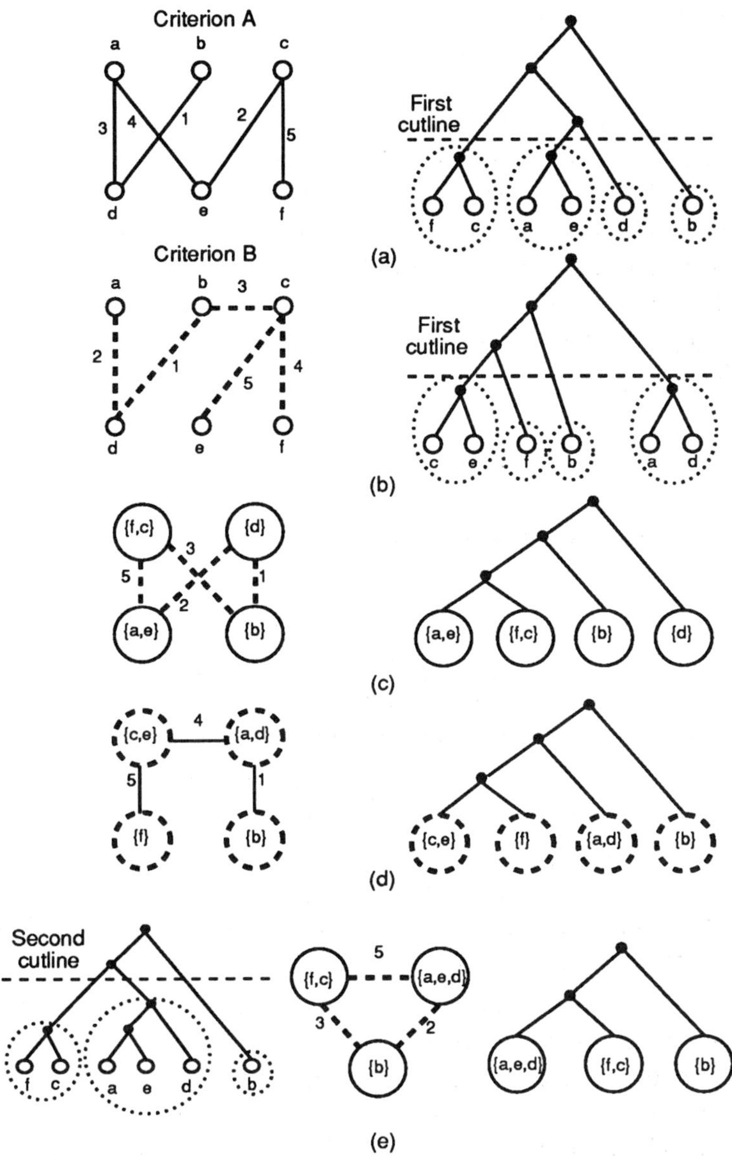

Figure 6.4: Clustering based on (a) criterion A only, (b) criterion B only, (c) criterion A below the first cut-line, then criterion B, (d) criterion B below the first cut-line, then criterion A, (e) criterion A below the second cut-line, then criterion B.

the clustering criteria can be decoupled over several stages. In the single-stage clustering, it is difficult to determine the proper weighting for various criteria for different design styles. However, multi-stage clustering provides a hierarchy of criteria that are applied in the order that satisfies given constraints. For example, if we use the single-stage clustering method to group a set of nodes based on three closeness measures (e.g., functional similarity, common connectivity and operation concurrency) at the same time, we have to decide a proper weighting for each closeness measure; this decision requires a lot of trial-and-error tests to ensure good results. On the other hand, if we use the multi-stage clustering method, we can apply different closeness measures at different levels of cluster granularity.

6.2.5   The Min-Cut Partitioning

The min-cut partitioning algorithm (also known as the Kernighan-Lin algorithm [KeLi70]) interchanges two groups of nodes between two partitions of equal size to produce a maximal partitioning improvement. The min-cut algorithm can be used also as the basis for solving general n-way partitioning problems. This algorithm is widely used in many applications since it produces good results in small amounts of CPU time.

The Kernighan-Lin algorithm partitions a given graph $G = (V, E)$ of $2n$ nodes into two equal subgraphs of n nodes minimizing the connections between the two subgraphs. The algorithm starts with an arbitrary partition of V into two subsets V_1 and V_2 (Figure 6.5). On each iteration the algorithm interchanges k pairs ($k \leq n$) of vertices between two sets. It stops when no further improvement is possible. This process can be executed repeatedly with an arbitrary number of initial partitions. Thus, the best solution can be selected among many generated partitions. In this way, the chances of finding a partition close to the global minimum are improved.

Consider the partitioning of a graph $G=(V,E)$ with $2n$ vertices into two subgraphs G_1 and G_2 of n vertices each. The cost of each edge $e_{ij} \in E$ is denoted by c_{ij}. For each vertex $v_i \in V_1$, we define the external cost as

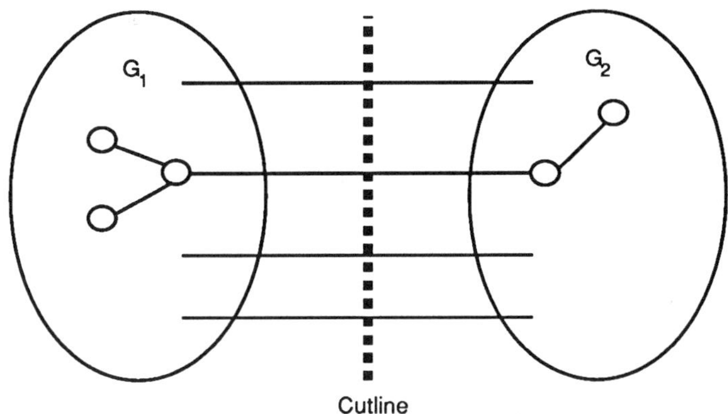

Figure 6.5: Two-way partition.

$$EC_i = \sum_{v_k \in V_2} c_{ik} \tag{6.1}$$

and the internal cost as

$$IC_i = \sum_{v_m \in V_1} c_{im}. \tag{6.2}$$

The difference between external and internal costs is denoted by $D_i = EC_i - IC_i$. Similarly, we can define EC_j, IC_j and D_j for each vertex $v_j \in V_2$.

Let c_{ij} define the number of edges between v_i and v_j. For any two vertices $v_i \in V_1$ and $v_j \in V_2$, we define the gain of interchanging v_i and v_j as

$$gain(v_i, v_j) = D_i + D_j - 2c_{ij}, \tag{6.3}$$

where D_i and D_j are given by Equations 6.1 and 6.2.

The cost c_{ij} contributes to both external costs EC_i and EC_j. After interchanging v_i and v_j, the contribution of those edges to the external cost remains the same. For example, the external and internal costs

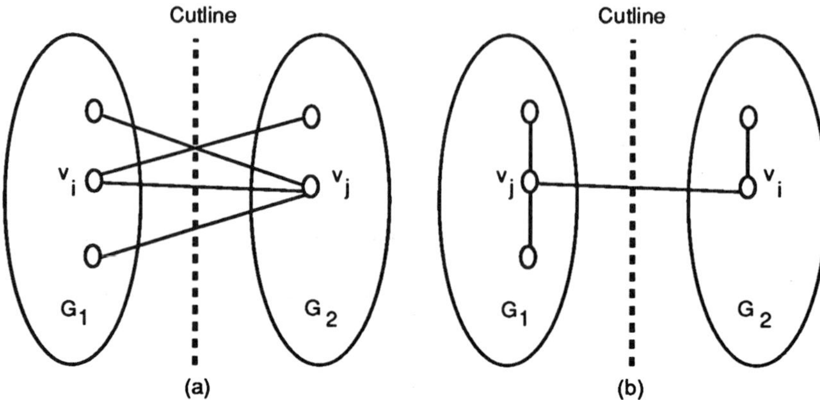

Figure 6.6: Interconnection reduction by node interchanging: (a) before interchanging v_i and v_j, (b) after interchanging v_i and v_j.

of v_i in Figure 6.6(a) are 2 and 0. For vertex v_j, $EC_j=3$ and $IC_j=0$. Thus, $D_i=2$ and $D_j=3$. Since there is a connection between v_i and v_j, c_{ij} is equal to 1. The gain of interchanging vertices v_i and v_j is equal to $2 + 3 - 2 = 3$. Hence, by interchanging v_i and v_j, the total number of connections between V_1 and V_2 is reduced by three (Figure 6.6(b)).

The Kernighan-Lin algorithm (Algorithm 6.3) interchanges a favorable group of vertices rather than interchanging one pair of vertices at a time. The algorithm first arbitrarily partitions vertices into two equal-sized groups. Then it computes the external costs, internal costs and the differences between these costs for all vertices. The algorithm finds a pair of vertices, one from each group, that generates the maximal gain through interchange. It stores the gain, readjusts the cost and locks the selected pair to prevent it from being considered for interchange again. This procedure continues until all n vertices in each subset are paired and a sequence of gains, $gain_1,...,gain_n$, is generated. The total gain of interchanging the first k-pair of vertices, where $1 \leq k \leq n$, is calculated as

$$GAIN(k) = \sum_{i=1}^{k} gain_i. \qquad (6.4)$$

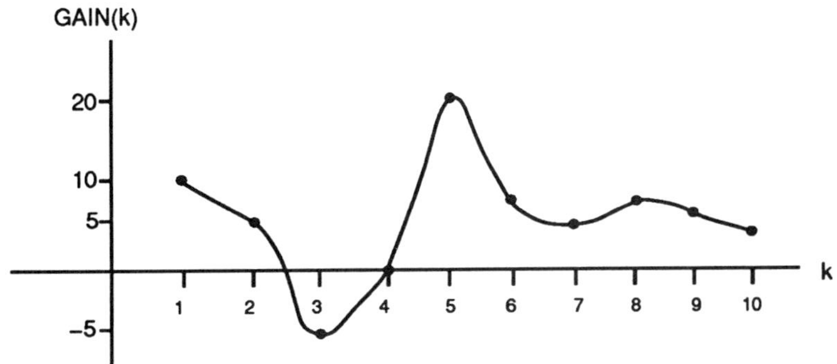

Figure 6.7: Min-cut search strategy.

The algorithm interchanges in reality only the first k pairs of vertices for which $GAIN(k)$ is maximal. If for all k, $GAIN(k)$ is equal to or less than zero the algorithm stops. For example, the sequence of gains in a 20-node graph is shown in Figure 6.7. By interchanging the first pair of vertices, the total gain is 10. By interchanging first and second vertex pairs, the total gain is 5, and so on. In this example, the maximum gain is 20 for $k = 5$. Thus the algorithm will interchange the first five pairs, although the first three interchanges produced a negative gain and the first four interchanges produced no improvement al all.

Let $G = (V, E)$ be a graph with $2n$ vertices. The array *Locked* stores the value 0 if a vertex is available for interchange and 1 if a vertex is locked. Procedure $CLUSTERGROWTH(G, n)$ selects a subgraph of size n. Procedure $EXCHANGE(V_1, V_2, v_i, v_j)$ interchanges vertices $v_i \in V_1$ and $v_j \in V_2$ in two subgraphs G_1 and G_2, respectively. The array *gain* stores the gain for each pair of vertices and arrays *max*1 and *max*2 store the indices of these vertices. Array $GAIN$ stores the accumulated gain for a sequence of n interchanges. Variables *bestk* and *bestGAIN* store the index and value of the maximal accumulated gain, respectively.

Algorithm 6.3: Min-Cut Partitioning.

```
G₁ = CLUSTERGROWTH(G, n);
G₂ = G − G₁;
bestGAIN = ∞;
while bestGAIN > 0 do
    bestGAIN = 0;
    for i = 1 to 2n do
       Locked(i) = 0;
    endfor
    for k = 1 to n do
       gain(k) = 0; GAIN(k) = 0;
       /*Find the best pair of vertices to interchange*/
       for all vᵢ ∈ G₁ AND Locked(i)=0 do
           for all vⱼ ∈ G₂ AND Locked(j)=0 do
               if Dᵢ + Dⱼ − 2cᵢⱼ > gain(k) then
                   gain(k) = Dᵢ + Dⱼ - 2cᵢⱼ;
                   max1(k) = i;
                   max2(k) = j;
               endif
           endfor
       endfor
       Locked(max1(k)) = Locked(max2(k)) = 1;
       /*Update the gain after tentative pairwise interchange*/
       for i = 1 to 2n do
           if vᵢ ∈ G₁ AND Locked(i)=0 then
               Dᵢ = Dᵢ − c_{i,max2(k)} + c_{i,max1(k)};
           endif
           if vᵢ ∈ G₂ AND Locked(i)=0 then
               Dᵢ = Dᵢ − c_{i,max1(k)} + c_{i,max2(k)};
           endif
       endfor
       /*Compute accumulated gain*/
       GAIN(k) = GAIN(k − 1) + gain(k);
       if GAIN(k) > bestGAIN then
           bestk = k;
           bestGAIN = GAIN(k);
       endif
    endfor
```

```
/*Interchange k pairs of vertices*/
for k = 1 to bestk do
    (G_1, G_2) = EXCHANGE(V_1, V_2, v_{max1(k)}, v_{max2(k)});
endfor
endwhile
```

The complexity of the Kernighan and Lin two-way partitioning algorithm is $O(n^2 logn)$ where n is the number of vertices. However, Fiduccia and Mattheyses [FiMa82] used a clever implementation to achieve linear complexity for each iteration of Algorithm 6.3. Dunlop and Kernighan [DuKe85] have compared the Kernighan and Lin algorithm with the Fiduccia and Mattheyses algorithm. They found that the results of Fiduccia and Mattheyses are inferior to those of Kernighan and Lin but that the execution time is substantially shorter.

Kernighan and Lin also extended their two-way partitioning technique to implement multi-way partitioning. Given the problem of partitioning a set S into m subsets, the multi-way partitioning algorithm executes two-way partitioning repeatedly to produce m subsets. For example, consider partitioning a set S with 20 nodes into four subsets of equal size. The algorithm first partitions S into two subsets with sizes $C_A=10$ and $C_B=10$, then further partitions subsets C_A and C_B into four subsets of five nodes each.

6.2.6 Simulated Annealing

Many iterative improvement algorithms use a greedy control strategy in which a pair of nodes is interchanged only if it results in a partitioning improvement. Such a greedy strategy may lead to locally optimal solutions. To avoid getting stuck in a local optimum, a hill-climbing mechanism is needed. The min-cut algorithm avoids a local minimum by accepting a sequence of pairwise exchanges in which some exchanges produce a negative gain as long as the total accumulated gain of the entire sequence is positive. Simulated annealing, proposed by Kirkpatrick et al. [KiGV83], is another control strategy that avoids locally minimal solutions. They observed that the combinatorial optimization process is analogous to the annealing process in physics, in which a material is melted and its minimal energy state is found by slowly lowering the temperature. The same idea can be applied to combinatorial optimization to move out of a locally minimal solution and on towards a globally minimal solution.

The simulated annealing algorithm starts with a random initial partition. Then the algorithm generates moves randomly. A move is defined as the relocation of an object from one partition to another. For each move, the algorithm calculates the new cost and checks whether this new cost satisfies the acceptance criterion based on the temperature T. Note that the acceptance of a move is proportional to the current simulated temperature. As the simulated temperature decreases, the probability of a move being accepted decreases.

Let the procedure $GENERATE$ generate a new partition (S_{new}) from the old partition (S_{old}) by relocating a node. The function F represents the acceptance strategy. In [KiGV83] it is defined as

$$F(\Delta c, T) = min(1, e^{-\frac{\Delta c}{T}}). \qquad (6.5)$$

When Δc is negative (i.e., the new partition is better than the old partition), F returns 1; otherwise, it returns a value in the range of [0,1]. The function RANDOM returns a random value in the range of [0,1]. Therefore, if S_{new} is better than S_{old}, it is definitely accepted; otherwise it is accepted with a probability determined by Δc, T and a random number. Since the random numbers are uniformly distributed, the acceptance rate is inversely proportional to Δc and directly proportional to T. The parameter T is called the annealing temperature. It is high in the beginning of the annealing process and low at the end. It can be updated by the function UPDATE in many different ways; for $0 < \alpha < 1$ a simple yet practical formula is

$$T_{new} = \alpha T_{old}. \qquad (6.6)$$

For a particular annealing temperature, the algorithm tries to improve the solution till the *inner loop criterion* is satisfied. The *inner loop criterion* is satisfied when the solution does not improve for a certain number of iterations. The annealing temperature is updated and the algorithm tries to improve the solution for the new annealing temperature until the *stopping criterion* is satisfied. The *stopping criterion* is satisfied when T is approximately zero. The simulated annealing algorithm is outlined in Algorithm 6.4.

Theoretical studies [RoSa85] have shown that the simulated-annealing algorithm can climb out of a local minimum and find the globally opti-

Algorithm 6.4: Simulated annealing.

```
T = initial_temperature;
S_old = initial_partition;
c_old = cost(S_old);
while "stopping criterion" is not satisfied do
    while "inner loop criterion" is not satisfied do
        S_new = GENERATE(S_old);
        c_new = cost(S_new);
        Δc = c_new - c_old;
        x = F(Δc, T);
        r = RANDOM(0,1);
        if r < x then
            S_old = S_new;
            c_old = c_new;
        endif
    endwhile
    T = UPDATE(T);
endwhile
```

mal solution if the stochastic process reaches an equilibrium state at each temperature, and if the temperature is lowered infinitely slowly. Both of the above conditions also imply that an infinite number of iterations has to be taken at each temperature step. Since it is impractical to find the optimal solution by performing an infinite number of iterations, several heuristic approaches [OtGi84, HuRS86] have been developed to control the simulated-annealing process. These heuristics determine when the equilibrium state is reached and how to lower the temperature. In general, simulated annealing usually produces very good results but suffers from very long run times.

6.3 Partitioning in High-Level Synthesis

Partitioning can be applied to a CDFG for scheduling, allocation and unit selection, or to the system description for decomposing it into chip and module descriptions.

Scheduling algorithms assign operations in a CDFG to control steps while preserving control and data dependencies between operations. Scheduling of operations into control steps can be viewed as a partitioning of operations into a set of clusters in which operations in the same cluster are executed in the same control step. Similarly, allocation can also be formulated as a general graph partitioning problem. For allocation, each cluster of operations represents a particular resource that is shared by all the operations assigned to this cluster. For example, a behavioral description may use an "add" operation 100 times, while the corresponding real design may have only two adders. Thus all the "add" operations must be partitioned into two groups, where the operations in each group execute on one adder. Basic algorithms for scheduling, unit selection and binding will be discussed in Chapters 7 and 8.

The most frequent application of partitioning algorithms to HLS is in unit selection, which defines the functional units to be used for scheduling and binding. Operators in the CDFG are partitioned into clusters according to their functional similarity, where each cluster defines a functional unit executing all the operations in the cluster. Selected functional units are used to obtain estimates of area and performance quality measures (described in Chapter 3) that are in turn used to guide scheduling and binding algorithms.

Another application of partitioning to HLS is in decomposing a behavioral description into a set of interconnected processes. These processes can be mapped to an FSMD or a set of communicating FSMDs as described in Chapter 2. Such a partitioning is similar to the scheduling of processes onto different processors in a multiprocessor. Partitioning algorithms group these processes into clusters so that each cluster is executed sequentially on one of the processors. The main goal of this grouping is to execute the entire cluster under one control unit while satisfying design constraints such as chip size, number of pins, clock rate, system delay and power consumption. Such a partitioning may require modification of the access protocols when extra delays are introduced by off-chip communication. Similarly, pin limitations may require merging of connections and hence necessitate multiplexing of communication channels between different components. Thus, such partitioning must be concurrently performed with interface synthesis.

In this section, we describe the application of partitioning to HLS for module selection and for chip-level partitioning.

6.3.1   Unit Selection for Scheduling and Binding

MacFarland and Kowalski [McFa86, McKo90] have proposed an approach for unit selection using a hierarchical clustering technique. Their algorithm partitions operations in a CDFG into clusters based on their similarity and concurrency. In general, the similarity of two objects is measured by the number of properties they have in common. If A_1 and A_2 are the sets of properties of two objects, then an estimate of the similarity of these two objects can be measured by $\mid A_1 \cap A_2 \mid / \mid A_1 \cup A_2 \mid$, that is, the ratio of the number of common properties to the number of all properties of these two objects. This measure of similarity can be used to define the functional and communication proximity of two operations in the CDFG. Let $fcost(o_1, o_2, .., o_n)$ be the cost of a functional unit executing the operations $o_1, o_2, .., o_n$. The functional similarity or functional proximity of two operations o_i and o_j is defined as

$$fp(o_i, o_j) = \frac{fcost(o_i) + fcost(o_j) - fcost(o_i, o_j)}{fcost(o_i, o_j)}. \qquad (6.7)$$

For example, if the cost of an adder or a subtracter bit-slice is six gates and the cost of a combined adder/subtracter bit-slice is eight gates, then the functional proximity of addition and subtraction, $fp(+, -) = (6 + 6 - 8)/8 = 0.5$.

The communication proximity of two operators can be defined similarly. Let $cconn(o_i, o_j)$ and $tconn(o_i, o_j)$ represent the number of common and total connections of the operations o_i and o_j. Then communication proximity is defined as

$$cp(o_i, o_j) = \frac{cconn(o_i, o_j)}{tconn(o_i, o_j)}. \qquad (6.8)$$

For example, if an addition operator and a subtraction operator have one common input then $cconn(+, -)=1$, $tconn(+, -)=6$ and $cp(+, -)=1/6$.

The potential concurrency or parallelism of two operations o_i and o_j is defined as

$$ppar(o_i, o_j) = \begin{cases} 1 & \text{if } o_i \text{ and } o_j \text{ can be executed in parallel,} \\ 0 & \text{otherwise.} \end{cases} \qquad (6.9)$$

The closeness metric used in operation clustering defines a distance between each pair of operations such that two operations are a short distance apart if there is a great advantage in having them share the same unit, while they have a large distance between them if there is little advantage in putting them together. Note that clustering similar operations into one unit reduces the potential parallelism and thus possibly increases execution time of the behavioral description. More precisely, the distance between two operations o_i and o_j is defined as a weighted sum given by

$$dist(o_i, o_j) = -(\alpha_1 \times fp(o_i, o_j)) - (\alpha_2 \times cp(o_i, o_j)) + (\alpha_3 \times ppar(o_i, o_j))$$
$$(6.10)$$

where α_1, α_2 and α_3 are the weights for each measure.

For example, consider the distance between o_1 and o_2 in Figure 6.8(a). Since both o_1 and o_2 are addition operations, we need at least one adder to execute both operations. Let C_{add} be the area of an adder. Hence, $fcost(o_1)=fcost(o_2)=fcost(o_1, o_2)=C_{add}$, and $fp(o_1, o_2)$ equals 1. The total number of connections for o_1 and o_2 is six, and the number of common interconnections between o_1 and o_2 is one. Thus, $cp(o_1, o_2)$ is 1/6. In addition, $ppar(o_1, o_2)$ is one because o_1 and o_2 can be executed in parallel. Hence, the distance between o_1 and o_2 is calculated as $dist(o_1, o_2)=\alpha_3-\alpha_1-(1/6)\alpha_2$.

As another example, consider the distance between o_1 and o_3. Since we need at least one adder and one multiplier to execute o_1 and o_3 separately or together, $fp(o_1, o_3)$ is zero. The total number of connections for o_1 and o_3 is six, and the number of common interconnections between o_1 and o_3 is one since the output of o_1 is the input of o_3. Thus, $cp(o_1, o_3)$ is 1/6. In addition, $par(o_1, o_3)$ is zero. Hence, $dist(o_1, o_3)=-(1/6)\alpha_2$.

BUD [McKo90] uses the hierarchical-clustering method described in Section 6.2.4. At the beginning of the search, the distance $dist(o_i, o_j)$ is computed for every two operations o_i and o_j. Then a hierarchical clustering tree is formed based on these distances. In such a tree, as mentioned in Section 6.2.4, the leaves represent the operations to be clustered, and the number of non-leaf nodes between two leaf nodes defines the relative distance between these two operations.

The cluster tree guides the search of the design space. A series of different module (i.e., functional unit) sets is generated by starting with a cut-line at the root and by moving the cut-line toward the leaves. Any cut across the tree partitions the tree into a set of subtrees, each of which represents a cluster. For each cut-line a different module set is formed and the resulting design is evaluated. Thus, the first design has all operations in the same cluster, sharing the same tightly coupled datapath. The second is made up of two clusters that have the greatest distance between them, and so on. This process continues until a stopping criterion, such as the lack of improvement in the design for a predetermined number of module sets, is reached.

Each time a new unit set is selected, the design is evaluated. First, the functional units required to execute all operations in each cluster are assigned to the design. Next, all operations are scheduled into control steps. After scheduling has been completed, the lifetimes of all values are computed and the maximum number of bits required to store those values between any two control steps is determined. The number of interconnections between each pair of clusters is also determined. With this information, the number and sizes of the registers, multiplexers and wires within each cluster is known so that the length, width and area of each cluster can be estimated. An approximate floorplanner based on a min-cut partitioning algorithm is used to estimate the total chip area. I/O pads are placed around the boundary of the chip for the global variables given in the behavioral description and for any off-chip memory. Finally, the datapath delays and the clock period are estimated. The total area and clock cycle time are used to compare the design with others that have been encountered in the search.

In the simple example in Figure 6.8(a), the operations are divided into two clusters, $G_1=\{o_1,o_2\}$ and $G_2=\{o_3\}$. Clusters G_1 and G_2 are mapped into the adder *add*1 and the multiplier *mult*1 respectively, with two buses represented by edges e_{13} and e_{23} connecting them. The design in Figure 6.8(a) requires three clock cycles to execute the three operations. Figure 6.8(b) shows a different partition of the same design. This partition consists of three clusters, $G_1=\{o_1\}$, $G_2=\{o_2\}$ and $G_3=\{o_3\}$ and requires two clock cycles to execute.

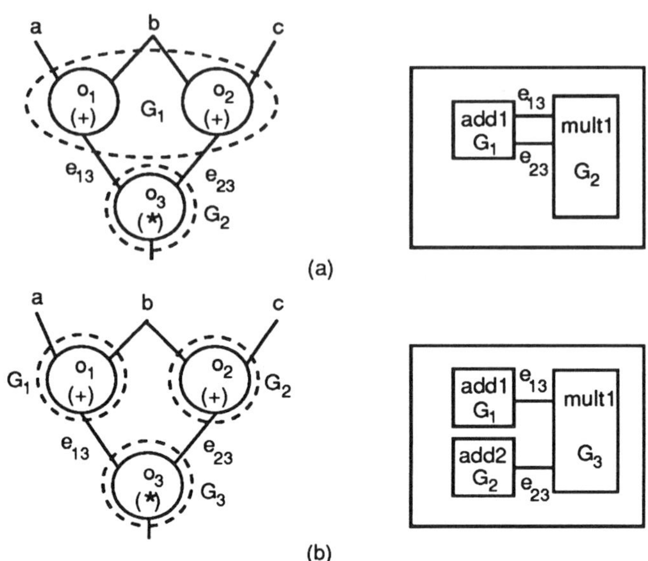

Figure 6.8: A clustering example: (a) two-cluster partition, (b) three-cluster partition.

The ideas of BUD have been extended in APARTY [LaTh91], an architectural partitioner that can choose a partitioning strategy for a behavior based on five different clustering criteria: control clustering, data clustering, procedure/control clustering, procedure/data clustering and operator clustering.

Control clustering attempts to group operators together into long threads of control in order to reduce the number of times that control is passed between clusters during the execution of the given description. Consider o_4 and o_5 in Figure 6.9, that are two operators executed in the same control sequence with no branches between them. If o_4 and o_5 are assigned to two separate clusters, control must passed from one cluster to the other. Thus, o_4 and o_5 are preferred to be assigned to the same cluster.

Data clustering reduces the number of data values that must be passed between clusters using the communication proximity measure given by Equation 6.8. Procedure/control clustering, similar to control

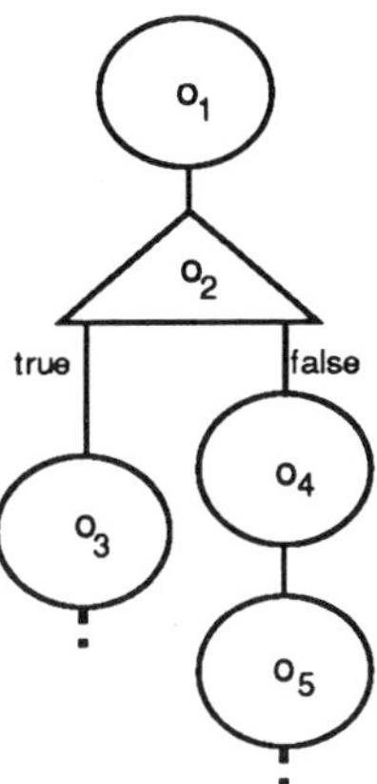

Figure 6.9: Control clustering.

clustering, minimizes the number of times that control is passed between clusters by grouping calling and called procedures. Procedure/data clustering groups together procedure calls in order to minimize the number of communication channels in the design. Operator clustering reduces the design cost by grouping similar operators together. The similarity measure combines functional proximity and concurrency measures defined by Equations 6.7 and 6.9.

APARTY does not combine all clustering criteria into a single closeness measure. Instead, it uses a multistage clustering algorithm which decouples the clustering criteria. The multistage algorithm allows better search of the design space since different criteria are applied at different times. In multistage clustering there are several clustering stages, with each stage having its own closeness criteria. In each stage, a complete tree of clusters is created. Then the cut-line is chosen at some level of the tree, and each subtree below that cut-line becomes an element to be considered for clustering in the next stage. The complete tree is built at each stage because this allows various cut lines to be evaluated and allows the best cut-line to be chosen. The cut-lines may be chosen automatically according to one of three physical criteria: area, connections and schedule length. Any criterion may be selected as the basis for choosing a cut-line for any stage.

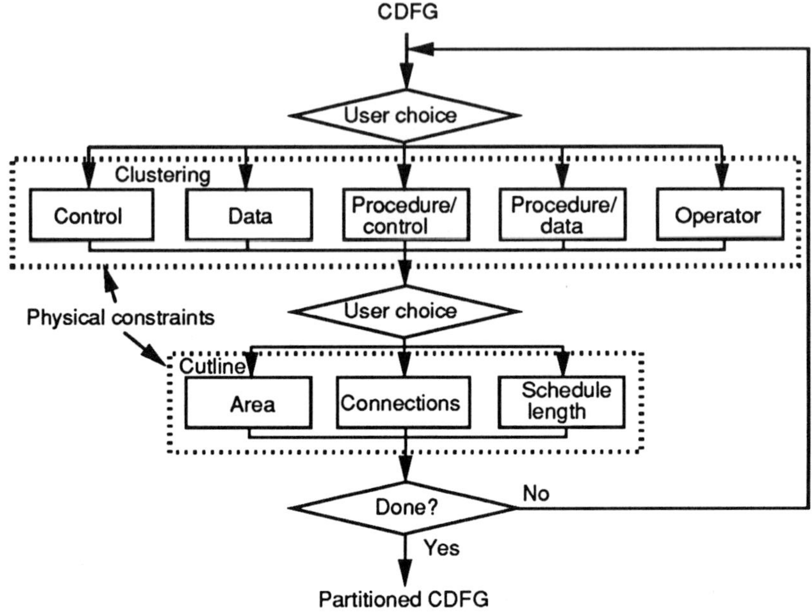

Figure 6.10: Possible partitioning scripts in APARTY.

The clustering stages and cut-line criteria may be thought of as building blocks for a variety of partitioners that produce different styles of partitions and thus different styles of design. All blocks of the APARTY partitioner are shown in Figure 6.10. The user provides a script with physical constraints for the design, such as area measures, estimates for operations, and chooses, for each stage, the type of clustering and the cut-line criterion to be used. These options allow the partitioner to produce different partitioning schemes. The resulting partitioning schemes are used to generate structures that can be passed along to other tools to guide synthesis.

6.3.2 Chip Partitioning

Chip-level partitioning can be performed at the CDFG or at the behavioral level. The first two approaches described in this section partition the CDFG, whereas the third approach partitions the behavioral description.

One approach for multi-chip partitioning is used in CHOP [KuPa91]. Here, the user first manually decomposes the nodes of a data-flow graph into partitions. A predictor then estimates the area and delay of each partition for various implementations. Finally, an iterative heuristic selects a feasible implementation of partitions that satisfies the given performance and physical constraints.

Gupta and De Micheli [GuDe90] have proposed a hypergraph partitioning technique to synthesize multi-chip systems from behavioral descriptions. Their approach assumed that each vertex of a CDFG will be implemented as a FSM or a combinational block. This assumption permits mapping of the CDFG directly to a hypergraph. In the hypergraph, each hypernode represents a functional unit or a combinational circuit. The operations or Boolean expressions assigned to the same hypernode share the same hardware. Each hyperedge represents the interconnections between hypernodes. The min-cut or simulated annealing algorithms are used to decompose the hypergraph into a set of sub-hypergraphs so that all constraints, such as the area-cost for each sub-hypergraph, the pin-count of each sub-hypergraph and the delay-cost between operations, are satisfied.

A different approach to partitioning is taken in SpecSyn [VaNG90]. A behavioral description is viewed as a set of behaviors, such as processes, procedures, and other code groupings imposed by the language, and a set of storage elements, including registers, memories, stacks, and queues. These behaviors and storage elements are then grouped into chips. This approach that groups behaviors is in contrast to the approach that converts a description to a CDFG, and then groups the data and control nodes; that is, the difference between CDFG and behavioral partitioning is in the granularity of the partitioned objects. In CDFG partitioning, the objects are data and control nodes while in behavioral partitioning the objects are code groupings that may include assignment statements, functions, procedures and processes. These behavioral objects are defined through a language that allows a system to be described as a hierarchy of concurrent and sequential behaviors which are entered and exited upon occurrence of events.

Each behavior in the description is mapped into a node of a hypergraph. Each behavior is assumed to be implemented by a single FSMD.

Area and performance measures for each behavior are estimated by using a FSMD implementation with a single functional unit for each type. The connections needed for communication between different behaviors are mapped into hyperedges. The number of wires in each connection is estimated from the declaration of ports and protocols used for communication, and the type of data being communicated. The hypergraph model of the description can be partitioned to satisfy physical or performance constraints. When physical constraints, such as the chip area or number of pins on a package, are the primary goal, the partitioning algorithms optimize component sharing at the expense of longer execution time caused by passing data and control to a unique component that performs required functions. When performance constraints are the primary goal, the partitioning algorithms minimize communication between different behaviors. The communication is minimized by using a single FSMD. Thus, the goal of performance optimization is to cluster non-concurrent behaviors and to map them onto the fewest number of FSMDs possibly communicating through a shared memory.

Once a satisfactory partitioning is obtained, the last step is to create a refined description reflecting the partitioning decisions. The refined description contains a set of interconnected chips, each consisting of behaviors and storage elements.

As an example, consider the behavioral description of a telephone answering machine shown in Figure 6.11(a). It consists of a hierarchical set of behaviors which respond to a phone line such that the machine answers the phone, plays an announcement, records a message, or is remotely programmed after checking a user identification number. The behavior of the leaf node *check_user_id* is shown in Figure 6.11(b).

Assume that nodes *main, respond_to_machine_button, monitor, check_user_id, respond_to_cmds, num_msgs,* and *user_id_ram* are the behaviors and storage elements selected as partitioning objects. Figure 6.12 shows the hypergraph for this set of objects. Note that the area estimate for *main* (3421) does not include estimates of any of its descendant behaviors that were also selected as partitioning objects. Instead, *main* merely controls (i.e., activates and deactivates) those behaviors. The connection estimates are calculated from declaration of storage elements and flow of data between behaviors. For example, consider the edge between *check_user_id* and *user_id_ram*. Since *user_id_ram* has four words

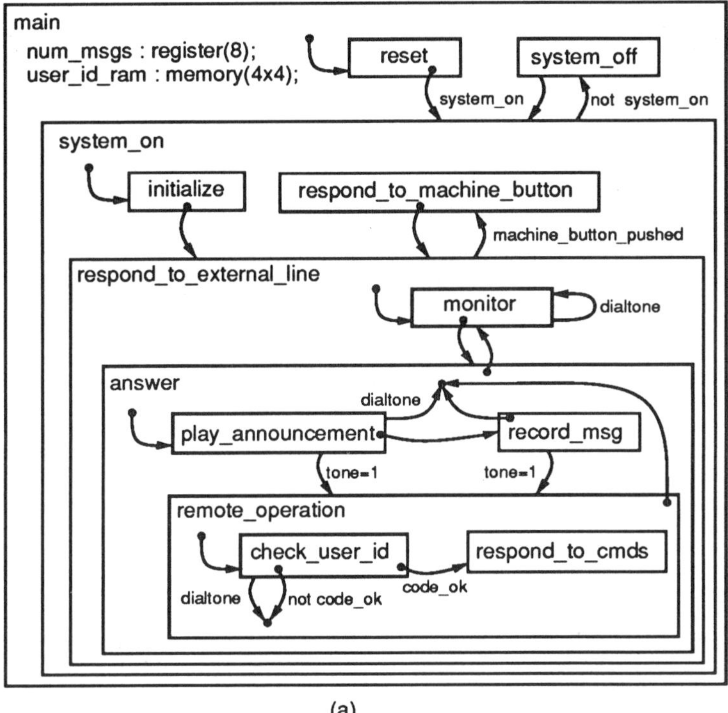

(a)

```
check_user_id          entered_code : memory(4x4);
                       i : integer range 1 to 5;
    i := 1;
    while i <= 4 loop
      wait until button_tone;
      entered_code[i] := button;
      i := i + 1;
    end loop;

    if (entered_code[1] = user_id_ram[1]) and
      ...
      (entered_code[4] = user_id_ram[4]) then
      code_ok <= true;
    else
      code_ok <= false;
    end if;
```

(b)

Figure 6.11: (a) A partial description of an answering machine, (b) a description of the *check_user_id* object.

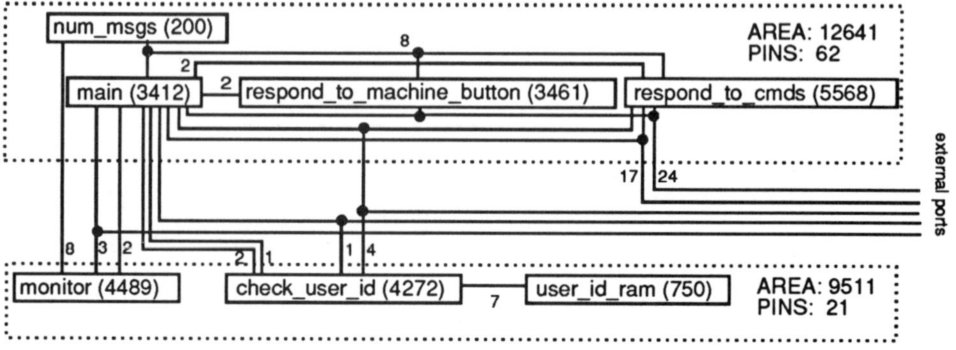

Figure 6.12: Partitioned hypergraph for the answering machine.

of four bits each, the estimator assumes that two bits are needed for an address, four for the data, and one for control. A sample partition of the hypergraph is indicated in Figure 6.12 by the dotted lines.

Partitioning at the behavioral description level rather than at the CDFG or structural level yields several advantages. First, the resulting partitioned description contains objects that are familiar to the designer from the original description. This simplifies the writing of functional test patterns for each partition. Second, the small number of familiar objects encourages designer interactivity, such as manual grouping of behaviors or overriding of partitioning decisions. In contrast, the complete answering machine example requires over 1500 nodes in a particular CDFG representation. This is an excessive amount for a designer to perform manual partitioning. Third, the behavior can be modified to make communication tradeoffs after the partitioning is known. For example, one might merge the address and data buses of a memory access, make the appropriate protocol modifications, and then repartition in order to reduce the number of pins at the expense of a slower memory access. Fourth, the focus is on considering performance using multiple processes and interprocess communication, rather than merely considering a DFG's input-to-output delay. Extending the latter to consider interprocess communication is cumbersome. Finally, note that systems which assume a pure DFG input are extremely limited. Most systems must be described with control and data operations, to which both description-level and CDFG-level partitioning can be applied.

Behavioral-description partitioning is therefore useful in cases where testability, designer interaction, communication tradeoffs (affecting pin counts) and overall system performance are given a high priority during the design process. One limitation, however, is that the sizes of objects considered for partitioning are greatly affected by the manner in which the original description is written. For example, if the entire description consists of only one large procedure, then it cannot be partitioned without redescribing it as a set of smaller procedures. In contrast, CDFG approaches can decompose a behavior to the granularity of operations, although they cannot easily partition control. A second limitation is that both behavioral and CDFG partitioning estimate area and performance with less accuracy than does structural-level partitioning, since assumptions about scheduling and allocation must be made. If the limitations concerning object sizes or estimation inaccuracy pose a problem for a particular example, behavioral partitioning can still provide a coarse initial partition, which can later be improved by CDFG or structural partitionings.

6.4 Summary and Future Directions

In this chapter, we introduced partitioning problems in HLS. We started with basic methods for solving general graph partitioning problems. We described in some detail two constructive (cluster growth and hierarchical partitioning) and two iterative-improvement (min-cut and simulated-annealing) methods. In the rest of the chapter, we described partitioning of CDFG and behavioral descriptions. CDFG partitioning is used generally for partitioning units and floorplans, whereas behavioral partitioning is used for partitioning systems into memories and processors.

Partitioning is important in HLS since many problems can be formulated as partitioning problems. One advantage of using partitioning in HLS is that physical information, including area and delay, can be provided to guide scheduling and allocation. However, most current high-level partitioning algorithms use an overly simplified design model and focus mostly on decomposing the given behavior to satisfy packaging constraints. The impact of partitioning on performance issues, such as inter-chip delays, need to be studied further. In addition, better estimation of quality measures from behavioral descriptions is needed to improve the quality of partitioning results.

6.5 Exercises

1. Formulate the multiple-chip partitioning problem as a general graph partitioning problem using a structural netlist as input and the following as cost functions: (a) chip capacity and (b) chip-pin counts.

2. Outline an algorithm for cluster growth and discuss the time complexity of your algorithm. Consider partitioning n nodes into two equal-size clusters and use the number of interconnections between two clusters as the objective function.

3. Using the cluster-growth method, partition the graph in Figure 6.13 into two subgraphs with equal number of vertices. Use vertex a as an initial seed and use
   (a) *Criterion A* closeness measures (Figure 6.13(a)),
   (b) *Criterion B* closeness measures (Figure 6.13(b)), and
   (c) the closeness measures of $0.3 \times$ *Criterion A* $+ 0.7 \times$ *Criterion B*.

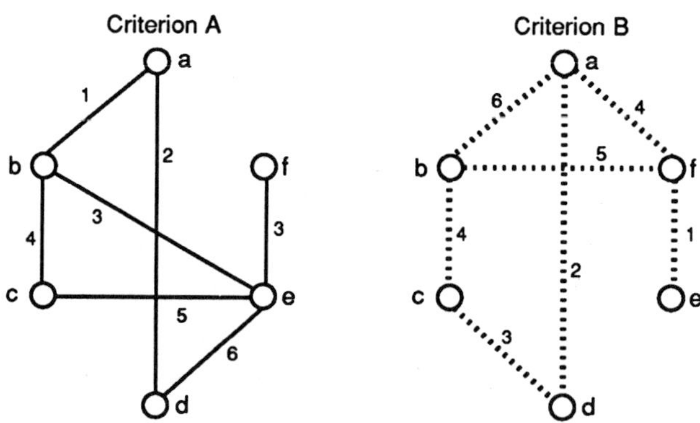

Figure 6.13: A sample graph (a) criterion A measures, (b) criterion B measures.

Also determine which initial seed produces the minimum interconnect costs between the two subgraphs using the closeness measure in (c) (the closeness measure is based on the sum of edge weights between two vertices or two clusters).

4. Build the cluster tree from the graphs in Figure 6.13 using the hierarchical clustering method.

5. Build the cluster tree from the graph in Figures 6.13 using:
(a) *Criterion A* closeness measures first; cut at the first-level of the cluster tree; then *Criterion B* closeness measures,
(b) *Criterion B* closeness measures first; cut at the first-level of the cluster tree; then *Criterion A* closeness measures, and finally,
(c) *Criterion A* closeness measures first; cut at the second-level of the cluster tree; then *Criterion B* closeness measures.
(The closeness score is based on the sum of edge weights between two vertices or two clusters.)

6. Address the differences between hierarchical clustering and multi-stage clustering methods, and the advantages and disadvantages of each.

7. Apply the min-cut partitioning algorithm to Figure 6.13(a) using the initial partitions {a,d,c} and {b,e,f}.

8. Justify why the term $(-2c_{ij})$ in Equation 6.3 is required to obtain the correct gain calculation when interchanging two vertices with a common connection.

9. Show that the time complexity of the min-cut algorithm is $O(n^2 log n)$, where n is the number of vertices.

10. Outline a multiple-way partitioning algorithm and discuss its time complexity.

11. *Outline a linear time complexity min-cut algorithm and justify its complexity. Hint: [FiMa82].

12. Build the closeness matrix for the graph in Figure 6.14 using the distance measures described in Section 6.3.1, with $\alpha_1=0.5$, $\alpha_2=0.4$, $\alpha_3=0.8$, where adder cost = 1 and subtracter cost = 1.

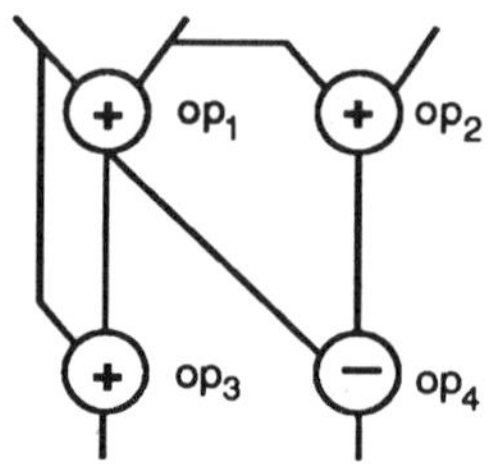

Figure 6.14: A sample data-flow graph.

13. Explain the difference between partitioning of behavioral descriptions and CDFGs.

14. Define several different quality measures for high-level partitioning.

15. *Outline an algorithm using a partitioning technique to guide the scheduling and allocation procedures. Hint: [McKo90].

16. Explain the differences between the five-clustering stages described in Section 6.3.1 and give an example of the loop relationship between clusters of operators. Hint: [LaTh91].

17. *Formulate a multiple-chip set partitioning into a hypergraph partitioning problem, define the objective functions and outline an algorithm. Hint: [GuDe90].

18. *Formulate a multiple-chip-set partitioning problem and outline an algorithm for chip partitioning as illustrated in Figures 6.11 and 6.12.

19. Explain the importance of estimation and interface synthesis in high-level partitioning. Hint: [NeTh86, Borr88, KuPa91].

20. *Discuss the interactions between estimation, interface synthesis and high-level partitioning.

21. **Develop a multiple-chip partitioning scheme including interface synthesis.

Chapter 7

Scheduling

7.1   Problem Definition

A behavioral description specifies the sequence of operations to be performed by the synthesized hardware. We normally compile such a description into an internal data representation such as the control/data flow graph (CDFG), which captures all the control and data-flow dependencies of the given behavioral description. Scheduling algorithms then partition this CDFG into subgraphs so that each subgraph is executed in one control step. Each control step corresponds to one state of the controlling finite-state machine in the FSMD model defined in Chapter 2. In the CDFG of the shift-and-add multiplier (Figure 5.3) dashed lines define the control step boundaries, and operations between two adjacent dashed lines are performed in the same control step.

Within a control step, a separate functional unit is required to execute each operation assigned to that step. Thus, the total number of functional units required in a control step directly corresponds to the number of operations scheduled in it. If more operations are scheduled into each control step, more functional units are necessary, which results in fewer control steps for the design implementation. On the other hand, if fewer operations are scheduled into each control step, fewer functional units are sufficient, but more control steps are needed. Scheduling is an important task in high-level synthesis because it impacts the tradeoff between design cost and performance.

Scheduling algorithms have to be tailored to suit the different target architectures discussed in Chapter 2. For example, a scheduling algorithm designed for a non-pipelined architecture would have to be reformulated for a target architecture with datapath or control pipelining. The types of functional and storage units and of interconnection topologies used in the architecture also influence the formulation of the scheduling algorithms.

The different language constructs also influence the scheduling algorithms. Behavioral descriptions that contain conditional and loop constructs require more complex scheduling techniques since dependencies across branch and loop boundaries have to be considered. Similarly, sophisticated scheduling techniques must be used when a description has multidimensional arrays with complex indices.

The tasks of scheduling and unit allocation are closely related. It is difficult to characterize the quality of a given scheduling algorithm without considering the algorithms that perform allocation (see Chapter 8). Two different schedules with the same number of control steps and requiring the same number of functional units may result in designs with substantially different quality metrics after allocation is performed.

In this chapter we discuss issues related to scheduling and present some well known solutions to the scheduling problem. We introduce the scheduling problem by discussing some basic scheduling algorithms on a simplified target architecture, using a simple design description. We then describe extensions to the basic algorithms for handling more sophisticated description models and realistic libraries of functional units.

7.2   Basic Scheduling Algorithms

In order to simplify the discussion of the basic scheduling algorithms, let us begin by assuming the following restrictions:

(a) behavioral descriptions do not contain conditional or loop constructs,

(b) each operation takes exactly one control step to execute,

(c) each type of operation can be performed by one and only one type
   of functional unit.

These assumptions will be relaxed later when more realistic models
are considered.

We can define two different goals for the scheduling problem, given
a library of functional units with known characteristics (e.g., size, delay,
power) and the length of a control step. First, we can minimize the number of functional units for a fixed number of control steps. We call this
fixed-control-step approach, time-constrained scheduling. Second, we
can minimize the number of control steps for a given design cost; the design cost can be measured in terms of the number of functional and storage units, the number of two-input NAND gates, or the chip-layout area.
This cost-minimizing approach is called resource-constrained scheduling.

In order to understand the scheduling algorithms discussed in this
chapter, we require a few simple definitions. A data-flow graph (DFG)
is a directed acyclic graph G(V, E), where V is a set of nodes and E a
set of edges. Each $v_i \in V$ represents an operation (o_i) in the behavioral
description. A directed edge e_{ij} from $v_i \in V$ to $v_j \in V$ exists in E
if the data produced by operation o_i (represented by v_i) is consumed
by operation o_j (represented by v_j). In this case we say that v_i is an
immediate predecessor of v_j. The set of all immediate predecessors of v_i
is denoted by $Pred_{v_i}$. Similarly v_j is an immediate successor of v_i. $Succ_{v_i}$
denotes the set of all immediate successors of v_i. Each operation o_i can
be executed in D_i control steps. Since we assume that each operation
takes exactly one control step, the value of D_i is 1 for every o_i.

We can illustrate the above definitions using the simple HAL example [PaKn89]. Figure 7.1(a) shows HAL's textual description, and
Figure 7.1(b) shows its DFG, which consists of eleven vertices, V = $\{v_1,$
$v_2, ..., v_{11}\}$ and eight edges, E = $\{e_{1,5}, e_{2,5}, e_{5,7}, e_{7,8}, e_{3,6}, e_{6,8}, e_{4,9},$
$e_{10,11}\}$.

DFGs expose parallelism in the design. Consequently, each DFG
node has some flexibility about the state into which it can be scheduled.
Many scheduling algorithms require the earliest and latest bounds within
which operations are to be scheduled. We call the earliest state to which
a node can possibly be assigned its ASAP value. This value is determined

```
while (x < a)   do
   x1 := x + dx;
   u1 := u − (3 * x * u * dx) − (3 * y * dx);
   y1 := y + (u*dx);
   x  := x1; u := u1; y := y1;
endwhile
```

(a)

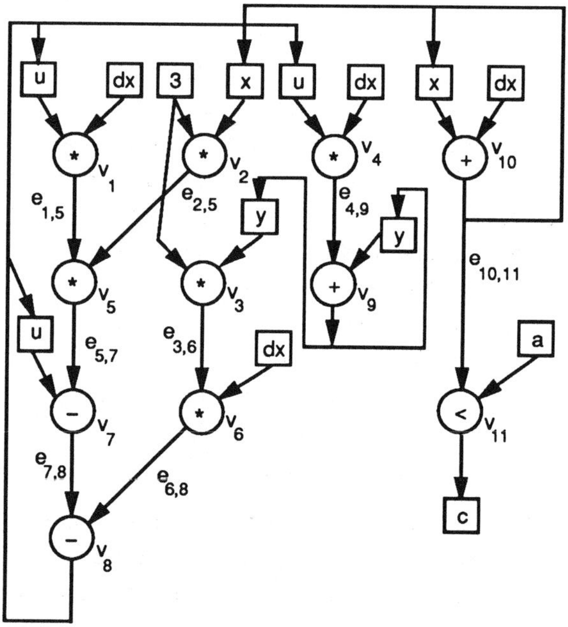

(b)

Figure 7.1: HAL example: (a) textual description, (b) DFG.

Algorithm 7.1: ASAP Scheduling.

```
for each node v_i ∈ V do
    if Pred_{v_i} = φ then
        E_i = 1;
        V = V - { v_i };
    else
        E_i = 0;
    endif
endfor

while V ≠ φ do
    for each node v_i ∈ V do
        if ALL_NODES_SCHED(Pred_{v_i}, E) then
            E_i = MAX(Pred_{v_i}, E) + 1;
            V = V - { v_i };
        endif
    endfor
endwhile
```

by the simple ASAP scheduling algorithm outlined in Algorithm 7.1.

The ASAP scheduling algorithm assigns an ASAP label (i.e., control-step index) E_i, to each node v_i of a DFG, thereby scheduling operation o_i into the earliest possible control step s_{E_i}. $Pred_{v_i}$ denotes all the nodes in the DFG that are immediate predecessors of the node v_i. The function ALL_NODES_SCHED($Pred_{v_i}$, E) returns true if all the nodes in set $Pred_{v_i}$ are scheduled (i.e., all immediate predecessors of v_i have a non-zero E label). The function MAX($Pred_{v_i}$, E) returns the index of the node with the maximum E value from the set of predecessor nodes for v_i.

The *for* loop in Algorithm 7.1 initializes the ASAP value of all the nodes in the DFG. It assigns the nodes which do not have any predecessors to state s_1 and the other nodes to state s_0. In each iteration, the *while* loop determines the nodes that have all their predecessors scheduled and assigns them to the earliest possible state. Since we assume that the delay of all operations is 1 control step, the earliest possible state is calculated using the equation $E_i = MAX(Pred_{v_i}, E) + 1$.

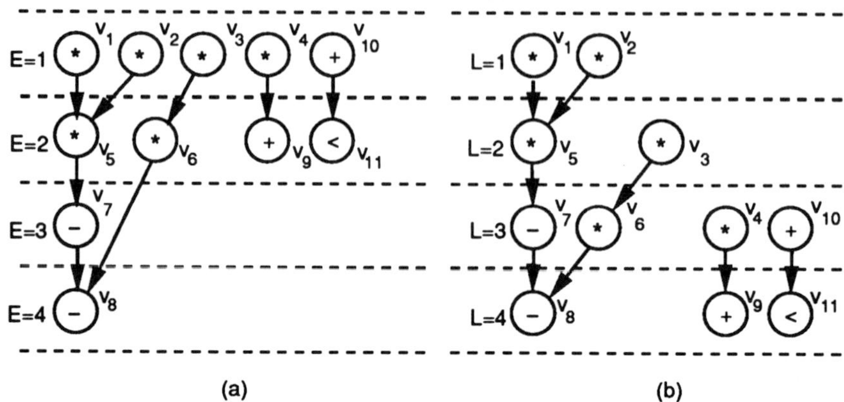

Figure 7.2: HAL example: (a) ASAP schedule, (b) ALAP schedule.

Figure 7.2(a) shows the results of the ASAP scheduling algorithm on the HAL example shown in Figure 7.1. Operations o_1, o_2, o_3, o_4 and o_{10} are assigned to control step s_1 since they do not have any predecessors. Operations o_5, o_6, o_9 and o_{11} are assigned to control step s_2 and operations o_7 and o_8 are assigned to control steps s_3 and s_4, respectively.

The ALAP value for a node defines the latest state to which a node can possibly be scheduled. We can determine the ALAP value using the scheduling algorithm detailed in Algorithm 7.2. Given a time constraint of T control steps, the algorithm determines the latest possible control step in which an operation must begin its execution. The ALAP scheduling algorithm assigns an ALAP label L_i to each node v_i of a DFG, thereby scheduling operation o_i in the latest possible control step s_{L_i}. The function ALL_NODES_SCHED($Succ_{v_i}$, L) returns true if all the nodes denoted by $Succ_{v_i}$ are scheduled (i.e., all immediate successors of v_i have a non-zero L label). The function MIN($Succ_{v_i}$, L) returns the index of the node with the minimum L value from the set of successor nodes for v_i.

The *for* loop in Algorithm 7.2 initializes the ALAP value of all the nodes in the DFG. It assigns the nodes which do not have any successors to the last possible state and the other nodes to state s_0. In each iteration, the *while* loop determines the nodes that have all their successors scheduled and assigns them to the latest possible state.

Algorithm 7.2: ALAP Scheduling.

```
for each node v_i ∈ V do
   if Succ_{v_i} = φ then
      L_i = T;
      V = V - { v_i };
   else
      L_i = 0;
   endif
endfor

while V ≠ φ do
   for each node v_i ∈ V do
      if ALL_NODES_SCHED(Pred_{v_i}, L) then
         L_i = MIN(Succ_{v_i}, L) - 1;
         V = V - { v_i };
      endif
   endfor
endwhile
```

Figure 7.2(b) shows the results of the ALAP scheduling algorithm (where $T = 4$) on the HAL example shown in Figure 7.1. Operations o_8, o_9 and o_{11} are assigned to the last control step s_4 since they do not have any successors. Operations o_4, o_6, o_7 and o_{10} are assigned to control s_3 and operations o_5 and o_3 are assigned to control step s_2. The remaining operations o_1 and o_2 are assigned to control step s_1.

Given a final schedule, we can easily compute the number of functional units that are required to implement the design. The maximum number of operations in any state denotes the number of functional units of that particular operation type. In the ASAP schedule (Figure 7.2(a)) the maximum number of multiplication operations scheduled in any control step is four (state s_1). Thus four multipliers are required. In addition, the ASAP schedule also requires an adder/subtracter and a comparator. In the ALAP schedule (Figure 7.2(b)) the maximum number of multiplication operations scheduled in any control step is two (states s_1, s_2 and s_3). Thus two multipliers are sufficient. In addition, the ALAP schedule also requires an adder, a subtracter and a comparator.

7.2.1 Time-Constrained Scheduling

Time-constrained scheduling is important for designs targeted towards applications in a real-time system. For example, in many digital signal processing (DSP) systems, the sampling rate of the input data stream dictates the maximum time allowed for carrying out a DSP algorithm on the present data sample before the next sample arrives. Since the sampling rate is fixed, the main objective is to minimize the cost of hardware. Given the control step length, the sampling rate can be expressed in terms of the number of control steps that are required for executing a DSP algorithm.

Time-constrained scheduling algorithms can use three different techniques: mathematical programming, constructive heuristics and iterative refinement. In this section we discuss the integer linear programming method (an example of mathematical programming), the force-directed scheduling method (an example of a constructive heuristic methodology) and an iterative rescheduling technique.

Integer Linear Programming Method

The integer linear programming (ILP) method [LeHL89] finds an optimal schedule using a branch-and-bound search algorithm that involves backtracking, i.e., some of the decisions made in the initial stages of the algorithm are revisited as the search progresses. Let s_{E_k} and s_{L_k} be the control steps into which operation o_k is scheduled by the ASAP and ALAP algorithms. Clearly, $E_k \leq L_k$. In a feasible schedule, o_k must begin its execution in a control step no sooner than s_{E_k} and no later than s_{L_k}.

The number of control steps between s_{E_k} and s_{L_k} is called the mobility range of operation o_k (i.e., mrange $(o_k) = \{s_j | E_k \leq j \leq L_k\}$). Figure 7.3(a) shows the range of every operation in the DFG for the HAL example, computed from the ASAP and ALAP labels in Figure 7.2. For example, the range of o_4 is $\{s_1, s_2, s_3\}$, since its ASAP and ALAP labels are $E_4 = 1$ and $L_4 = 3$.

We can now use the ASAP, ALAP and range values of the operations to formulate the scheduling problem using ILP. We use the following notations for the ILP formulation:

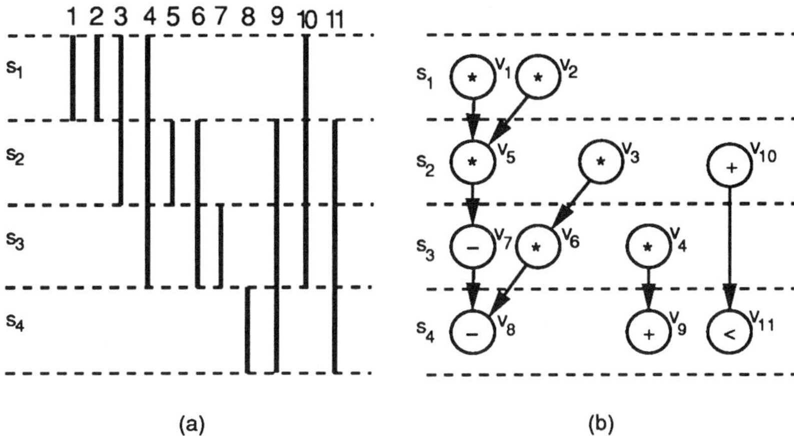

Figure 7.3: An ILP scheduling example: (a) ranges for operations, (b) final schedule.

Let OP = $\{o_i \mid 1 \leq i \leq n\}$ be the set of operations in the flowgraph and t_i = type(o_i) be the type of each operation o_i. Let T = $\{t_k \mid 1 \leq k \leq m\}$ be the set of possible operation types. The set OP_{t_k} consists of operations from the set OP which are of type t_k (i.e., OP_{t_k} = $\{ o_i \mid o_i \in \text{OP} \wedge \text{type}(o_i) = t_k\}$). Let $INDEX_{t_k}$ define the set of all operation indices in OP_{t_k}. (i.e., $INDEX_{t_k} = \{ i \mid o_i \in OP_k \}$). Let N_{t_k} be the number of units performing operation t_k and let C_{t_k} be the cost of such an unit. Let S = $\{s_j \mid 1 \leq j \leq r\}$ be the set of control steps available for scheduling the operations. Let x_{ij} be the 0-1 integer variables, which assume a value of 1 if operation o_i is scheduled into step s_j and 0 otherwise.

The scheduling problem can be formulated as follows:

$$minimize \quad \sum_{k=1}^{m}(C_{t_k} \times N_{t_k}) \qquad (7.1)$$

under the assumptions,

$$\forall i, 1 \leq i \leq n, (\sum_{E_i \leq j \leq L_i} x_{i,j} = 1), \qquad (7.2)$$

$$\forall j, 1 \leq j \leq r, \forall k, \ 1 \leq k \leq m, (\sum_{i \in INDEX_{t_k}} x_{i,j} \leq N_{t_k}), \qquad (7.3)$$

and

$$\forall i, j, o_i \in Pred_{o_j}(\sum_{E_i \leq k \leq L_i} (k \times x_{i,k}) - \sum_{E_j \leq l \leq L_j} (l \times x_{j,l}) \leq -1) \qquad (7.4)$$

The objective function (7.1) minimizes the total cost of the required functional units. Condition 7.2 requires that every operation o_i be scheduled into one and exactly one control step no sooner than E_i and no later than L_i. Condition 7.3 ensures that no control step contains more than N_{t_k} operations of type t_k. Finally, Condition 7.4 guarantees that for an operations o_j, all its predecessors (i.e., $Pred_{o_j}$)) are scheduled in an earlier control step. In other words, if $x_{i,k} = x_{j,l} = 1$ then $k < l$.

Let us now use the DFG from Figure 7.1(b) to illustrate the ILP formulation for scheduling the flowgraph into four control steps. Since the DFG contains four different types of operations (i.e., multiplication, addition, subtraction and comparison), we need four types of functional units from the library. Let C_m, C_a, C_s and C_c be the costs of a multiplier, an adder, a subtracter and a comparator, respectively. Let N_m, N_a, N_s and N_c be the number of multipliers, adders, subtracters and comparators needed in the final schedule. The ILP formulation for scheduling the DFG of Figure 7.1(b) into four control steps is

$$minimize \quad C_m \times N_m + C_a \times N_a + C_s \times N_s + C_c \times N_c \qquad (7.5)$$

subject to

$$
\begin{aligned}
x_{1,1} &= 1 \\
x_{2,1} &= 1 \\
x_{3,1} + x_{3,2} &= 1 \\
x_{4,1} + x_{4,2} + x_{4,3} &= 1 \\
x_{5,2} &= 1 \\
x_{6,2} + x_{6,3} &= 1 \\
x_{7,3} &= 1 \\
x_{8,4} &= 1 \\
x_{9,2} + x_{9,3} + x_{9,4} &= 1 \\
x_{10,1} + x_{10,2} + x_{10,3} &= 1 \\
x_{11,2} + x_{11,3} + x_{11,4} &= 1 \\
x_{1,1} + x_{2,1} + x_{3,1} + x_{4,1} &\leq N_m \\
x_{3,2} + x_{4,2} + x_{5,2} + x_{6,2} &\leq N_m \\
x_{4,3} + x_{6,3} &\leq N_m \\
x_{7,3} &\leq N_s \\
x_{8,4} &\leq N_s \\
x_{10,1} &\leq N_a \\
x_{9,2} + x_{10,2} &\leq N_a \\
x_{9,3} + x_{10,3} &\leq N_a \\
x_{9,4} &\leq N_a \\
x_{11,2} &\leq N_c \\
x_{11,3} &\leq N_c \\
x_{11,4} &\leq N_c \\
1x_{3,1} + 2x_{3,2} - 2x_{6,2} - 3x_{6,3} &\leq -1 \\
1x_{4,1} + 2x_{4,2} + 3x_{4,3} - 2x_{9,2} - 3x_{9,3} - 4x_{9,4} &\leq -1 \\
1x_{10,1} + 2x_{10,2} + 3x_{10,3} - 2x_{11,2} - 3x_{11,3} - 4x_{11,4} &\leq -1.
\end{aligned}
$$

If we assume that $C_m = 2$ and $C_a = C_b = C_c = 1$, the cost function is minimized and all inequalities are satisfied when the values for the variables are set as follows: $N_m = 2$, $N_a = N_s = N_c = 1$, $x_{1,1} = x_{2,1} = x_{3,2} = x_{4,3} = x_{5,2} = x_{6,3} = x_{7,3} = x_{8,4} = x_{9,4} = x_{10,2} = x_{11,4} = 1$, and other xs $= 0$. This solution to the ILP formulation of the scheduling problem is shown in Figure 7.3(b). In the figure, an operation o_i is scheduled into control step s_j, if and only if $x_{i,j}$ is set to 1.

The size of the ILP formulation increases rapidly with the number of control steps. For example, if we increase the number of control steps by 1, we will have n additional x variables in the inequalities since we have to consider an extra control step for every operation. Moreover, the number of inequalities due to Condition 7.3 will increase by an amount that depends on the structure of the given DFG. Since the execution time of the algorithm grows rapidly with the number of variables or the number of inequalities, in practice the ILP approach is applicable only to very small problems.

Since the ILP method is impractical for large descriptions, heuristic methods that run efficiently at the expense of the design optimality have been developed. The improved efficiency of the heuristic methods is obtained by eliminating the backtracking in the ILP method. Backtracking can be eliminated by scheduling the operations one at a time, based on a criterion that produces the right decisions most of the time. In these heuristic approaches, the cost of the scheduled design depends largely on the selection of the next operation to be scheduled and the assignment of that operation to the best control step.

Force-Directed Heuristic Method

The force-directed scheduling (FDS) heuristic [PaKn89] is a well known heuristic for scheduling with a given timing constraint. In this section we present a simplified version of the FDS algorithm. The main goal of the algorithm is to reduce the total number of functional units used in the implementation of the design. The algorithm achieves its objective by uniformly distributing operations of the same type (e.g., multiplication) into all the available states. This uniform distribution ensures that functional units allocated to perform operations in one control step are used efficiently in all other control steps, which leads to a high unit utilization rate.

Like the ILP formulation, the FDS algorithm relies on both the ASAP and the ALAP scheduling algorithms to determine the range of control steps for every operation (i.e., mrange (o_i)). It also assumes that each operation o_i has a uniform probability of being scheduled into any of the control steps in the range, and probability zero of being scheduled in any other control step. Thus, for a given state s_j, such that $E_i \leq j \leq L_i$, the probability that operation o_i will be scheduled in that state is given by $p_j(o_i) = 1/(L_i - E_i + 1)$.

We can illustrate these probability calculations using the HAL example from Figure 7.1, with the ASAP (E_i) and ALAP (L_i) values from Figure 7.2 used in the calculations. The operation probabilities for the example are shown in Figure 7.4(a). Operations o_1, o_2, o_5, o_7 and o_8 have probability values of "1" for being scheduled in steps s_1, s_1, s_2, s_3 and s_4 respectively, because the s_{E_i} value is equal to the s_{L_i} value for these operations. The width of a rectangle in this figure represents the probability $(1/(L_i - E_i + 1))$ of an operation getting scheduled in that particular control step. For example, operation o_3 has a probability of "0.5" of being assigned to either s_1 or s_2. Therefore, the value of $p_1(o_3) = p_2(o_3) = 0.5$.

A set of probability distribution graphs (i.e., bar graphs) are created from the probability values of each operation with a separate bar graph being constructed for each operation type. A bar graph for a particular operation type (e.g., multiplication), represents the expected operator cost (EOC) in each state. The expected operator cost in state s_j for operation type k is given by $EOC_{j,k} = c_k * \sum_{i, s_j \in mrange(o_i)} p_j(o_i)$, where o_i is an operation of type k and c_k is the cost of the functional unit performing the operation of type k. Figure 7.4(b) is a bar graph of expected operation costs for the multiplication operation in each control step. We can calculate the value for $EOC_{1,multiplication}$ as $c_{multiplication} \times (p_1(o_1) + p_1(o_2) + p_1(o_3) + p_1(o_4))$, which is $c_k \times (1.0 + 1.0 + 0.5 + 0.33)$ or $2.83 \times c_k$. Consequently, the bar graph in state s_1 for the multiplication operation shows an EOC value of 2.83.

The bar graph in Figure 7.4(b) shows that the EOC for multiplication in the four states are 2.83, 2.33, 0.83 and 0.0. Since the functional units can be shared across states, the maximum of the expected operator costs over all states gives a measure of the total cost of implementing all operations of that type. Bar graphs similar to Figure 7.4(b) are constructed for all other operation types.

Since the main goal of the FDS algorithm is efficient sharing of functional units across all states, it attempts to balance the EOC value

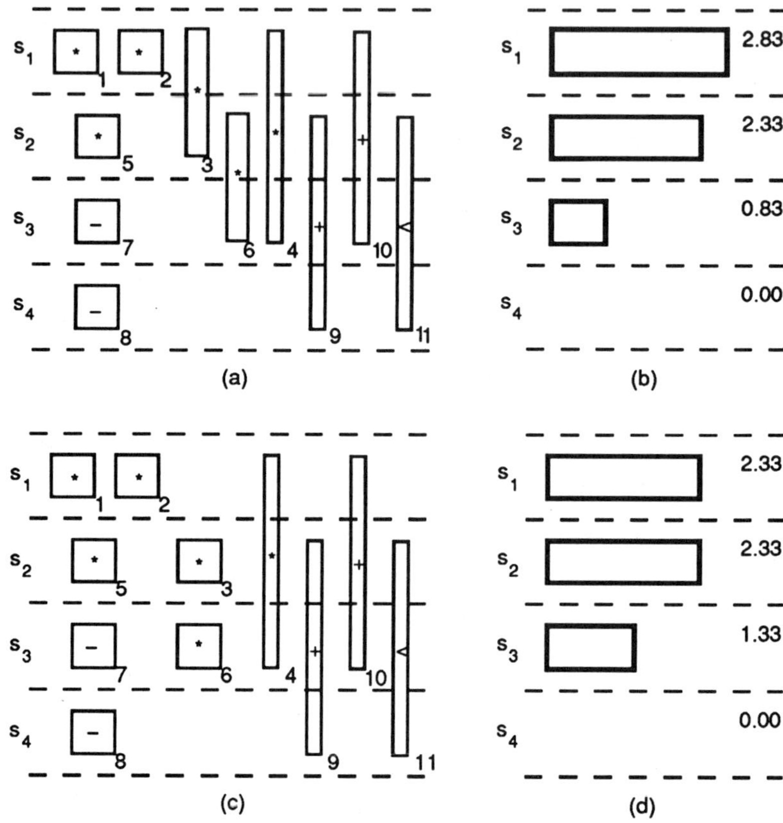

Figure 7.4: Force-directed scheduling example: (a) probability of scheduling operations into control steps, (b) operator cost for multiplications in a, (c) probability of scheduling operations into control steps after operation o_3 is scheduled to step s_2, (d) operator cost for multiplications in c.

Algorithm 7.3: Force-Directed Scheduling.

```
Call ASAP (V);
Call ALAP (V);
while there exists oᵢ such that Eᵢ ≠ Lᵢ do
    MaxGain = −∞;
    /* Try scheduling all unscheduled operations to every */
    /* state in its range */
    for each oᵢ, Eᵢ ≠ Lᵢ do
        for each j, Eᵢ ≤ j ≤ Lᵢ do
            S_work = SCHEDULE_OP(S_current, oᵢ, sⱼ);
            ADJUST_DISTRIBUTION(S_work, oᵢ, sⱼ);
            if COST(S_current) - COST(S_work) > MaxGain then
                MaxGain = COST(S_current) - COST(S_work);
                BestOp = oᵢ; BestStep = sⱼ;
            endif
        endfor
    endfor
    S_current = SCHEDULE_OP(S_current, BestOp, BestStep);
    ADJUST_DISTRIBUTION(S_current, BestOp, BestStep);
endwhile
```

for each operation type. Algorithm 7.3 (Force-Directed Scheduling) describes a method to achieve this uniform value of expected operator costs. During the execution of the algorithm, $S_{current}$ denotes the most recent partial schedule. S_{work} is a copy of the schedule on which temporary scheduling assignments are attempted. In each iteration, the variables *BestOp* and *BestStep* maintain the best operation to schedule and the best control step for scheduling the operation. When the *BestOp* and *BestStep* are determined for a given iteration, the $S_{current}$ schedule is changed appropriately using the function SCHEDULE_OP($S_{current}$, o_i, s_j) which returns a new schedule, after scheduling operation o_i into state s_j on $S_{current}$. Scheduling a particular operation into a control step affects the probability values of other operations because of data dependencies. The function ADJUST_DISTRIBUTION scans through the set of vertices and adjusts the probability distributions of the successor and predecessor nodes in the graph. The FDS the algorithm is shown in Algorithm 7.3.

The function COST(S) evaluates the cost of implementing a partial schedule S based on any given cost function. A simple cost function could add the EOC values for each operation type as shown in Equation 7.6.

$$COST(S) = \sum_{1 \leq k \leq m} \max_{1 \leq j \leq s} EOC_{j,k}. \qquad (7.6)$$

This cost is calculated using the ASAP (E_i) and the ALAP (L_i) values for all the nodes.

During each iteration, the cost of assigning each unscheduled operation to possible states within its range (i.e., mrange(o_i)) is calculated using S_{work}. The assignment that leads to the minimal cost is accepted and the schedule $S_{current}$ is updated. Therefore, during each iteration an operation o_i gets assigned into control step s_k where $E_i \leq k \leq L_i$. The probability distribution for operation o_i is changed to $p_k(o_i) = 1$ and $p_j(o_i) = 0$ for all j not equal to k. The operation o_i remains fixed and does not get moved in later iterations.

Returning to our example, from the initial probability distribution for the multiplication operation shown in Figure 7.4(b), the costs for assigning each unscheduled operation into possible control steps are calculated. The assignment of o_3 to control step s_2 results in the minimal expected operator costs for multiplication, because Max(P_j) falls from 2.83 to 2.33. This assignment is accepted. When operation o_3 is assigned to control step s_2, the probability values from operation o_6 also change as shown in Figure 7.4(c). The operation o_3 is never moved while the iterations for scheduling other unscheduled operations continue.

In each iteration of the FDS algorithm, one operation is assigned to its control step based on the minimum expected operator costs. If there are two possible control-step assignments with close or identical operator costs, then the above algorithm cannot estimate the best choice accurately. Alternatively, we can invert the strategy by pruning only one control step from an operation's possible destinations during each iteration [VAKL91]. By doing so, we effectively postpone the decision of choosing one of the many feasible control step assignments to a later stage when the feasible assignment set is smaller.

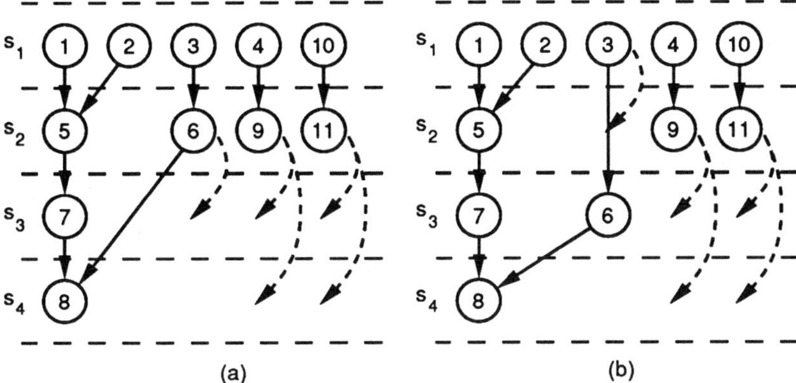

Figure 7.5: Rescheduling: (a) initial schedule where three operations are movable to five control steps (dashed arrows), (b) after operation 6 is moved and locked.

Iterative Refinement Method

We call algorithms similar to FDS "constructive" because they construct a solution without performing any backtracking. The decision to schedule an operation into a control step is made on the basis of a partially scheduled DFG; it does not take into account future assignments of operators to the same control step. Most likely, the resulting solution will not be optimal, due to the lack of a look-ahead scheme and the lack of compromises between early and late decisions. We can cope with this weakness by rescheduling some of the operations in the given schedule.

The essential issues in rescheduling are: the choice of a candidate for rescheduling, the rescheduling procedure and the control of the improvement process. We describe a method [PaKy91] based on the paradigm originally proposed for the graph-bisection problem by Kernighan and Lin (KL) [KeLi70]. Chapter 6 gives a detailed description of the KL algorithm.

Any scheduling algorithm can be used to compute the initial schedule. New schedules can be obtained by simply rescheduling one operation at a time. An operation can be rescheduled into an earlier or a later step, as long as it does not violate the data dependence constraints. For example, there are five possible moves (o_6 to s_3, o_9 to s_3, o_9 to s_4, o_{11} to s_3 and o_{11} to s_4) as indicated by the dashed arcs in the schedule depicted in Figure 7.5(a). Each move leads to a new schedule. For example, moving operation o_6 to state s_3 will result in the new schedule shown in Figure 7.5(b). Possible next moves are indicated by dashed arcs in Figure 7.5(b). An operation that has been moved once, (i.e., o_6 in Figure 7.5(b)) is locked and not allowed to move again. After a sequence of $m \le n$ move and locks, all operations become locked. We can find a subsequence of $k \le m$ move and locks that maximizes the cumulative gain, where gain is defined as the decrease in the cost of implementing the new schedule. If the gain is non-negative (i.e., the cost decreases), we change the initial schedule according to the first k moves, unlock all operations and iterate the refinement process. If no non-negative cumulative gain is attainable, the process stops.

In the iterative rescheduling algorithm (Algorithm 7.4), the variable $S_{current}$ maintains the most recently updated copy of the schedule. S_{move} and S_{work} are temporary versions of the schedule used to evaluate various rescheduling options. The function SCHEDULE_OP(S_{work}, o_i, s_j) returns a new schedule, after scheduling operation o_i into state s_j in S_{work}. The function COST(S) calculates the cost of a particular schedule. This cost can be determined from equation 7.6. The set UnlockOps is a set of nodes that remains unlocked in a particular iteration and hence can be rescheduled in future iterations. Array $O[m]$ maintains the sequence of operations that are moved in each iteration and array $S[m]$ maintains the states into which each operation from $O[m]$ is to be moved. The function MAX_CUM_GAIN scans through the array G[m] and returns the maximum possible cumulative gain. The variable Max-GainIndex stores the number of moves required to attain the maximum cumulative gain.

The *while* loop in the algorithm determines the sequence of best possible moves until all the operations become locked. The largest gain for each move is stored in the array G[m]. The two functions MAX_CUM_GAIN and MAX_CUM_GAIN_INDEX scan this array G[m] and extract the sub-sequence of moves that results in the largest cumulative gain. The schedule $S_{current}$ is changed to reflect the modifications induced by these moves.

The KL algorithm has been studied extensively and has proven to be

Algorithm 7.4: Iterative Rescheduling.

repeat
    $S_{work} = S_{current}$; m = 0;
    UnlockOps = {All operations in S_{work}};
    while UnlockOps $\neq \phi$ **do**
        m = m + 1; G[m] = $-\infty$;
        for each operation $o_i \in S_{work}$ **do**
            for each possible destination s_j of o_i **do**
                S_{move} = SCHEDULE_OP(S_{work}, o_i, s_j);
                gain = COST(S_{work}) - COST(S_{move});
                if gain > G[m] **then**
                    O[m] = i; S[m] = j; G[m] = gain;
                endif
            endfor
        endfor
        S_{work} = SCHEDULE_OP(S_{work}, $o_{O[m]}$, $s_{S[m]}$);
        UnlockOps = UnlockOps - {$o_{O[m]}$};
    endwhile

    MaxGain = MAX_CUM_GAIN(G);
    MaxGain_index = MAX_CUM_GAIN_INDEX(G);
    if MaxGain $\geq$ 0 **then**
        for j = 1 to MaxGain_index **do**
            $S_{current}$ = SCHEDULE_OP($S_{current}$, $o_{O[j]}$, $s_{S[j]}$);
        endfor
    endif
until MaxGain < 0.

useful in the physical design domain. Its usefulness for scheduling can be increased by incorporating some enhancements proposed for the graph bipartitioning problem. Two of them are outlined below.

1. The first enhancement is probabilistic in nature and exploits the fact that the quality of a result produced by the KL method depends greatly on the initial solution. Since the algorithm is computationally efficient, we can run the algorithm many times, each with a different initial solution, and then pick the best solution.

2. The second enhancement, called the look-ahead scheme [Kris84], relies on a more sophisticated strategy of move selection. Instead of just evaluating the gain of a move, the algorithm looks ahead and evaluates the present move on the basis of possible future moves. The depth of the look-ahead determines the tradeoff between the computation time and the design quality: the deeper the look ahead, the more accurate the move selection and more expensive the computation.

7.2.2   Resource-Constrained Scheduling

The resource-constrained scheduling problem is encountered in many applications where we are limited by the silicon area. The constraint is usually given in terms of either a number of functional units or the total allocated silicon area. When total area is given as a constraint, the scheduling algorithm determines the type of functional units used in the design. The goal of such an algorithm is to produce a design with the best possible performance but still meeting the given area constraint.

In resource-constrained scheduling, we gradually construct the schedule, one operation at a time, so that the resource constraints are not exceeded and data dependencies are not violated. The resource constraints are satisfied by ensuring that the total number of operations scheduled in a given control step does not exceed the imposed constraints. The area constraints are satisfied by ensuring that the area of all units or the floorplan area does not exceed the constraints. A unit constraint can be checked for each control step, but the total area constraint can only be checked for the whole design. The dependence constraints can be satisfied by ensuring that all predecessors of a node are scheduled before the node is scheduled. Thus, when scheduling operation o_i into a control state s_j, we have to ensure that the hardware requirements for o_i and other operations already scheduled in s_j do not exceed the given constraint and that all predecessors of node o_i have already been scheduled.

We describe two resource-constrained scheduling algorithms: the list-based scheduling method and the static-list scheduling method.

List-Based Scheduling Method

List scheduling is one of the most popular methods for scheduling operations under resource constraints. Basically, it is a generalization of the ASAP scheduling technique since it produces the same result in the absence of resource constraints.

The list based scheduling algorithm maintains a priority list of ready nodes. A ready node represents an operation that has all predecessors already scheduled. During each iteration the operations in the beginning of the ready list are scheduled till all the resources get used in that state. The priority list is always sorted with respect to a priority function. Thus the priority function resolves the resource contention among operations. Whenever there are conflicts over resource usage among the ready operations (e.g., three additions are ready but only two adders are given in the resource constraint), the operation with higher priority gets scheduled. Operations with lower priority will be deferred to the next or later control steps. Scheduling an operation may make some other non-ready operations ready. These operations are inserted into the list according to the priority function. The quality of the results produced by a list-based scheduler depends predominantly on its priority function.

Algorithm 7.5 outlines the list scheduling method. The algorithm uses a priority list $PList$ for each operation type ($t_k \in T$). These lists are denoted by the variables $PList_{t_1}$, $PList_{t_2}$, ..., $PList_{t_m}$. The operations in these lists are scheduled into control steps based on N_{t_k} which is the number of functional units performing operation of type t_k. The function INSERT_READY_OPS scans the set of nodes, V, determines if any of the operations in the set are ready (i.e., all its predecessors are scheduled), deletes each ready node from the set V and appends it to one of the priority lists based on its operation type. The function

Algorithm 7.5: List Scheduling.

```
INSERT_READY_OPS(V, PList_{t_1}, PList_{t_2}, ... PList_{t_m});
Cstep = 0;
while((PList_{t_1} ≠ φ) or ... or (PList_{t_m} ≠ φ)) do
    Cstep = Cstep + 1;
    for k = 1 to m do
        for funit = 1 to N_k do
            if Plist_{t_k} ≠ φ then
                SCHEDULE_OP(S_{current}, FIRST(PList_{t_k}),Cstep);
                PList_{t_k} = DELETE(PList_{t_k}, FIRST(PList_{t_k}));
            endif
        endfor
    endfor
    INSERT_READY_OPS(V, PList_{t_1}, ..., PList_{t_m} );
endwhile
```

SCHEDULE_OP($S_{current}$, o_i, s_j) returns a new schedule after scheduling the operation o_i in control step s_j. The function DELETE($Plist_{t_k}$, o_i), deletes the indicated operation o_i from the specified list.

Initially, all nodes that do not have any predecessors are inserted into the appropriate priority list by the function INSERT_READY_OPS, based on the priority function. The *while* loop extracts operations from each priority list and schedules them into the current step until all the resources are exhausted in that step. Scheduling an operator in the current step makes other successor-operations ready. These ready operations are scheduled during the next iterations of the loop. These iterations continue till all the priority lists are empty.

We will illustrate the list scheduling process with an example (Figure 7.6). Suppose the available resources are two multipliers, one adder, one subtracter and one comparator. (Figure 7.6(c)). Each operation o_i in the DFG in Figure 7.6(a) is labeled with its mobility range (i.e., mrange(o_i)). Nodes with lower mobility must be scheduled first since delaying their assignment to a control step increases the probability of extending the schedule. Consequently, the mobility value is a good priority function. For each operator type, a priority list (Figure 7.6(b)) is constructed in which priority is given to ready nodes with lower mobility.

If two operations have the same mobility, then the one with a smaller index is given a higher priority.

During the first iteration, when ready operations are scheduled into control step s_1, there are five ready operations: o_1, o_2, o_3, o_4 and o_{10}. Operation o_{10}, which is the only addition operation, is scheduled into control step s_1 without considering any other factors. Since only two multipliers are available, only two out of the four ready multiplications can be scheduled into control step s_1; o_1 and o_2 are chosen for scheduling since they have lower mobilities than o_3 or o_4.

After the first iteration, operations o_1, o_2 and o_{10} are scheduled into control step s_1. For the second iteration, operations o_5 and o_{11} are added to the ready list because these two operations have all their input nodes scheduled. The ready list during the second iteration consists of o_5, o_{11}, o_3 and o_4. This ready list is sorted by mobilities and the whole process is repeated again. After four iterations, all the operations are scheduled into the appropriate control steps (Figure 7.6(d)).

As we stated previously, the success of a list-scheduler depends mainly on its priority function. Mobility is a good priority function because a smaller value of mobility indicates a higher urgency for scheduling an operation since the schedule will run out of alternative control steps earlier. Mobility is just one of the many priority functions that have been proposed as a priority function for list-scheduling. An alternative priority function uses the length of the longest path from the operation node to a node with no immediate successor. This longest path is proportional to the number of additional steps needed to complete the schedule if the operation is not scheduled into the current step. Therefore, an operation with a longer path label gets a higher priority. Yet another scheme uses the number of immediate successor nodes for an operation as a priority function: an operation node with more immediate successors is scheduled earlier because it makes more of these operations ready than a node with fewer successors.

Static-List Scheduling Method

Instead of dynamically building a priority list while scheduling every control step, we can create a single large list before starting scheduling [JMSW91]. This approach is different from the ordinary list-based

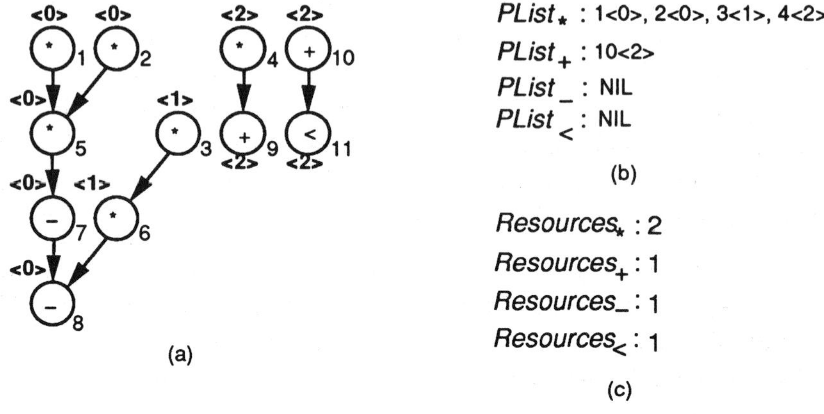

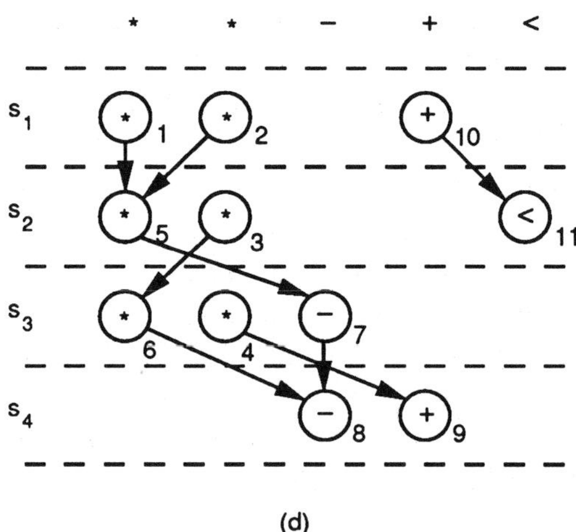

Figure 7.6: List scheduling: (a) a DFG with mobility labeling (inside <>), (b) the list of ready operations for control state s_1, (c) the resource constraints, (d) the scheduled DFG.

scheduling not only in the selection of candidate operations but also in the assignment of the target steps for the candidates. The algorithm sorts all operations using the ALAP labels in ascending order as the primary key and the ASAP labels in descending order as the secondary key. If both the keys have the same value, an arbitrary ordering is used. This sorted list is maintained as the priority list that determines the order in which operations are scheduled.

Figure 7.7(a) shows the HAL dataflow graph, and Figure 7.7(b) shows the ASAP and ALAP values for each node and the complete priority list for the example. Operations o_8, o_9 and o_{11} have the lowest ALAP value of 1 and so fall into the first three slots in the priority list. Among these three nodes, o_8 has a higher ASAP labeling than both o_9 and o_{11}. Therefore it falls in the lowest priority slot. Operations o_9 and o_{11} have the same ASAP and ALAP values and tie for the second and third slots on the list. Operation o_9 is selected randomly to precede operation o_{11}. The rest of the list is formed in a similar manner.

Once the priority list is created, the operations are scheduled sequentially starting with the last operation (i.e., the highest priority) in the list. An operation is scheduled as early as possible, subject only to available resources and operator dependencies. Thus, operation o_2 is the first one to be scheduled. It is scheduled into the first control step, since both the multipliers are available. The next operation to be scheduled is o_1 and it is assigned to the second multiplier in the same control step. Operation o_3 cannot be scheduled into state s_1 since there are no available multipliers, so it is scheduled into state s_2. Operation o_5 cannot be scheduled into state s_1 because its input data is available only in state s_2, so it is scheduled into state s_2. Although operation o_{10} has a lower priority compared to operations o_3 and o_5, it is scheduled into control step s_1 because an adder is available in state s_1 and it does not depend on any other previously scheduled operations. The final schedule for the example is shown in Figure 7.7(d).

7.3   Scheduling with Relaxed Assumptions

In Section 7.2 we outlined some basic scheduling algorithms that used a set of simplifying assumptions. We now extend the scheduling algo-

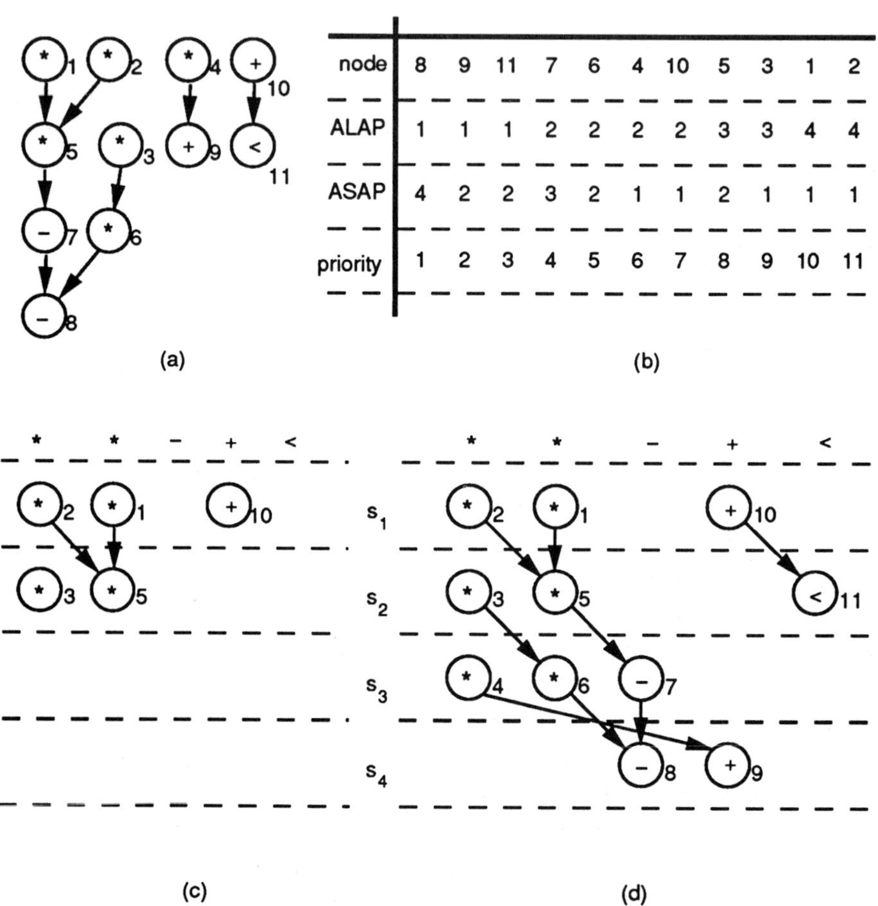

Figure 7.7: Static-list scheduling: (a) a DFG, (b) priority list, (c) partial schedule of five nodes, (d) the final schedule.

rithms to handle more realistic design models, such as functional units with varying execution times, multi-function units, and behavioral descriptions that are not limited to straight-line code.

7.3.1 Functional Units with Varying Delays

Real functional units have different propagation delays based on their design. For example, a 32-bit floating-point multiplier is much slower than a fixed-point adder of the same word length. Therefore we should not assume that every operation finishes in one control step. This assumption would lead to a clock cycle that is unusually lengthened to accommodate the slowest unit in the design, as shown in Figure 7.8(a), where the multiplier takes approximately three times longer than an adder or a subtracter. As a result, units with delay shorter than the clock cycle remain idle during part of the cycle and so are underutilized. Allowing functional units with arbitrary delays will improve the utilization of the functional units. Since functional units with arbitrary delays may no longer all execute within a single clock cycle, we have to generalize the execution model of operation to include multicycling, chaining and pipelining.

If the clock cycle period is shortened to allow the fast operations to execute in one clock cycle, then the slower operations take multiple clock cycles to complete execution. These slower operations, called multicycle operations, have to be scheduled across two or more control steps (Figure 7.8(b)). This increases the utilization of the faster functional units since two or more operations can be scheduled on them during a single operation on the multicycle units. However, input latches are needed in front of the multicycle functional units to hold its operands until the result is available a number of steps later. Also, a shorter clock cycle results in a larger number of control steps to execute the CDFG, which in turn increases the size of the control logic.

Another method of increasing the functional unit utilization is to allow two or more operations to be performed serially within one step, i.e., chaining. We can feed the result of one functional unit directly to the input of some other functional units (Figure 7.8(c)). This method requires direct connections between functional units in addition to the connections between the functional and storage units.

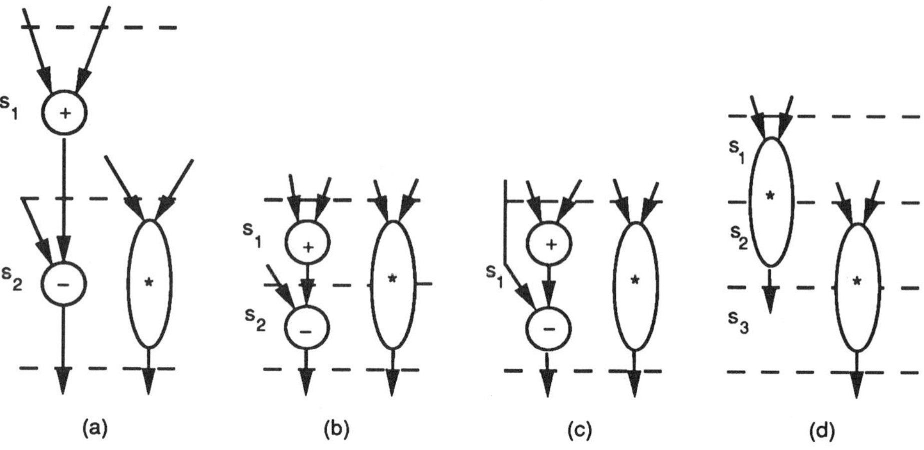

Figure 7.8: Scheduling with arbitrary-delay units: (a) schedule in which every operation is executed in one step, (b) schedule with a 2-cycle multiplier, (c) schedule with a chained adder and subtracter, (d) schedule with a 2-stage pipelined multiplier.

As described in Chapter 2, pipelining is a simple yet effective technique for increasing parallelism. When a pipelined functional unit is used, the scheduler has to calculate the resource requirements differently. For example, in Figure 7.8(d), the two multiplications can share the same two-stage pipelined multiplier despite the fact that the two operations are executing concurrently. This sharing is possible because each multiplication operation uses a different stage of the pipelined multiplier. Therefore, only one pipelined multiplier is needed instead of two non-pipelined multipliers.

7.3.2 Multi-functional Units

In the previous section, we extended the scheduling problem to functional units with non-uniform propagation delays and different number of pipeline stages. However, we have still assumed that each functional unit performs exactly one operation. This assumption is unrealistic since in practice multi-function units cost less than a set of uni-functional units that perform the same operations. For example, although an adder-subtracter component costs twenty percent more than an adder or subtracter separately, it is considerably cheaper than using one adder and one subtracter. Therefore, it is reasonable to expect that designers will use as many multi-function units as possible. The scheduling algorithms have to be reformulated to handle these multi-function units.

We have also been assuming single physical implementations for each functional unit. In reality, the component library has multiple implementations of the same component, with each implementation having a different area/delay characteristic. For example an addition can be done either quickly with a large (hence, costly) carry-look-ahead adder or slowly with a small (hence, inexpensive) ripple-carry adder. Given a library with multiple implementations of the same unit, the scheduling algorithms must simultaneously perform two important tasks : operation scheduling, in which operations are assigned to control steps, and component selection, in which the algorithm selects the most efficient implementation of the functional unit for each operation in the library.

We can exploit the range of component implementations in a library by using a technology-based scheduling algorithm [RaGa91]. Given a library that has multiple implementations of functional units, the main goal of the algorithm is to implement the design within the specified time constraint, with minimal costs. For each operation in the flowgraph an efficient component is selected so that the operations on the critical path are implemented with faster components, and the operations not on the critical path are implemented with slower components. Simultaneously, the algorithm ensures that the operations are scheduled into appropriate control steps that enable sharing the functional units across the various states.

7.3.3   Realistic Design Descriptions

In addition to blocks of straight line code, behavioral descriptions generally contain both conditional and loop constructs. Thus, a realistic scheduling algorithm must be able to deal with both of these modeling constructs.

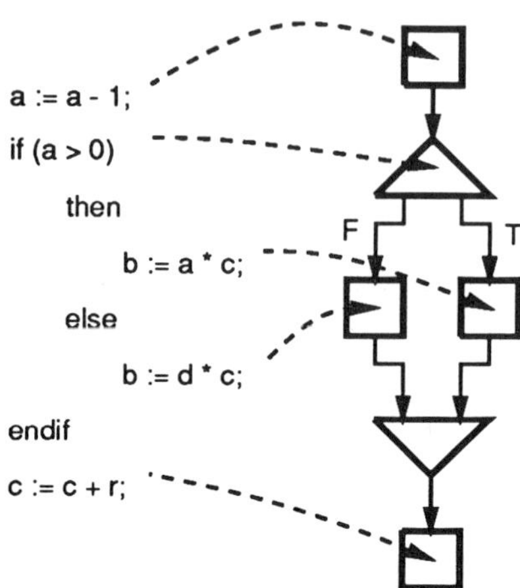

Figure 7.9: CDFG for a behavioral description with a conditional construct.

Conditional Constructs

A conditional construct is analogous to an "if" or "case" statement in programming languages. It results in several branches that are mutually exclusive. During one execution path of the design, only one branch gets executed based on the outcome of an evaluated condition. Figure 7.9 shows a segment of a behavioral description that contains a conditional *if* statement. It is not possible to represent this behavior with a pure data-flow graph (see Chapter 5). Instead, a flow graph with control and data dependencies (e.g., a CDFG) is needed. In this example, the control flow determines the execution of the four DFGs represented by boxes.

An effective scheduling algorithm shares the resources among mutually exclusive operations. For example, only one of the two multiplications in Figure 7.9 gets executed during any execution instance of the given behavior. Therefore, the scheduling algorithm can schedule both the multiplication operations in the same control step, even if only one multiplier is available in each step.

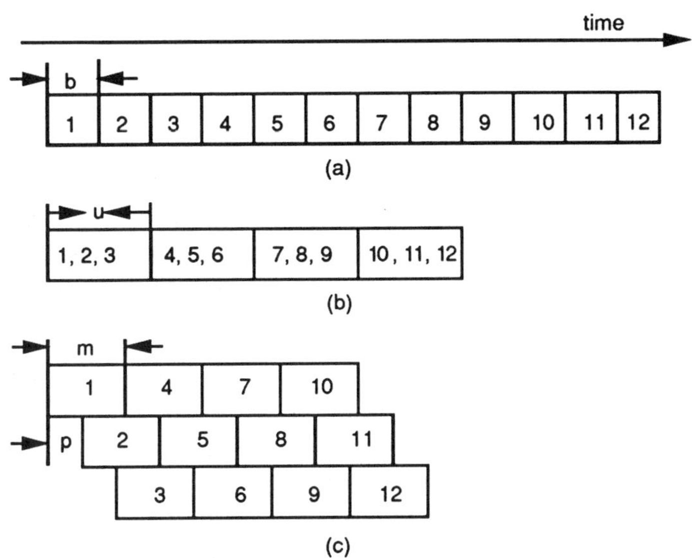

Figure 7.10: Loop scheduling: (a) sequential execution, (b) partial loop unrolling, (c) loop folding.

Loop Constructs

A behavioral description sometimes contains loop constructs. For example, in a filter for digital signal processing, a set of actions is repeatedly executed for every sample of the input data stream. This repeated action is modeled using a loop construct. In these descriptions, optimization of the loop body improves the performance of the design.

Loop constructs exhibit potential parallelism between different iterations of the loop. This parallelism can be exploited by executing several iterations of the same loop concurrently. Scheduling a loop body differs from scheduling a pure data flow graph, in that we need to consider potential parallelism across different loop iterations.

We use Figure 7.10 to illustrate three different ways of scheduling a loop. Suppose the loop is finite and consists of n iterations ($n = 12$ in Figure 7.10). The first and most simple approach assumes sequential execution of the loop and schedules each loop iteration into b control steps. If the twelve iterations are executed sequentially, the total execution time is $12b$ control steps as shown in Figure 7.10(a). The other two approaches exploit parallelism across loop iterations.

The second technique is called loop unrolling since a certain number of loop iterations are unrolled. This action results in a loop with a larger loop body but with fewer iterations. The larger loop body provides a greater flexibility for compacting the schedule. In Figure 7.10(b) three iterations are unrolled into a super-iteration resulting in four super-iterations. Each super-iteration is scheduled over u control steps. Therefore, if $u < 3b$, the total execution time is less than $12b$ control steps.

The third method exploits intraloop parallelism using loop folding, by which successive iterations of the loop are overlapped in a pipelined fashion. Figure 7.10(c) shows a loop with a body that takes m control steps to execute. In loop folding, we initiate a new iteration every p control steps, where $p < m$; that is, we overlap successive iterations. The total execution time is $m+(n-1)\times p$ control steps (Figure 7.10(c)). Loop unrolling is applicable only when the loop count is known in advance, but loop folding is applicable to both fixed and unbounded loops.

To illustrate these three methods of loop execution, we introduce an example DFG that contains multiple iterations. Figure 7.11(a) shows the DFG of a loop body [Cytr84] that contains 17 identical operations, with each taking one control step to execute. The dashed arcs indicate data dependencies across loop boundaries. For instance, the arc from node P to node J indicates that the result produced by operation P during one iteration is used by operation J of the next iteration. We schedule this DFG using the three loop scheduling techniques. In order to evaluate the quality of the resulting schedules, we use two performance measures: functional unit utilization, i.e., the percentage of states in which the functional units are used to perform some operation, and control cost, which is measured by the number of unique control words in the control unit.

In Figure 7.11(b) we show a schedule with six control steps using three functional units. Since a datapath with three functional units needs at least six control steps to execute seventeen operations and the length of the critical path of the DFG is six, this schedule is optimal. The functional unit utilization rate is $17/(3 \times 6) = 17/18$. The control cost is six words assuming that one word is needed in the control unit

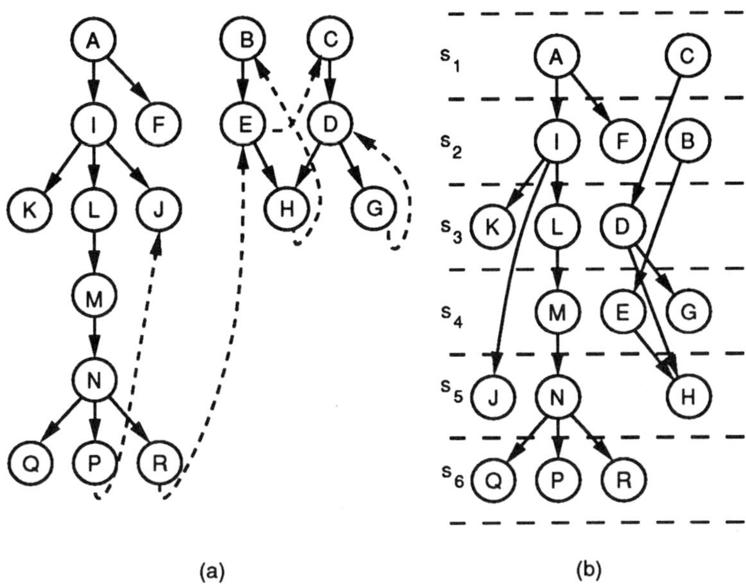

Figure 7.11: Standard loop scheduling: (a) DFG with dependencies across iterations, (b) sequential schedule using three functional units.

for each scheduled control step. We can increase the performance of this design only by exploiting parallelism across loop boundaries.

Figure 7.12(a) shows a schedule of the example in Figure 7.11 using loop unrolling. Two iterations, i.e., two copies of the DFG, are scheduled into nine control steps using four functional units. The hardware utilization rate remains the same: $(17 \times 2)/(4 \times 9) = 17/18$. However, the average time required to execute an iteration is reduced from 6 to $9/2 = 4.5$ control steps. The control cost increases from six to nine control words.

Figure 7.12(b) shows a schedule of the same example using loop folding. Successive iterations begin their execution three control steps apart. Six functional units are used and their utilization rate is $(5+6+6)/(6 \times 3) = 17/18$, which is same as in the two previous schedules. But the average time for executing an iteration is approximately three control steps

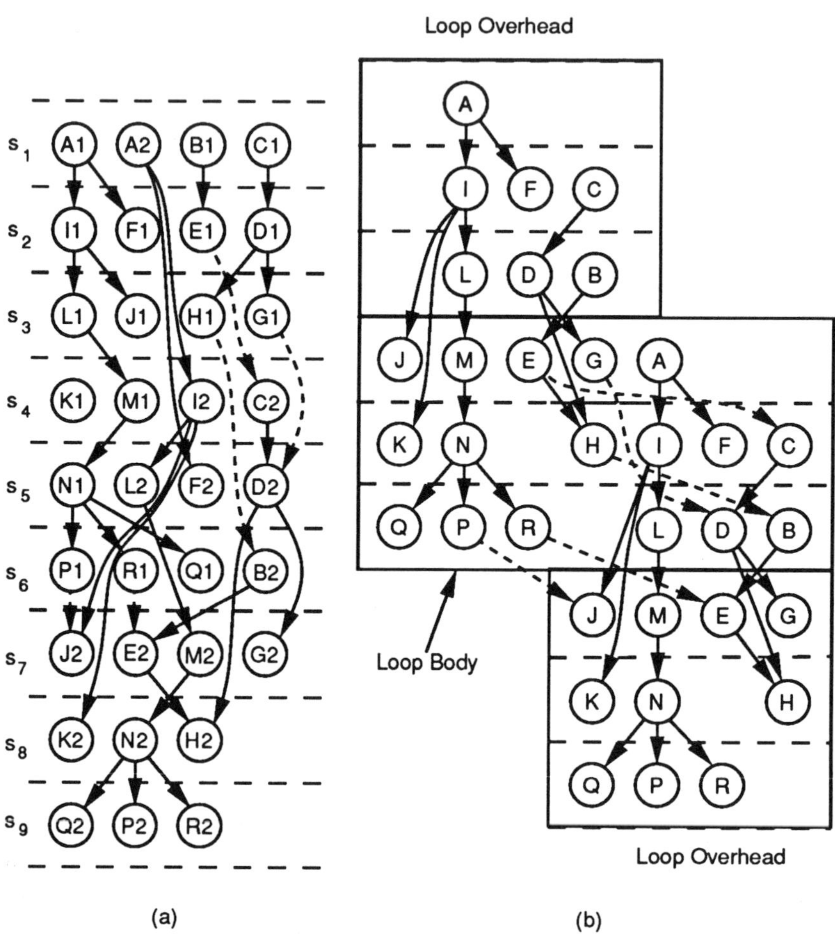

Figure 7.12: Scheduling with iteration overlapping: (a) loop unrolling, (b) loop folding.

when the number of iterations is very large. The control cost is nine (e.g., three each for the head, body and tail of the loop).

In the context of high-level synthesis, we must consider the change in control costs when performing these optimizations. Both loop unrolling and loop folding increase the control cost. When unrolling a loop, the scheduler can explore more parallelism by expanding more iterations, but the control cost increases as well. Therefore, it performs tradeoffs between the execution speed and the control cost. When folding a loop, there are two important constituents to the control cost: the loop overhead and the loop body. The loop overhead constituent is proportional to the execution time of an iteration. The loop body constituent is proportional to the product of the execution time of an iteration and the number of control steps between successive iterations, i.e., the latency.

Scheduling an unrolled loop is fairly easy since the basic scheduling algorithms described in Section 7.2 can be applied to the unrolled DFG. Problem size is the only concern. Some algorithms that have high computational complexity (e.g., the ILP method) will perform poorly on deep unrolling of loops. On the other hand, scheduling a folded loop is more complicated. For a fixed latency k, we can extend the list-scheduling algorithm to schedule operations under a resource constraint. In a control-step q where $q > k$, the operations from various iterations of the loop are executed concurrently. Consequently, the resources required to execute different operations from all the concurrent iterations would have to be provided.

7.4 Other Scheduling Formulations

In addition to the basic scheduling algorithms described in Section 7.2, a variety of other techniques can be adopted. We discuss three approaches: simulated annealing, path-based scheduling and DFG restructuring. We will use the simplifying assumptions discussed in Section 7.2 in order to explain these algorithms clearly.

7.4.1 Simulated Annealing

Simulated annealing (described in Algorithm 6.4) can also be used to schedule a dataflow graph [DeNe89]. The schedules are represented using a two-dimensional table where one dimension corresponds to the control steps and the other dimension corresponds to the the available functional units. We view scheduling as a placement problem in which operations are placed in table entries. By placing the operation nodes on the table, we define the control step in which the operation is executed and the functional unit that will carry out the operation. Since a functional unit can perform only a single operation in a given control step, no table entry can be occupied by more than one operation. For example, Figure 7.13(a) shows a schedule of five operations into three control steps using three ALUs.

Beginning with an initial schedule, the annealing procedure improves upon the schedule by iteratively modifying it and evaluating the modification. In [DeNe89] the quality measure for a given schedule is computed as the weighted sum of the number of functional units, registers, busses and the number of control steps.

The simulated annealing algorithm modifies a schedule by displacing an arbitrary operation from one table entry to another or by swapping two arbitrary operations. For example, Figure 7.13(b) shows the schedule resulting from swapping two operations of the schedule depicted in Figure 7.13(a): $(v_1 = v_2 + v_3)$ and $(v_4 = v_2 * v_3)$. Figure 7.13(c) shows the schedule resulting from displacing an operation (i.e., $v_2 = v_4 * v_5$). A modification is accepted for the next iteration if it results in a better schedule. If the modification does not result in a better schedule, its acceptance is based on a randomized function of the quality improvement and the annealing temperature.

Although the simulated annealing approach is robust and easy to implement, it suffers from long computation times and requires complicated tuning of the annealing parameters.

7.4.2 Path-Based Scheduling

Path-based scheduling algorithm [Camp91] minimizes the number of control steps needed to execute the critical paths in the CDFG. It first ex-

	ALU1	ALU2	ALU3
s1	v1 = v2 + v3	v4 = v2 * v3	
s2	v5 = v1 + v4		v6 = v4 / v1
s3		v2 = v4 * v5	

(a)

	ALU1	ALU2	ALU3
s1	v4 = v2 * v3	v1 = v2 + v3	
s2	v5 = v1 + v4		v6 = v4 / v1
s3		v2 = v4 * v5	

(b)

	ALU1	ALU2	ALU3
s1	v4 = v2 * v3	v1 = v2 + v3	
s2	v5 = v1 + v4	v2 = v4 * v5	v6 = v4 / v1
s3			

(c)

Figure 7.13: Scheduling using simulated annealing: (a) an initial schedule, (b) after swapping two operations, (c) after displacing an operation.

tracts all possible execution paths from a given CDFG and schedules them independently. The schedules for the different paths are then combined to generate the final schedule for the design.

Consider the CDFG shown in Figure 7.14(a). In order to extract the execution paths, the CDFG is made acyclic by deleting all feedback edges of loops. A path starts with either the first node of the CDFG or the first node of a loop, and ends at a node with no successors. Figure 7.14(b) shows a path of the original CDFG in Figure 7.14(a).

We then partition each path in the CDFG into control steps in such a way that:

(a) no variable is assigned more than once in each control step,

(b) no I/O port is accessed more than once in each control step,

(c) no functional unit is used more than once in each control step,

(d) the total delay of operations in each control step is not greater than the given control-step length,

(e) all designer imposed constraints for scheduling particular operations in different control steps are satisfied.

In order to satisfy any of the above conditions, the path-based scheduling algorithm generates constraints between two nodes that must be scheduled into two different control steps. Such a constraint is represented by an interval starting and ending at the two conflicting nodes. For example, operations indicated by the nodes 5 and 9 in Figure 7.14(b), cannot be in the same control step since both the operations assign values to the same variable. This constraint is represented by the interval i_1 in Figure 7.14(b). Similarly operations indicated by nodes 3 and 9 use the same adder and must occur in separate control steps. This constraint is represented by interval i_2 in Figure 7.14(b). If two constraints have overlapping intervals, both the constraints can be simultaneously satisfied by introducing a control step between any two nodes that belong to the overlapping part of the two intervals. For example, introducing a new control step between nodes 5 and 9 will satisfy both constraints i_1 and i_2.

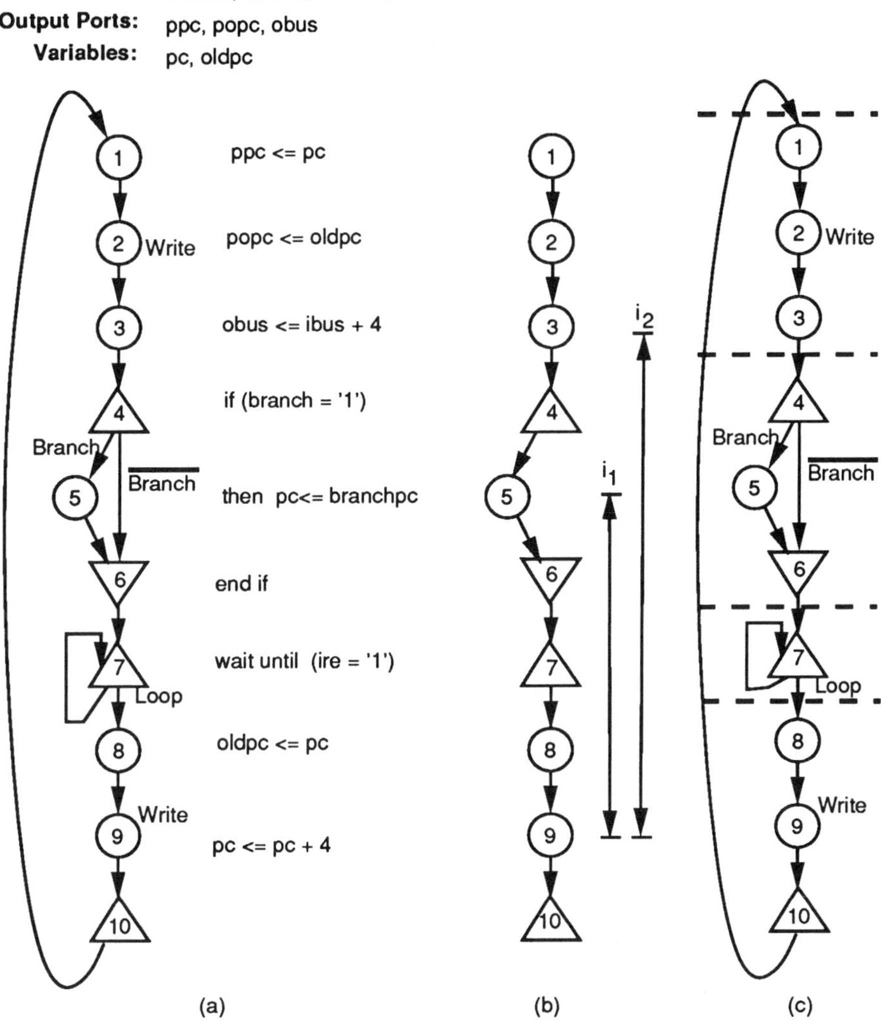

Figure 7.14: Path-based scheduling: (a) an example CDFG, (b) a path in the CDFG with constraint intervals, (c) scheduled CDFG.

In order to schedule the entire path efficiently, it is important to introduce the minimum number of control steps. The problem of introducing the minimum number of control steps that satisfy all constraints can be transformed into a clique-partitioning problem for a graph. A graph is constructed in which each node represents a constraint interval. An edge exists between two nodes if the two corresponding intervals overlap with each other.

The clique-partitioning solution indicates a set of minimum non-overlapping intervals for the given path. Similar intervals are obtained for each path in the graph. The clique partitioning technique is used again to combine the intervals generated from different paths. The results of this clique partitioning represent a final set of intervals for the entire CDFG. Introducing a new control step between any two nodes in this final set of intervals generates an unique schedule for the original CDFG. Figure 7.14(c) shows the final schedule for the example in Figure 7.14(a).

7.4.3   DFG Restructuring

The performance (in terms of the number of control steps) attainable by the scheduling techniques described so far is bounded by the length of the critical path of the DFG. In Figure 7.15(a), the critical-path length is four (i.e., operations 1, 2, 3 and 4). Therefore, none of the techniques we described can do better than scheduling into four control steps. However, we could lower this bound by modifying the structure of the DFG. Two techniques, tree-height reduction [NiPo91] and redundant-operation introduction [LoPa91], can modify the structure of the DFG while preserving its behavior.

Tree-height reduction, also discussed in Chapter 5, restructures a DFG represented as a tree by exploiting the associativity of some operations, e.g., addition. For example, in Figure 7.15(a), the DFG, which represents the computation $(((a + b) + c) + d) + (e + f)$, has a critical-path length of four; the DFG in Figure 7.15(b), which represents the computation $((a + b) + c) + (d + (e + f))$, has a critical-path length of three. Both DFGs have the same behavior since they compute $a + b + c + d + e + f$. However, the second DFG has a shorter tree height, resulting in a shorter critical path, and therefore a possibly

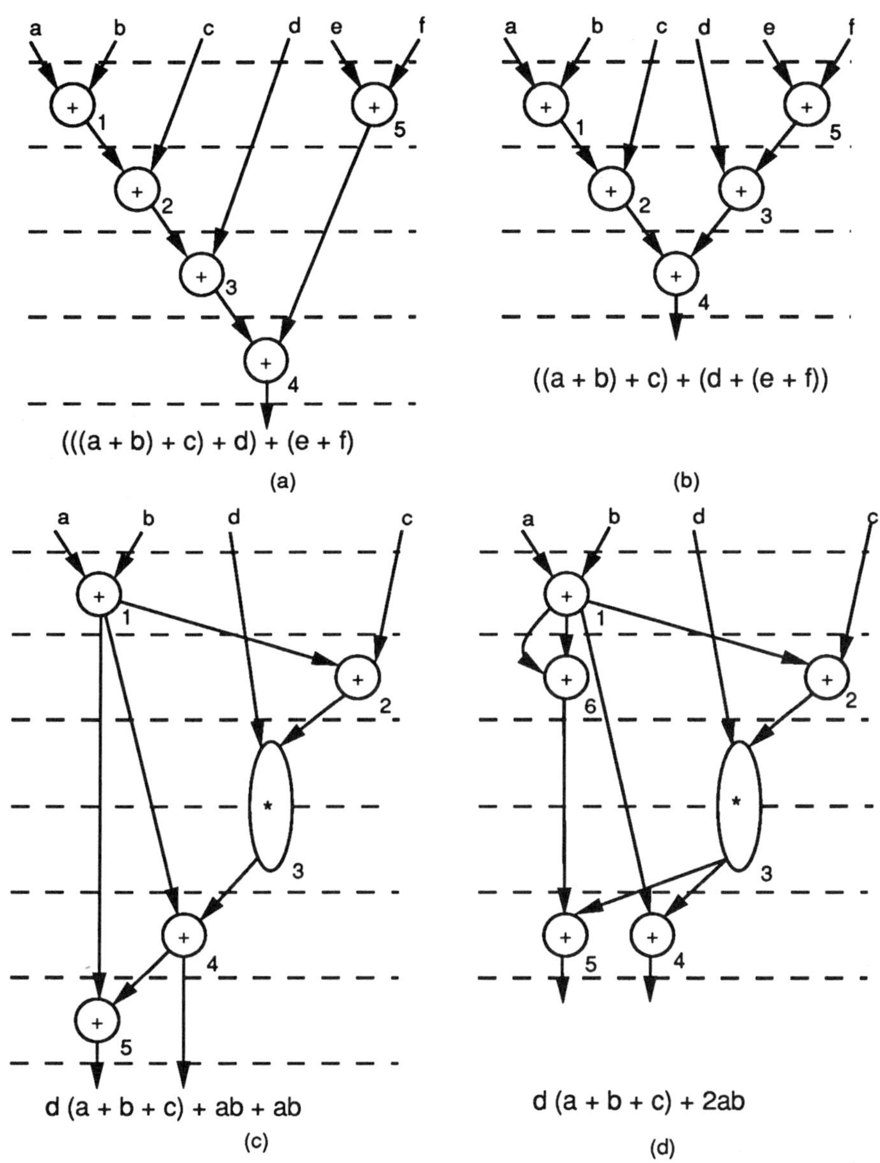

Figure 7.15: DFG restructuring: (a) DFG1, (b) DFG1 after tree-height reduction, (c) DFG2, (d) DFG2 after redundant operation insertion.

better schedule.

Redundant-operation insertion also tries to decrease the critical-path length by introducing additional operators to offload computations on the critical path. For example, Figure 7.15(c) shows a DFG with a critical-path length of six (operation 3 takes two control steps). After introducing a new operation 6, the last operation, 5, on the critical path can be scheduled earlier. Therefore, the critical-path length is reduced from six to five (Figure 7.15(d)).

7.5   Summary and Future Directions

In this chapter, we described several algorithms for scheduling operations into control steps. We first discussed two types of scheduling problems: time-constrained and resource-constrained scheduling. The integer linear programming approach solves the time-constrained problem optimally but suffers from long execution times. On the other hand, the force-directed heuristic produces schedules quickly, but the optimality of the solution cannot be guaranteed. The iterative refinement approach improves the design quality of an initial schedule generated by any scheduling algorithm. We described list scheduling with various priority functions for the resource-constrained formulation of the scheduling problem. We initially used restrictive models for input descriptions and target architectures to explain basic algorithms. We later discussed realistic extensions to the basic scheduling formulations. Finally, we discussed techniques for improving performance by exploiting parallelism beyond the loop boundaries and by restructuring input descriptions.

In scheduling, work is needed on utilizing more realistic libraries, target architectures and cost functions. Given a library of functional units, the scheduling algorithm should be able to select the functional units to optimize both the schedule and the cost. Thus, scheduling algorithms that combine both scheduling and module selection must be developed. Similarly, scheduling algorithms must be combined with allocation, since scheduling and allocation are interdependent.

Furthermore, the cost functions used in scheduling must be realistic. Simple quality measures, such as the number of operators or the number of RT components will not be sufficient. The cost functions must include layout parameters, particularly information about possible floorplans and routing delays for performance-driven scheduling.

Many scheduling algorithms are intended for straight-line code. Few scheduling algorithms exist for scheduling arbitrary descriptions with conditional and loop constructs. Techniques used in optimizing compilers, such as loop-pipelining and tree-height reduction, can be integrated into scheduling for datapath design. Scheduling algorithms must also be extended to include arrays and other data structures in the input description as well as memories with multiclock access times in the target architecture.

Scheduling algorithms should be expanded to incorporate different target architectures, for example a RISC architecture with a large register file and a few functional units, for which minimization of load-and-store instructions from the main memory is the primary goal. Such an architecture could be extended to very large instruction word (VLIW) architectures, in which several RISC datapaths execute in parallel. Similarly, we need scheduling algorithms for other specific architectures, such as those found in DSP applications.

7.6 Exercises

1. Write the constraint inequalities for scheduling the DFG shown in Figure 7.16 into four control steps.

2. Derive a formula for the number of inequalities in the ILP formulation for the DFG in Figure 7.16 given a time constraint of five control steps. Assume that there are two types of functional units: a multiplier for multiplication and an ALU for all other operations.

3. Extend the ILP formulation for multicycled and chained operations.

4. Extend the ILP formulation to minimize a linear cost function of storage, bus and functional units. Estimate the growth in the size of your formulation as compared to the original formulation in Section 7.2.1.

5. **Extend the ILP formulation to handle resource constrained scheduling and module selection simultaneously. Hint: For each type of operation, the component library can contain a number of functional units with different speed/cost attributes. Moreover, the units may be multi-functional.

6. Schedule the DFG in Figure 7.16 into five control steps using the FDS algorithm.

7. Demonstrate that the FDS algorithm can generate non-optimal schedules by using an example with less than ten operations.

8. Calculate the operation costs of the read-data transfers for the DFG in Figure 7.16 for a five control-step schedule.

9. *Extend the concept of balancing operation cost to that of balancing the data-transfer cost, resulting in a high utilization of the bus units. Hint: Read [PaKn89].

10. Improve the effectiveness of the iterative rescheduling method by implementing a look-ahead scheme. Write a pseudo-code description of your algorithm.

11. Given one multiplier and two ALUs, schedule the DFG shown in Figure 7.16 using:
    (a) static-list scheduling, and
    (b) list scheduling based on the critical-path length.
    Can you decrease the number of control steps with two multipliers and two ALUs?

12. Is there a difference between static-list scheduling and list scheduling that uses the same priority function as static-list scheduling?

13. Suppose the DFG in Figure 7.16 is to be executed endlessly where variables r and s carry the data dependence across iterations. Fold the loop using two multipliers and two ALUs. Calculate the average execution time per iteration, the functional unit utilization rate and the control cost.

14. Fold the loop in Figure 7.11(a) using no more than five functional units. Calculate the average execution time per iteration, the functional unit utilization rate and the control cost.

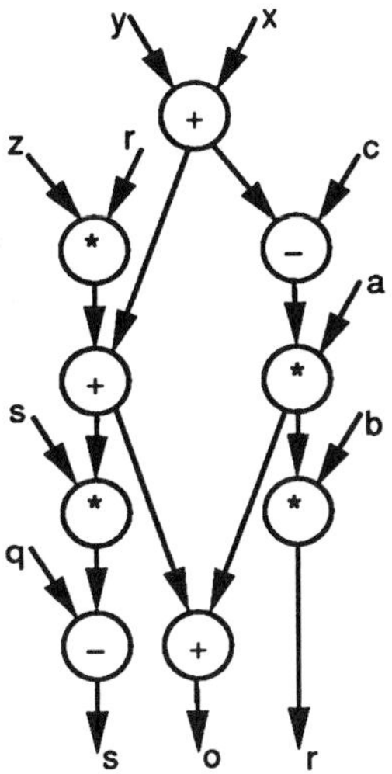

Figure 7.16: A sample DFG.

15. Can you fold the loop in Figure 7.11 so that a new iteration can start its execution every two control steps? Is there a lower bound on the achievable latency (i.e., initiation interval) when folding a loop?

16. Under what conditions can a loop be folded to fully (i.e., 100%) to utilize the functional units?

17. Derive a formula for the control cost in terms of the execution time of an iteration and the number of control steps between the initiation of successive iterations.

18. **Design a folding algorithm for a loop that contains branch constructs. Hint: Your algorithm should be able to explore the parallelism achievable by moving a subset of the branch's subconstructs across the loop boundary.

19. Extract all possible execution paths for the CDFG shown in Figure 5.3.

20. Define a set of rules for adding operations to a DFG in order to increase the degree of parallelism and decrease the number of control

steps in the schedule.

21. *Assume that you have a library that contains multiple implementations of functional units with different sizes and speeds. Devise an algorithm that would perform scheduling in combination with unit selection. Hint: An intuitive strategy for scheduling is to use the fast components only for the operations that are critical to the performance of the overall design while implementing the non-critical operations with the slower components.

22. Compute the cost of implementing the control logic for the schedules given in Figure 7.12(b), Figure 7.13(a) and Figure 7.13(b) assuming a datapath with 3, 4 and 6 functional units and a shared register file with 6, 8 and 12 read ports and 3, 4 and 6 write ports. Assume that the register file has 256 words.

Chapter 8

Allocation

8.1 Problem Definition

As described in the previous chapter, scheduling assigns operations to control steps and thus converts a behavioral description into a set of register transfers that can be described by a state table. A target architecture for such a description is the FSMD given in Chapter 2. We derive the control unit for such a FSMD from the control-step sequence and the conditions used to determine the next control step in the sequence. The datapath is derived from the register transfers assigned to each control step; this task is called datapath synthesis or datapath allocation.

A datapath in the FSMD model is a netlist composed of three types of register transfer (RT) components or units: functional, storage and interconnection. Functional units, such as adders, shifters, ALUs and multipliers, execute the operations specified in the behavioral description. Storage units, such as registers, register files, RAMs and ROMs, hold the values of variables generated and consumed during the execution of the behavior. Interconnection units, such as buses and multiplexers, transport data between the functional and storage units.

Datapath allocation consists of two essential tasks: unit selection and unit binding. Unit selection determines the number and types of RT components to be used in the design. Unit binding involves the mapping of the variables and operations in the scheduled CDFG into the functional, storage and interconnection units, while ensuring that the

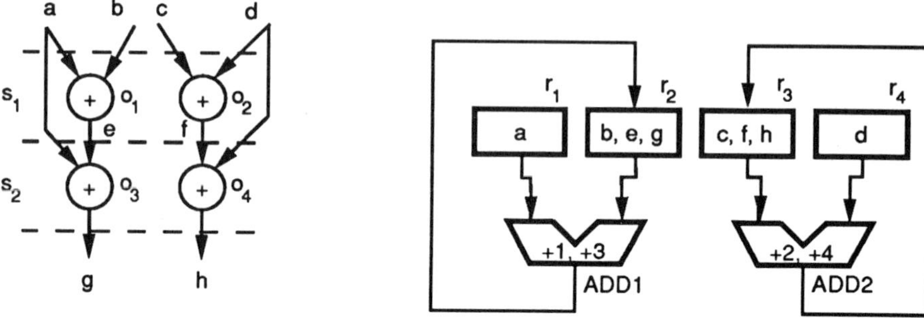

Figure 8.1: Mapping of behavioral objects into RT components.

design behavior operates correctly on the selected set of components. For every operation in the CDFG, we need a functional unit that is capable of executing the operation. For every variable that is used across several control steps in the scheduled CDFG, we need a storage unit to hold the data values during the variable's lifetime. Finally, for every data transfer in the CDFG, we need a set of interconnection units to effect the transfer. Besides the design constraints imposed on the original behavior and represented in the CDFG, additional constraints on the binding process are imposed by the type of hardware units selected. For example, a functional unit can execute only one operation in any given control step. Similarly, the number of multiple accesses to a storage unit during a control step is limited by the number of parallel ports on the unit.

We illustrate the mapping of variables and operations in the DFG of Figure 8.1 into RT components. Let us assume that we select two adders, *ADD1* and *ADD2*, and four registers, r_1, r_2, r_3 and r_4. Operations o_1 and o_2 cannot be mapped into the same adder because they must be performed in the same control step s_1. On the other hand, operation o_1 can share an adder with operation o_3 because they are carried out during different control steps. Thus, operations o_1 and o_3 are both mapped into *ADD1*. Variables a and e must be stored separately because their values are needed concurrently in control step s_2. Registers r_1 and r_2, where variables a and e reside, must be connected to the input ports of $ADD1$; otherwise, operation o_3 will not be able to execute in $ADD1$. Similarly, operations o_2 and o_4 are mapped to $ADD2$. Note that there are several different ways of performing the binding. For example, we can map o_2 and o_3 to $ADD1$ and o_1 and o_4 to $ADD2$.

Besides implementing the correct behavior, the allocated datapath must meet the overall design constraints in terms of the metrics defined in Chapter 3 (e.g., area, delay, power dissipation, etc.). To simplify the allocation problem, we use two quality measures for datapath allocation: the total size (i.e., the silicon area) of the design and the worst-case register-to-register delay (i.e., the clock cycle) of the design.

We can solve the allocation problem in three ways: greedy approaches, which progressively construct a design while traversing the CDFG; decomposition approaches, which decompose the allocation problem into its constituent parts and solve each of them separately; and iterative methods, which try to combine and interleave the solution of the allocation subproblems.

We begin this chapter with a brief discussion of typical datapath architectural features and their effects on the datapath-allocation problem. Using a simple design model, we then outline the three techniques for datapath allocation: the greedy constructive approach, the decomposition approach and the iterative refinement approach. Finally, we conclude this chapter with a brief discussion of future trends.

8.2 Datapath Architectures

In Chapter 2, we discussed basic target architectures and showed how pipelined datapaths are used for performance improvements with negligible increase in cost. In Chapter 3, we presented formulas for calculating the clock cycle for such architectures. In this section we will review some basic features of real datapaths and relate them to the formulation of the datapath-allocation problem.

A datapath architecture defines the characteristics of the datapath units and the interconnection topology. A simple target architecture may greatly reduce the complexity of the synthesis problems since the number of alternative designs is greatly reduced. On the other hand, a less constrained architecture, although more difficult to synthesize, may result in higher quality designs. While an oversimplified datapath architecture leads to elegant synthesis algorithms, it also usually results in unacceptable designs.

The interconnection topology that supports data transfers between the storage and functional units is one of the factors that has a significant influence on the datapath performance. The complexity of the interconnection topology is defined by the maximum number of interconnection units between any two ports of functional or storage units. Each interconnection unit can be implemented with a multiplexer or a bus. For example, Figure 8.2 shows two datapaths, using multiplexer and bus interconnection units respectively, which implement the following five register transfers:

$$s_1: r_3 \Leftarrow ALU1\,(r_1, r_2); \quad r_1 \Leftarrow ALU2\,(r_3, r_4);$$
$$s_2: r_1 \Leftarrow ALU1\,(r_5, r_6); \quad r_6 \Leftarrow ALU2\,(r_2, r_5);$$
$$s_3: r_3 \Leftarrow ALU1\,(r_1, r_6); \; .$$

We call the interconnection topology "point-to-point" if there is only one interconnection unit between any two ports of the functional and/or storage units. The point-to-point topology is most popular in high-level synthesis since it simplifies the allocation algorithms. In this topology, we create a connection between any two functional or storage units as needed. If more than one connection is assigned to the input of a unit, a multiplexer or a bus is used. In order to minimize the number of interconnections, we can combine registers into register files with multiple ports. Each port may support read and/or write accesses to the data. Some register files allow simultaneous read and write accesses through different ports. Although register files reduce the cost of interconnection units, each port requires dedicated decoder circuitry inside the register file, which increases the storage cost and the propagation delay.

To simplify the binding problem, in this section we assume that all register transfers go through functional units and that direct interconnections of two functional units are not allowed. Therefore, we only need interconnection units to connect the output ports of storage units

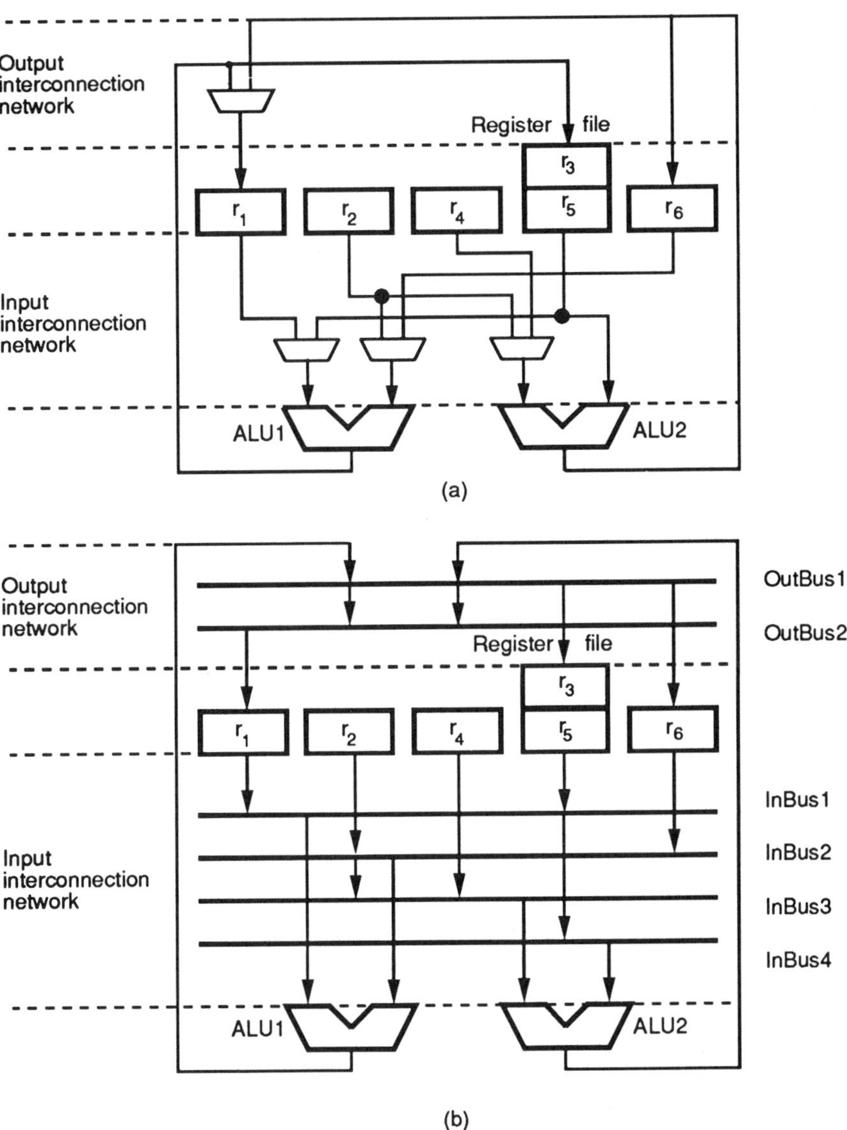

Figure 8.2: Datapath interconnections: (a) a multiplexer-oriented datapath, (b) a bus-oriented datapath.

to the input ports of functional units (i.e., the input interconnection network) and the output ports of functional units to the input ports of storage units (i.e., the output interconnection network).

The complexity of the input and output interconnection networks need not be the same. One can be simplified at the expense of the other. For example, selectors may not be allowed in front of the input ports of storage units. This results in a more complicated input interconnection network, and, hence, an imbalance between the read and write times of the storage units. In addition to affecting the allocation algorithms, such an architecture increases testability [GrDe90]. Furthermore, the number of buses that each register file drives can be constrained. If a unique bus is allocated for each register-file output, tri-state bus drivers are not needed between the registers and the buses [GeEl90]. This restriction on register-file outputs produces more multiplexers at the input ports of functional units. Moreover, some variables may need to be duplicated across different register files in order to simplify the selector circuits between the buses and the functional units.

Another interconnection scheme commonly used in processors has the buses partitioned into segments so that each bus segment is used by one functional unit at a time [Ewer90]. The functional and storage units are arranged in a single row with bus segments on either side, and so looks like a 2-bus architecture. A data transfer between the segments is achieved through switches between bus segments. Thus, we accomplish interconnection allocation by controlling the switches between the bus segments to allow or disallow data transfers.

Let us analyze the delays involved in register-to-register transfers for the five-transfer example in Figure 8.2. The relative timing of read, execute and write micro-operations in the first two clock cycles of the example are shown in Figure 8.3. Let t_r be the time delay involved for reading the data out of the registers and then propagating through the input interconnection network; t_e, the propagation delay through a functional unit; and t_w, the delay for data to propagate from the functional units through the output interconnection network and be written to the registers. In the target architectures described so far, all the components along the path from the registers' output ports back to the registers' input ports (i.e., the input interconnection network, the ALUs and the output interconnection network) are combinational. Thus, the clock

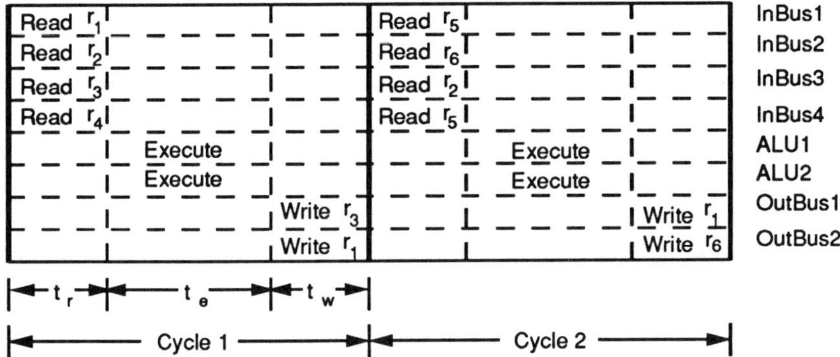

Figure 8.3: Sequential execution of three micro-operations in the same clock period.

cycle will be equal to or greater than $t_r + t_e + t_w$.

As described in Chapter 2, latches may be inserted at the input and/or output ports of functional units to improve the datapath performance. When latches are inserted only at the outputs of the functional units (Figure 8.4), the read accesses and functional-unit execution for operations scheduled into the current control step can be performed at the same time as the write accesses of operations scheduled into the previous control step (Figure 8.5). The cycle period is reduced to $\max(t_r + t_e, t_w)$. But the register transfers are not well balanced: reading of the registers and execution of an ALU operation are performed in the first cycle, while only writing of the result back into the registers is performed during the second cycle. Similarly, if only the inputs of the functional units are latched, the read accesses for operations scheduled into the next control step can be performed at the same time as the functional unit-execution and the write accesses for operations scheduled into the current control step. The cycle period in that case will be $\max(t_r, t_e + t_w)$. In either case, the register files and latches are controlled by a single-phase clock.

Operation execution and the reading/writing of data can take place

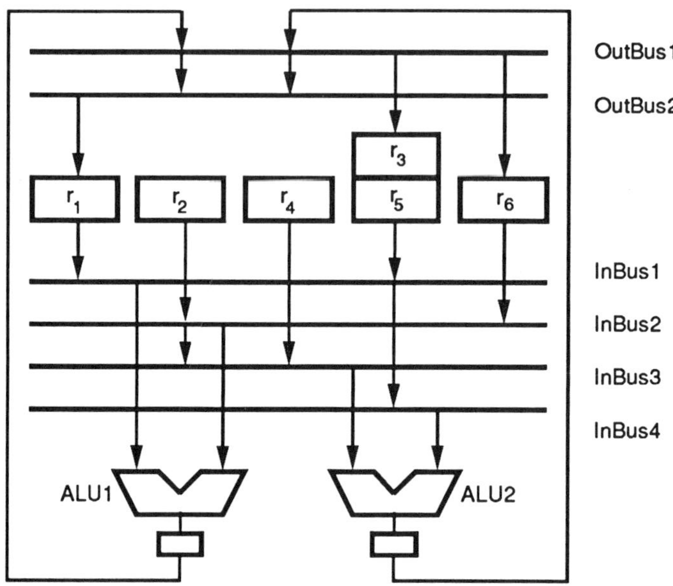

Figure 8.4: Insertion of latches at the output ports of the functional units.

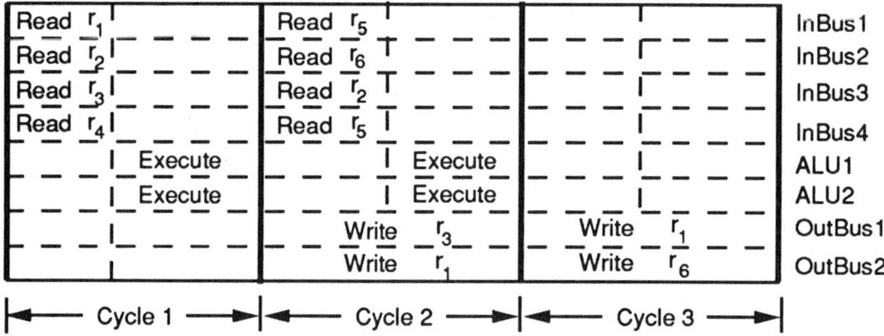

Figure 8.5: Overlapping read and write data transfers.

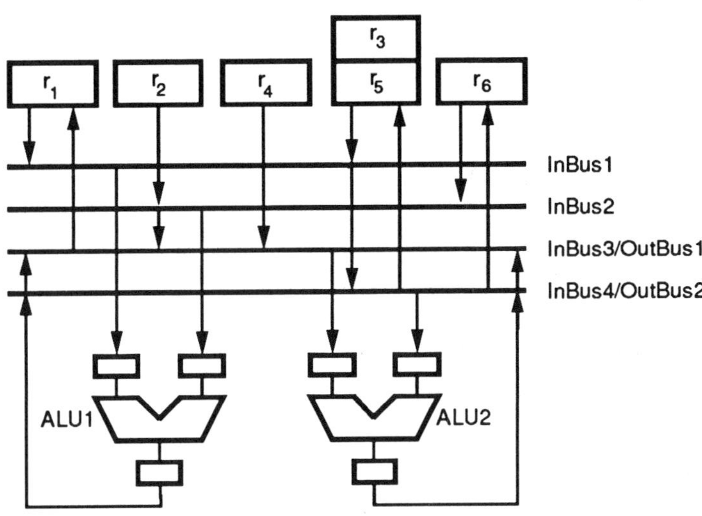

Figure 8.6: Insertion of latches at both the input and output ports of the functional units.

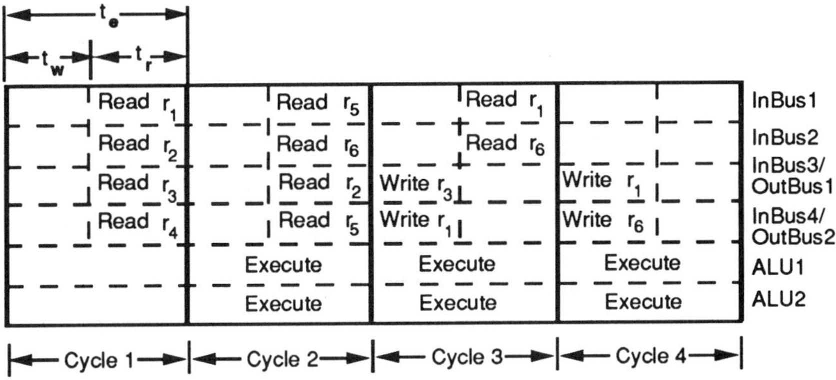

Figure 8.7: Overlapping data transfer with functional-unit execution.

concurrently when both the inputs and outputs of a functional unit are latched (Figure 8.6). The three combinational components, namely, the input-interconnection units, the functional units and the output-interconnection units, can all be active concurrently. Figure 8.7 shows how this pipelining scheme works. The clock cycle consists of two minor cycles. The execution of an operation is spread across three consecutive clock cycles. The input operands for an operation are transferred from the register files to the input latches of the functional units during the second minor cycle of the first cycle. During the second cycle, the functional unit executes the operation and writes the result to the output latch by the end of the cycle. The result is transferred to the final destination, the register file, during the first minor cycle of the third cycle. A two-phase non-overlapping clocking scheme is needed. Both the input and output latches are controlled by one phase since the end of the read access and the end of operation execution occur simultaneously. The other phase is used to control the write accesses to the register files.

By overlapping the execution of operations in successive control steps, we can greatly increase the hardware utilization. The cycle period is reduced to $\max(t_e,\ t_r + t_w)$. Moreover, the input and output networks can share some interconnection units. For example, *OutBus1* is merged with *InBus3* and *OutBus2* with *InBus4* in Figure 8.6.

Thus, inserting input and output latches makes available more interconnection units for merging, which may simplify the datapath design. By breaking the register-to-register transfers into micro-operations executed in different clock cycles, we achieve a better utilization of hardware resources. However, this scheme requires binding algorithms to search through a larger number of design alternatives.

Operator chaining was introduced in Section 7.3.1 as the execution of two or more operations in series during the same control step. To support operation chaining, links are needed from the output ports of some functional units directly to the input ports of other functional units. In an architecture with a shared bus (e.g., Figure 8.6), this linking can be accomplished easily by using the path from a functional unit's output port through one of the buses to some other functional unit's input port. Since such a path must be combinational, bypass circuits have to be added around all the latches along the path of chaining.

8.3 Allocation Tasks

Datapath synthesis consists of four different yet interdependent tasks:
module selection, functional-unit allocation, storage allocation and interconnection allocation. In this section, we define each task and discuss
the nature of their interdependence.

8.3.1 Unit Selection

A simple design model may assume that we have only one particular
type of functional unit for each behavioral operation. However, a real
RT component library contains multiple types of functional units, each
with different characteristics (e.g., functionality, size, delay and power
dissipation) and each implementing one or several different operations
in the register-transfer description. For example, an addition can be
carried out by either a small but slow ripple adder or by a large but
fast carry look-ahead adder. Furthermore, we can use several different
component types, such as an adder, an adder/subtracter or an entire
ALU, to perform an addition operation. Thus, unit selection selects
the number and types of different functional and storage units from the
component library. A basic requirement for unit selection is that the
number of units performing a certain type of operation must be equal
to or greater than the maximum number of operations of that type to
be performed in any control step. Unit selection is frequently combined
with binding into one task called allocation.

8.3.2 Functional-Unit Binding

After all the functional units have been selected, operations in the behavioral description must be mapped into the the set of selected functional
units. Whenever we have operations that can be mapped into more
than one functional unit, we need a functional-unit binding algorithm to
determine the exact mapping of the operations into the functional units.
For example, operations o_1 and o_3 in Figure 8.1 have been mapped into
adder *ADD1*, while the operations o_2 and o_4 have been mapped into
adder *ADD2*.

8.3.3 Storage Binding

Storage binding maps data carriers (e.g., constants, variables and data structures like arrays) in the behavioral description to storage elements (e.g., ROMs, registers and memory units) in the datapath. Constants, such as coefficients in a DSP algorithm, are usually stored in a read-only memory (ROM). Variables are stored in registers or memories. Variables whose lifetime intervals do not overlap with each other may share the same register or memory location. The lifetime of a variable is the time interval between its first value assignment (the first variable appearance on the left-hand side of an assignment statement) and its last use (the last variable appearance on the right-hand side of an assignment statement). After variables have been assigned to registers, the registers can be merged into a register file with a single access port if the registers in the file are not accessed simultaneously. Similarly, registers can be merged into a multiport register file as long as the number of registers accessed in each control step does not exceed the number of ports.

8.3.4 Interconnection Binding

Every data transfer (i.e., a read or write) needs an interconnection path from its source to its sink. Two data transfers can share all or part of the interconnection path if they do not take place simultaneously. For example, in Figure 8.1, the reading of variable b in control step s_1 and variable e in control step s_2 can be achieved by using the same interconnection unit. However, writing to variables e and f, which occurs simultaneously in control step s_1, must be accomplished using disjoint paths. The objective of interconnection binding is to maximize the sharing of interconnection units and thus minimize the interconnection cost, while still supporting the conflict-free data transfers required by the register-transfer description.

8.3.5 Interdependence and Ordering

All the datapath synthesis tasks (i.e., scheduling, unit selection, functional unit binding, storage binding and interconnection binding) depend on each other. In particular, functional-unit, storage and inter-

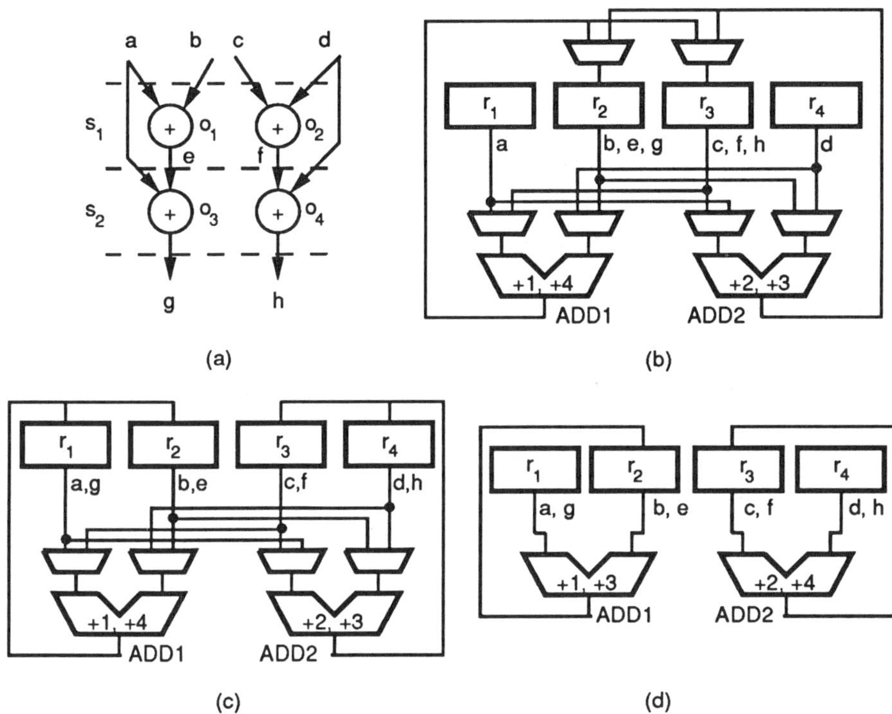

Figure 8.8: Interdependence of functional-unit and storage binding: (a) a scheduled DFG, (b) a functional-unit binding requiring six multiplexers, (c) improved design with two fewer multiplexers, obtained by register reallocation, (d) optimal design with no multiplexers, obtained by modifying functional-unit binding.

connection binding are tightly related to each other. For example, Figure 8.8 shows how both functional unit and storage binding affect interconnection allocation. Suppose the eight variables (a through g) in the DFG of Figure 8.8(a) have been partitioned into four registers as follows: $r_1 \leftarrow \{a\}$, $r_2 \leftarrow \{b, e, g\}$, $r_3 \leftarrow \{c, f, h\}$, $r_4 \leftarrow \{d\}$. Given two adders, *ADD1* and *ADD2*, there are two ways of grouping the four addition operations, o_1, o_2, o_3, and o_4, in Figure 8.8(a) so that each group is assigned to one adder:

(1) $ADD1 \leftarrow \{o_1, o_4\}$, $ADD2 \leftarrow \{o_2, o_3\}$, or

(2) $ADD1 \leftarrow \{o_1, o_3\}$, $ADD2 \leftarrow \{o_2, o_4\}$.

For the given register binding, we need six 2-to-1 multiplexers for unit interconnection in case (1) (Figure 8.8(b)). However, we can eliminate two 2-to-1 multiplexers (Figure 8.8(c)), by modifying the register binding to be : $r_1 \leftarrow \{a, g\}$, $r_2 \leftarrow \{b, e\}$, $r_3 \leftarrow \{c, f\}$, $r_4 \leftarrow \{d, h\}$. If we then modify the functional-unit binding to case (2) above, no multiplexers are needed (Figure 8.8(d)). Thus, this design is optimal if the interconnection cost is measured by the number of multiplexers needed. Clearly, both functional-unit and storage binding potentially affect the optimization achievable by interconnection allocation.

The previous example also raises the issue of ordering among the allocation tasks. The requirements on interconnection become clear after both functional-unit and storage allocation have been performed. Furthermore, functional-unit allocation can make correct decisions if storage allocation is done beforehand, and vice versa. To break this deadlock situation, we choose one task ahead of the other. Unfortunately, in such a ordering, the first task chosen cannot use the information from the second task, which would have been available had the second task been performed first.

8.4   Greedy Constructive Approaches

We can construct a datapath using a greedy approach in which RT components are assigned to operations in a step-by-step fashion [KuPa90].

A constructive algorithm starts with an empty datapath and builds the datapath gradually by adding functional, storage and interconnection units as necessary. For each operation, it tries to find a functional unit on the partially designed datapath that is capable of executing the operation and is idle during the control step in which the operation must be executed. In case there are two or more functional units that meet these conditions, we choose the one which results in a minimal increase in the interconnection cost. On the other hand, if none of the functional units on the partially designed datapath meet the conditions, we add a new functional unit from the component library that is capable of carrying out the operation. Similarly, we can assign a variable to an available register only if its lifetime interval does not overlap with those of variables already assigned to that register. A new register is allocated only when no allocated register meets the above condition. Again, when multiple alternatives exist for assignment of a variable to a register, we select the one that minimally increases the datapath cost.

Figures 8.9(b)-(g) show several modified datapaths obtained by adding interconnection and/or functional units to the partial design of Figure 8.9(a). We discuss each case below:

1. In Figure 8.9(b) we add a new connection from each of r_3 and *Bus1* to the left input of *ALU1*.

2. In Figure 8.9(c) we add a connection from r_3 to the right input of *ALU2* through *Bus1*. Since the connection from *Bus1* to the right input of *ALU2* already exists, we add only a tri-state buffer.

3. Figure 8.9(d) is similar to Figure 8.9(c), except that we add the multiplexer *Mux2* instead of a tri-state buffer.

4. In Figure 8.9(e) we add a new functional unit, *ALU3*, and a new connection from each of r_3 and *Bus1* to the right input of *ALU3*.

5. Figure 8.9(f) is similar to Figure 8.9(e), except that we add the multiplexer *Mux2* instead of a tri-state buffer.

6. In Figure 8.9(g) we merge *Mux1* and *Bus1* into a single bus, *Bus1*.

Interconnection allocation follows immediately after both the source and sink of a data transfer have been bound. For example, the partial

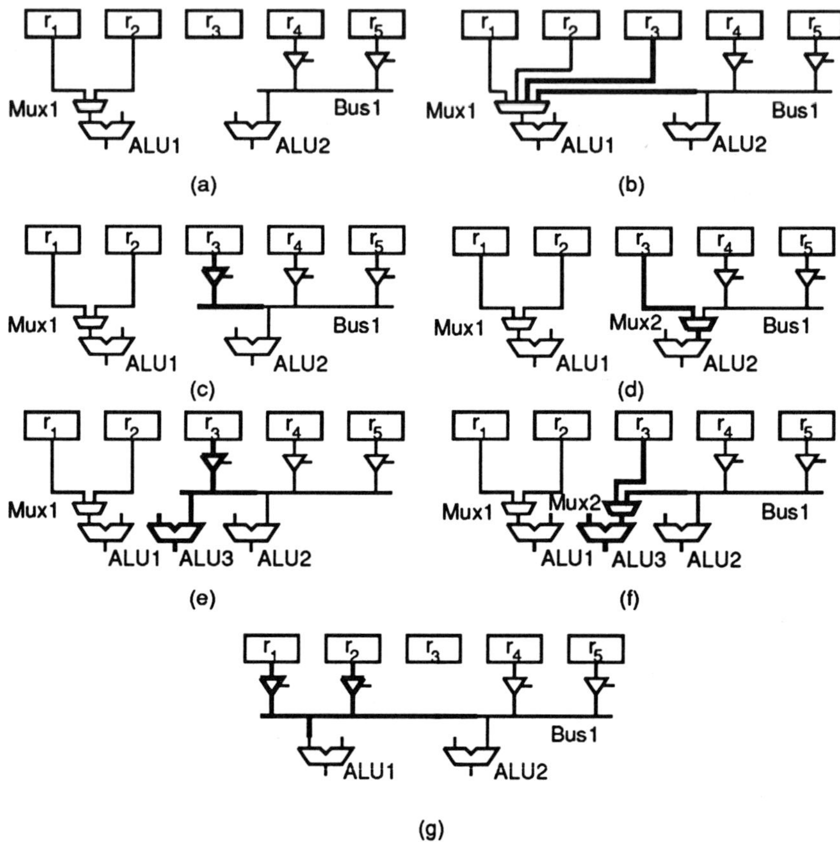

Figure 8.9: Datapath construction: (a) an initial partial design, (b) addition of two more inputs to a multiplexer, (c) addition of a tri-state buffer to a bus, (d) addition of a multiplexer to the input of a functional unit, (e) addition of a functional unit and a tri-state buffer to a bus, (f) addition of a functional unit and a multiplexer, (g) conversion of a multiplexer to a shared bus.

datapath in Figure 8.9(a) does not have a link between register r_3 and *ALU2*. Suppose a variable that is one of the inputs to an operation that has been assigned to *ALU2* is just assigned to register r_3. Then, an interconnection link has to be established as shown by the bold wires in Figure 8.9(c). Each interconnection unit that is required by a data transfer in the behavioral description contributes to the datapath cost. For example, in Figure 8.9(b), two more connections are made to the left input port of *ALU1*, where a two-input multiplexer already exists. Therefore, the cost of this modification is the difference between the cost of a two-input multiplexer and that of a four-input multiplexer. However, at least two additional data transfers are now supported with this modification.

Algorithm 8.1 describes the greedy constructive allocation method. Let UBE be the set of unallocated behavioral entities and $DP_{current}$ be the partially designed datapath. The behavioral entities being considered could be variables that have to be mapped into registers, operations that have to be mapped into functional units, or data transfers that have to be mapped into interconnection units. $DP_{current}$ is initially empty. The procedure ADD(DP, *ube*) structurally modifies the datapath DP by adding to it the components necessary to support the behavioral entity *ube*. The function COST(DP) evaluates the area/performance cost of a partially designed datapath DP. DP_{work} is a temporary datapath which is created in order to evaluate the cost c_{work} of performing each modification to $DP_{current}$.

Starting with the set UBE, the inner *for* loop determines which unallocated behavioral entity, *BestEntity*, requires the minimal increase in the cost when added to the datapath. This is accomplished by adding each of the unallocated behavioral entities in UBE to $DP_{current}$ individually and then evaluating the resulting cost. The procedure ADD then modifies $DP_{current}$ by incorporating *BestEntity* into the datapath. *BestEntity* is deleted from the set of unallocated behavioral entities. The algorithm iterates in the outer *while* loop until all behavioral entities have been allocated (i.e., $UBE = \phi$).

In order to use the greedy constructive approach, we have to address two basic issues: the cost-function calculation and the order in which the unallocated behavioral entities are mapped into the datapath. The costs can be computed as explained in Chapter 3. For example, the cost

Algorithm 8.1: Constructive Allocation.

```
DP_current = φ;
while UBE ≠ φ do
    LowestCost = ∞;

    for all ube ∈ UBE do
        DP_work = ADD(DP_current, ube);
        c_work = COST(DP_work);
        if c_work < LowestCost then
            LowestCost = c_work;
            BestEntity = ube;
        endif
    endfor

    DP_current = ADD(DP_current, BestEntity);
    UBE = UBE - BestEntity;
endwhile
```

of converting the datapath in Figure 8.9(a) to the one in Figure 8.9(g) depends on the difference between the cost of one 2-to-1 multiplexer and that of three tri-state buffers. Since the buses of Figure 8.9(a) and Figure 8.9(g) are of different length, the two datapaths may also have different wiring costs.

The order in which unallocated entities are mapped into the datapath can be determined either statically or dynamically. In a static approach, the objects are ordered before the datapath construction begins. The ordering is not changed during the construction process. By contrast, in a dynamic approach no ordering is done beforehand. To select an operation or variable for binding to the datapath, we evaluate every unallocated behavioral entity in terms of the cost involved in modifying the partial datapath, and the entity that requires the least expensive modification is chosen. After each binding, we revaluate the costs associated with the remaining unbound entities. Algorithm 8.1 uses the dynamic strategy.

In hardware sharing, a previously expensive binding may become inexpensive after some other bindings are done. Therefore, a good strategy incorporates a look-ahead factor into the cost function. That is, the cost of a modification to the datapath should be lower if it decreases the cost of other bindings done in the future.

8.5   Decomposition Approaches

The intuitive approach taken by the constructive method falls into the category of greedy algorithms. Although greedy algorithms are simple, the solutions they find can be far from optimal. In order to improve the quality of the results, some researchers have proposed a decomposition approach, where the allocation process is divided into a sequence of independent tasks; each task is transformed into a well-defined problem in graph theory and then solved with a proven technique.

While a greedy constructive approach like the one described in Algorithm 8.1 might interleave the storage, functional-unit, and interconnection allocation steps, decomposition methods will complete one task before performing another. For example, all variable-to-register assignments might be completed before any operation-to-functional-unit assignments are performed, and vice versa. Because of interdependencies among these tasks, no optimal solution is guaranteed even if all the tasks are solved optimally. For example, a design using three adders may need one fewer multiplexer than the one using only two adders. Therefore, an allocation strategy that minimizes the usage of adders is justified only when an adder costs more than a multiplexer.

In this section we describe allocation techniques based on three graph-theoretical methods: clique partitioning, left-edge algorithm and the weighted bipartite matching algorithm. For the sake of simplicity, we illustrate these allocation techniques by applying them to behavioral descriptions consisting of straight-line code without any conditional branches.

8.5.1   Clique Partitioning

The three tasks of storage, functional-unit and interconnection allocation can be solved independently by mapping each task to the well known problem of graph clique-partitioning [TsSi86].

We begin by defining the clique-partitioning problem. Let $G = (V, E)$

denote a graph, where V is the set of vertices and E the set of edges. Each edge $e_{i,j} \in E$ links two different vertices v_i and $v_j \in V$. A subgraph SG of G is defined as (SV, SE), where $SV \subseteq V$ and $SE = \{e_{i,j} \mid e_{i,j} \in E, v_i, v_j \in SV\}$. A graph is complete if and only if for every pair of its vertices there exists an edge linking them. A clique of G is a complete subgraph of G. The problem of partitioning a graph into a minimal number of cliques such that each node belongs to exactly one clique is called clique partitioning. The clique-partitioning problem is a classic NP-complete problem for which heuristic procedures are usually used.

Algorithm 8.2 describes a heuristic proposed by Tseng and Siewiorek [TsSi86] to solve the clique-partitioning problem. A super-graph $G'(S, E')$ is derived from the original graph $G(V, E)$. Each node $s_i \in S$ is a super-node that can contain a set of one or more vertices $v_i \in V$. E' is identical to E except that the edges in E' now link super-nodes in S. A super-node $s_i \in S$ is a common neighbor of the two super-nodes s_j and $s_k \in S$ if there exist edges $e_{i,j}$ and $e_{i,k} \in E'$. The function COMMON_NEIGHBOR(G', s_i, s_j) returns the set of super-nodes that are common neighbors of s_i and s_j in G'. The procedure DELETE_EDGE(E', s_i) deletes all edges in E' which have s_i as their end super-node.

Initially, each vertex $v_i \in V$ of G is placed in a separate super-node $s_i \in S$ of G'. At each step, the algorithm finds the super-nodes s_{Index1} and s_{Index2} in S such that they are connected by an edge and have the maximum number of common neighbors. These two super-nodes are merged into a single super-node, $s_{Index1Index2}$, which contains all the vertices of s_{Index1} and s_{Index2}. The set $CommonSet$ contains all the common neighbors of s_{Index1} and s_{Index2}. All edges originating from s_{Index1} or s_{Index2} in G' are deleted. New edges are added from $s_{Index1Index2}$ to all the super-nodes in $CommonSet$. The above steps are repeated until there are no edges left in the graph. The vertices contained in each super-node $s_i \in S$ form a clique of the graph G.

Figure 8.10 illustrates the above algorithm. In the graph of Figure 8.10(a), $V = \{v_1, v_2, v_3, v_4, v_5\}$ and $E = \{e_{1,3}, e_{1,4}, e_{2,3}, e_{2,5}, e_{3,4}, e_{4,5}\}$. Initially, each vertex is placed in a separate super-node (labeled s_1 through s_5 in Figure 8.10(b)). The three edges, $e'_{1,3}$, $e'_{1,4}$ and $e'_{3,4}$, of the super-graph G' have the maximum number of common neighbors among all edges (Figure 8.10(b)). The first edge, $e'_{1,3}$, is selected and the

Algorithm 8.2: Clique Partitioning.

```
/* create a super graph G'(S, E') */
S = φ;     E' = φ;
for each vᵢ ∈ V do   sᵢ = {vᵢ}; S = S ∪ {sᵢ};   endfor
for each eᵢ,ⱼ ∈ E do   E' = E' ∪ {e'ᵢ,ⱼ};   endfor

while E' ≠ φ do

    /* find s_Index1, s_Index2 having most common neighbors */
    MostCommons = -1;
    for each e'ᵢ,ⱼ ∈ E' do
       cᵢ,ⱼ = | COMMON_NEIGHBOR(G', sᵢ, sⱼ) |;
       if cᵢ,ⱼ > MostCommons then
          MostCommons = cᵢ,ⱼ;
          Index1 = i;     Index2 = j;
       endif
    endfor

    CommonSet = COMMON_NEIGHBOR(G', s_Index1, s_Index2);

    /* delete all edges linking s_Index1 or s_Index2 */
    E' = DELETE_EDGE(E', s_Index1);
    E' = DELETE_EDGE(E', s_Index2);

    /* merge s_Index1 and s_Index2 into s_Index1Index2 */
    s_Index1Index2 = s_Index1 ∪ s_Index2;
    S = S - s_Index1 - s_Index2;
    S = S ∪ {s_Index1Index2};

    /* add edge from s_Index1Index2 to super-nodes in CommonSet */
    for each sᵢ ∈ CommonSet do
       E' = E' ∪ {e'ᵢ,Index1Index2};
    endfor

endwhile
```

following steps are carried out to yield the graph of Figure 8.10(c).

(1) s_4, the only common neighbor of s_1 and s_3 is put in *CommonSet*.

(2) All edges are deleted that link either super-nodes s_1 or s_3 (i.e., $e'_{1,3}$, $e'_{1,4}$, $e'_{2,3}$ and $e'_{3,4}$).

(3) Super-nodes s_1 and s_3 are combined into a new super-node s_{13}.

(4) An edge is added between s_{13} and each super-node in *CommonSet*; i.e., the edge $e_{13,4}$ is added.

On the next iteration, s_4 is merged into s_{13} to yield the super-node s_{134} (Figure 8.10(d)). Finally, s_2 and s_5 are merged into the super-node s_{25} (Figure 8.10(e)). The cliques are $s_{134} = \{v_1, v_3, v_4\}$ and $s_{25} = \{v_2, v_5\}$ (Figure 8.10(f)).

In order to apply the clique partitioning technique to the allocation problem, we have to first derive the graph model from the input description. Consider register allocation as an example. The primary goal of register allocation is to minimize the register cost by maximizing the sharing of common registers among variables. To solve the register allocation problem, we construct a graph $G = (V, E)$, in which every vertex $v_i \in V$ uniquely represents a variable v_i and there exists an edge $e_{i,j} \in E$ if and only if variables v_i and v_j can be stored in the same register (i.e., their lifetime intervals do not overlap). All the variables whose representative vertices are in a clique of G can be stored in a single register. A clique partitioning of G provides a solution for the datapath storage-allocation problem that requires a minimal number of registers. Figure 8.11 shows a solution of the register-allocation problem using the clique-partitioning algorithm.

Both functional-unit allocation and interconnection allocation can be formulated as a clique-partitioning problem. For functional-unit allocation, each graph vertex represents an operation. An edge exists between two vertices if two conditions are satisfied:

(1) the two operations are scheduled into different control steps, and

(2) there exists a functional unit that is capable of carrying out both operations.

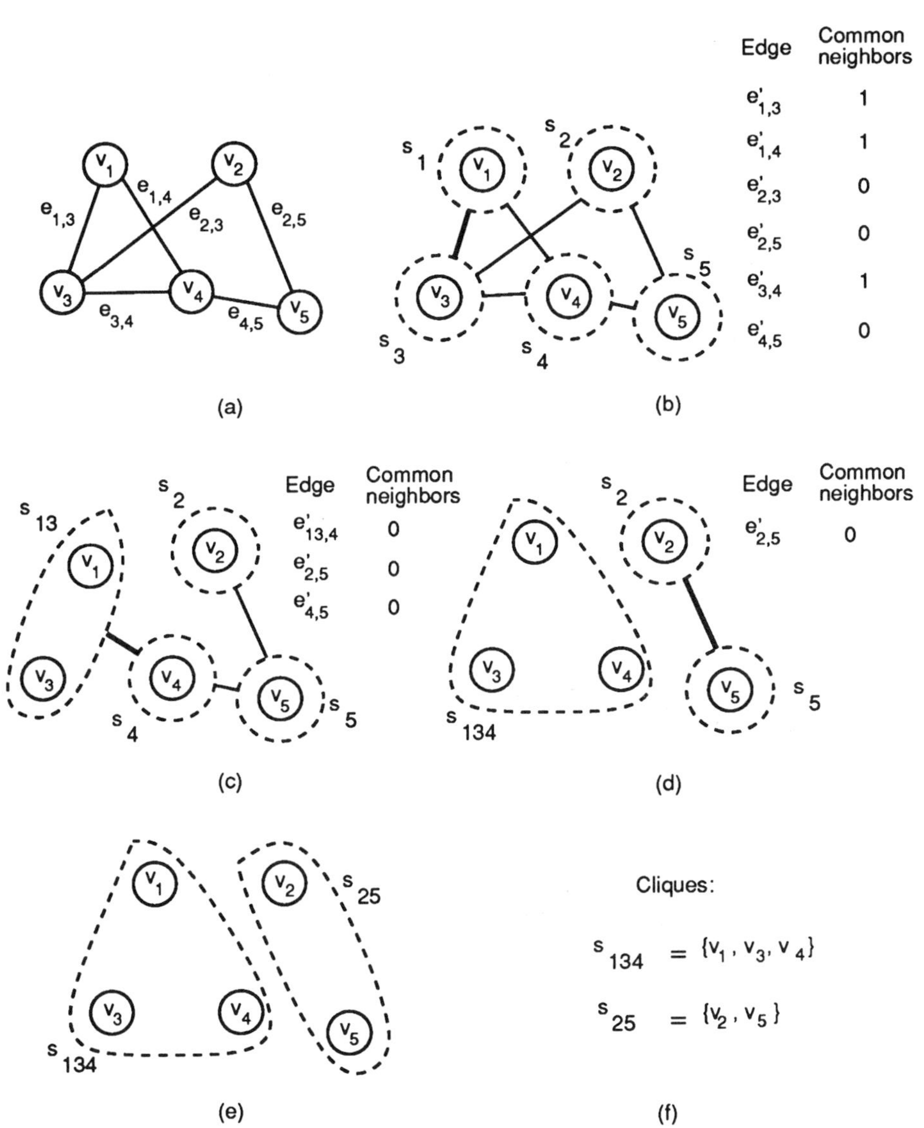

Figure 8.10: Clique partitioning: (a) given graph G, (b) calculating the common neighbors for the edges of graph G', (c) super-node s_{13} formed by considering edge $e'_{1,3}$, (d) super-node s_{134} formed by considering edge $e'_{13,4}$, (e) super-node s_{25} formed by considering edge $e'_{2,5}$, (f) resulting cliques s_{134} and s_{25}.

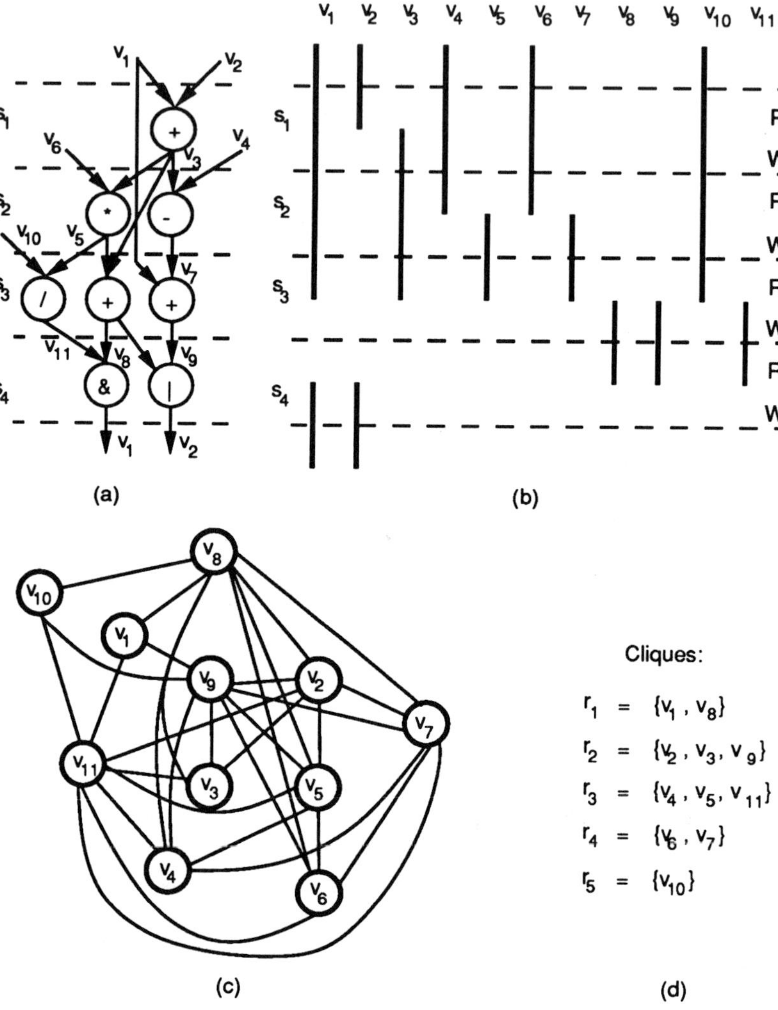

Figure 8.11: Register allocation using clique partitioning: (a) a scheduled DFG, (b) lifetime intervals of variables, (c) the graph model for register allocation, (c) a clique-partitioning solution.

A clique-partitioning solution of this graph would yield a solution for the functional-unit allocation problem. Since a functional unit is assigned to each clique, all operations whose representative vertices are in a clique are executed in the same functional unit.

For interconnection-unit allocation, each vertex corresponds to a connection between two units, whereas an edge links two vertices if the two corresponding connections are not used concurrently in any control step. A clique-partitioning solution of such a graph implies partitioning of connections into buses or multiplexers. In other words, all connections whose representative vertices are in the same clique use the same bus or multiplexer.

Although the clique-partitioning method when applied to storage allocation can minimize the storage requirements, it totally ignores the interdependence between storage and interconnection allocation. Paulin and Knight [PaKn89] extend the previous method by augmenting the graph edges with weights that reflect the impact on interconnection complexity due to register sharing among variables. An edge is given a higher weight if sharing of a register by the two variables corresponding to the edge's two end vertices reduces the interconnection cost. On the other hand, an edge is given a lower weight if the sharing causes an increase in the interconnection cost. The modified algorithm prefers cliques with heavier edges. Hence, variables that share a common register are more likely to reduce the interconnection cost.

8.5.2 Left-Edge Algorithm

The left-edge algorithm [HaSt71] is well known for its application in channel-routing tools for physical-design automation. The goal of the channel routing problem is to minimize the number of tracks used to connect points on the channel boundary. Two points on the channel boundary are connected with one horizontal (i.e., parallel to the channel) and two vertical (i.e., orthogonal to the channel) wire segments. Since the channel width depends on the number of horizontal tracks used, the channel-routing algorithms try to pack the horizontal segments into as few tracks as possible. Kurdahi and Parker [KuPa87] apply the left-edge algorithm to solve the register-allocation problem, in which variable lifetime intervals correspond to horizontal wire segments and registers to wiring tracks.

The input to the left-edge algorithm is a list of variables, L. A lifetime interval is associated with each variable. The algorithm makes several passes over the list of variables until all variables have been assigned to registers. Essentially, the algorithm tries to pack the total lifetime of a new register allocated in each pass with as many variables whose lifetimes do not overlap, by using the channel-routing analogy of packing horizontal segments into as few tracks as possible.

Algorithm 8.3 describes register allocation using the left-edge algorithm. If there are n variables in the behavioral description, we define L to be the list of all variables v_i, $1 \leq i \leq n$. Let the 2-tuple $<Start(v), End(v)>$ represent the lifetime interval of a variable v where $Start(v)$ and $End(v)$ are respectively the start and end times of its lifetime interval. The procedure SORT(L) sorts the variables in L in ascending order with their start times, $Start(v)$, as the primary key and in descending order with their end times, $End(v)$, as the secondary key. The procedure DELETE (L, v) deletes the variable v from list L, FIRST(L) returns the first variable in the sorted list L and NEXT(L, v) returns the variable following v in list L. The array MAP keeps track of the registers assigned to each variable. The value of reg_index represents the index of the register being allocated in each pass. The end time of the interval of the most recently assigned variable in that pass is contained in *last*.

Initially, the variables are not assigned to any of the registers. During each pass over L, variables are assigned to a new register, r_{reg_index}. The first variable from the sorted list whose lifetime does not overlap with the lifetime of any other variables assigned to r_{reg_index} is assigned to the same register. When a variable is assigned to a register, the register is entered in the array MAP for that variable. On termination of the algorithm, the array MAP contains the registers assigned to all the variables and reg_index represents the total number of registers allocated.

Figure 8.12(a) depicts the sorted list of the lifetime intervals of the variables of the DFG in Figure 8.11(a). Note that variables v_1 and v_2 in Figure 8.11(a) are divided into two variables each (v_1, v_1' and v_2, v_2') in order to obtain a better packing density. Figure 8.12(b) illustrates how the left-edge algorithm works on the example. Starting with a new empty register r_1, the first variable in the sorted list, v_1, is put into r_1. Traveling

Algorithm 8.3: Register Allocation using Left-Edge Algorithm.

```
for all v ∈ L do   MAP[v] = 0;   endfor
SORT(L);

reg_index = 0;
while L ≠ φ do
    reg_index = reg_index + 1;
    curr_var = FIRST(L);
    last = 0;

    while curr_var ≠ null do
        if Start(curr_var) ≥ last then

            MAP[curr_var] = r_reg_index;
            last = End(curr_var);

            temp_var = curr_var;
            curr_var = NEXT(L, curr_var);
            DELETE(L, temp_var);
        else
            curr_var = NEXT(L, curr_var);
        endif
    endwhile

endwhile
```

down the list, no variables can be packed into r_1 before v_8 is encountered. After packing v_8 into r_1, the next packable variable down the list is v_1'. No more variables can be assigned to r_1 without overlapping variable lifetimes. Hence the algorithm allocates a new register (r_2) and starts from the beginning of the list again. The sorted list now has three fewer variables than it had in the beginning (i.e., v_1, v_8 and v_1' have been removed). The list becomes empty after five registers have been allocated.

Unlike the clique-partitioning problem, which is NP-complete, the left-edge algorithm has a polynomial time complexity. Moreover, this algorithm allocates the minimum number of registers [KuPa87]. However, it cannot take into account the impact of register allocation on the interconnection cost, as can the weighted version of the clique-partitioning

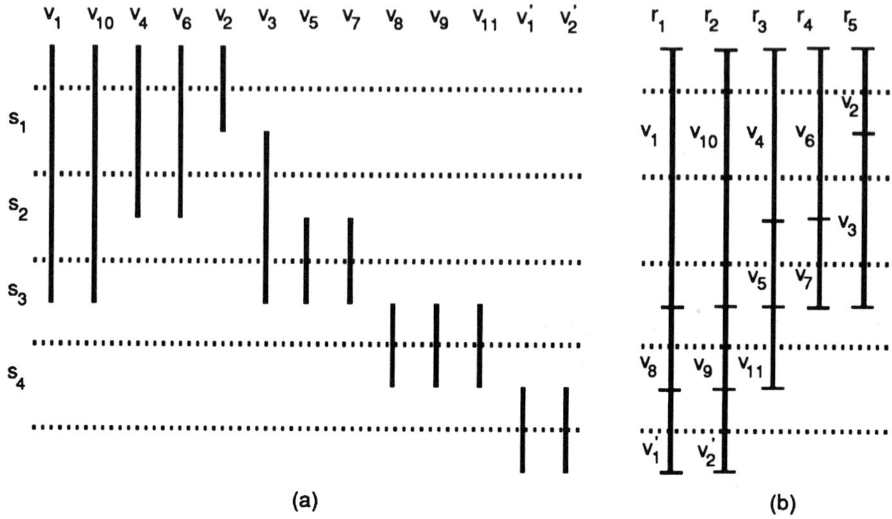

Figure 8.12: Register allocation using the left-edge algorithm: (a) sorted variable lifetime intervals, (b) five-register allocation result.

algorithm.

8.5.3 Weighted Bipartite-Matching Algorithm

Both the register and functional-unit allocation problems can be transformed into a weighted bipartite-matching algorithm [HCLH90]. Unlike the left-edge algorithm, which binds variables sequentially to one register at a time, the bipartite-matching algorithm binds multiple variables to multiple registers simultaneously. Moreover, it takes interconnection cost into account during allocation of registers and functional units.

Algorithm 8.4 describes the weighted bipartite-matching method. Let L be the list of all variables as defined in the description of the left-edge algorithm. The number of required registers, num_reg, is the maximum density of the lifetime intervals (i.e., the maximum number of overlapping variable lifetimes). Let R be the set of nodes representing the registers $\{r_1, r_2, ..., r_{num\_reg}\}$ into which the set of variables, V, will be mapped. The set of variables is partitioned into clusters of variables with mutually overlapping lifetimes. The function OVERLAP($Cluster_i$, v) returns the value true if either $Cluster_i$ is empty or if the lifetime of variable v overlaps with the lifetimes of all the variables already assigned to $Cluster_i$. The function BUILD_GRAPH (R, V) returns the set of edges E containing edges between nodes of the two sets R, representing registers, and V, representing variables in a particular cluster. An edge $e_{i,j} \in E$ will represents feasible assignment of variable v_j to a register r_i if and only if the lifetime of v_j does not overlap with the lifetime of any variable already assigned to r_i. The procedure MATCHING($G(R \cup V, E)$) finds the subset $E' \subseteq E$, which represents a bipartite-matching solution of the graph G. Each edge $e_{i,j}$ in E' represents the assignment of a variable $v_j \in V$ to register $r_i \in R$. No two edges in E' share a common end-node (i.e., no two variables in the same cluster can be assigned to the same register). The array MAP, procedures SORT and DELETE, and the function FIRST are identical to that of the left-edge algorithm.

After sorting the list of variables according to their lifetime intervals in the same way as the left-edge algorithm, this algorithm divides them into clusters of variables. The lifetime of each variable in a cluster overlaps with the lifetimes of all the other variables in the same cluster (Figure 8.13(a)). In the figure, the maximum number of overlapping lifetimes is five. Thus, the set of registers, R, will have five registers (r_1, r_2, r_3, r_4 and r_5) into which all the variables will be mapped. A cluster of variables is assigned to the set of registers simultaneously. For example, the first cluster of five variables v_1, v_{10}, v_4, v_6 and v_2, shown in Figure 8.13(b), have been assigned to the registers r_1, r_2, r_3, r_4 and r_5 respectively. The algorithm then tries to assign the second cluster of three variables, v_3, v_5 and v_7, to the registers.

A variable can be assigned to a register only if its lifetime does not overlap with the lifetimes of all the variables already assigned to that register. In Figure 8.13(b), each graph edge represents a possible variable-to-register assignment. As in the clique-partitioning algorithm, weights can be associated with the edges. An edge $e_{i,j}$ is given a higher weight if the assignment of v_j to r_i results in a reduced interconnection cost. For example, let variable v_m be bound to register r_n. If another variable

Algorithm 8.4: Register Allocation using Bipartite Matching.

> **for** all $v \in L$ **do** $MAP[v] = 0$;   **endfor**
> SORT(L);
>
> /* divide variables into clusters */
> $clus\_num = 0$;
> **while** $L \neq \phi$ **do**
> $clus\_num = clus\_num + 1$;
> $Cluster_{clus\_num} = \phi$;
>
>   **while** $(L \neq \phi)$ and OVERLAP($Cluster_{clus\_num}$, FIRST(L)) **do**
> $Cluster_{clus\_num} = Cluster_{clus\_num} \cup \{\text{FIRST}(L)\}$;
> $L = \text{DELETE}(L, \text{FIRST}(L))$;
>   **endwhile**
> **endwhile**
>
> /* allocate registers for one cluster of variables at a time */
> **for** $k = 1$ to $clus\_num$ **do**
>
> $V = Cluster_k$;
> $E = \text{BUILD\_GRAPH}(R, V)$;
> $E' = \text{MATCHING}(G(R \cup V, E))$;
>
>   **for** each $e_{i,j} \in E'$, where $v_j \in V$ and $r_i \in R$ **do**
> $MAP[v_j] = r_i$;
>   **endfor**
> **endfor**

v_k is to be used by the same functional units that also use variable v_m, then since the two variables can share the same interconnections, it is desirable that v_k also be assigned to r_n.

In a bipartite graph, the node set is partitioned into two disjoint subsets and every edge connects two nodes in different subsets. The graph depicted in Figure 8.13(b) is bipartite. It has two sets, the set of registers, $R = \{r_1, r_2, r_3, r_4, r_5\}$ and the set of variables, $V = \{v_3, v_5, v_7\}$. The graph also has a set of edges $E = \{e_{3,5}, e_{3,7}, e_{4,5}, e_{4,7}, e_{5,3}, e_{5,5}, e_{5,7}\}$ returned by the function BUILD_GRAPH. The problem of matching each variable to a register is equivalent to the classic job-assignment problem. The largest subset of edges that do not have any common end-nodes is

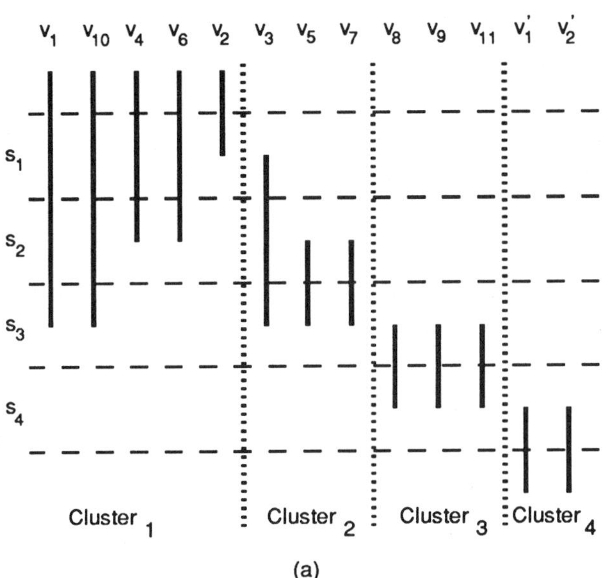

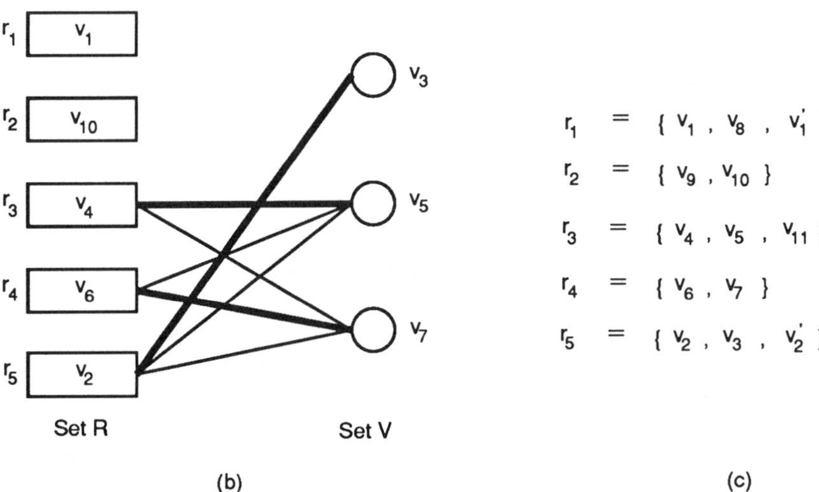

Figure 8.13: Weighted bipartite-matching for register allocation: (a) sorted lifetime intervals with clusters, (b) bipartite graph for binding of variables in $Cluster_2$ after $Cluster_1$ has been assigned to the registers, (c) final binding of all variables to the set of registers.

defined as the maximal matching of the graph. The maximal edge-weight matching of a graph is the maximal matching with the largest sum of the weights of its edges. A polynominal time algorithm for obtaining the maximum weight matching is presented in [PaSt82].

The set E' indicates the assignments of variables to registers as determined by the maximum matching algorithm in function MATCH-ING. $E' = \{e_{5,3}, e_{3,5}, e_{4,7}\}$ is indicated in the graph of Figure 8.13(b) with bold lines. After binding the second cluster of variables to the registers according to the matching solution, namely, v_3 to r_5, v_5 to r_3 and v_7 to r_4, the algorithm proceeds to allocate the third cluster of variables, v_8, v_9 and v_{11}, and so on. The final allocation of variables to registers is given in Figure 8.13(c).

The matching algorithm, like the left-edge algorithm, allocates a minimum number of registers. It also takes partially into consideration the impact of register allocation on interconnection allocation since it can associate weights with the edges.

8.6   Iterative Refinement Approach

Given a datapath synthesized by constructive or decomposition methods, its quality can be improved by reallocation. As an example, consider functional-unit reallocation. It is possible to reduce the interconnection cost by just swapping the functional-unit assignments for a pair of operations. For instance, if we start with the datapath shown in Figure 8.8(c), swapping the functional unit assignments for operations o_3 and o_4 will reduce the interconnection cost by four 2-to-1 multiplexers (Figure 8.8(d)).

Changing some variable-to-register assignments can be beneficial too. In Figure 8.8(b), if we move variable g from register r_2 to register r_1 and h from r_3 to r_4, we get an improved datapath with two fewer multiplexers (Figure 8.8(c)).

The main issues in the iterative refinement approach are the types of modifications to be applied to a datapath, the selection of a modification type during an iteration and the termination criteria for the refinement process.

The most straightforward approach could be a simple assignment exchange. In this approach, the modification to a datapath is limited to a swapping of two assignments (i.e., variable pairs or operation pairs). Assume that only operation swapping is used for the iterative refinement. The pairwise exchange algorithm performs a series of modifications to the datapath in order to decrease the datapath cost. First, all possible swappings of operation assignments scheduled into the same control step are evaluated in terms of the gain in the datapath cost due to a change in the interconnections. Then, the swapping that results in the largest gain is chosen and the datapath is updated to reflect the swapped operations. This process is repeated until no amount of swapping results in a positive gain (i.e., a further reduction in the datapath cost).

Algorithm 8.5 describes the pairwise exchange method. Let $DP_{current}$ represent the current datapath structure and DP_{work} represent a temporary datapath created to evaluate the cost of each operation assignment swap. The function $COST(DP)$ evaluates the cost of the datapath DP. The datapath costs of $DP_{current}$ and DP_{work} are represented by $c_{current}$ and c_{work}. The procedure $SWAP(DP, o_i, o_j)$ exchanges the assignments for operations o_i and o_j of the same type and updates the datapath DP accordingly. In each iteration of the inner-most loop, $CurrentGain$ represents the reduction in datapath cost due to the swapping of operations in that iteration. $BestGain$ keeps track of the largest reduction in the cost attainable by any single swapping of operations evaluated so far in the current iteration.

This approach has two weaknesses. First, using a simple pairwise swapping alone may be inadequate for exploring all possible refinement opportunities. For instance, no amount of variable-swapping can change the datapath of Figure 8.8(b) to that of Figure 8.8(d). Second, the greedy strategy of always going for the most profitable modification is likely to lead the refinement process into a local minimum. While a probabilistic method, such as simulated annealing [KiGV83], can be used to cope with the second problem at the expense of more computation time, the first one requires a more sophisticated solution, such as swapping several assignments at the same time.

Suppose operation o_i has been assigned to functional unit fu_j and one of its input variables has been bound to register r_k. The removal of o_i from fu_j will not eliminate the interconnection from r_k to fu_j unless no other operation that has been previously assigned to fu_j has its input

Algorithm 8.5: Pairwise Exchange.

repeat
    $BestGain = -\infty$;
    $c_{current} = \text{COST}(DP_{current})$;

    for all control steps, s **do**
       for each o_i, o_j of the same type scheduled into s, $i \neq j$ **do**
         $DP_{work} = \text{SWAP}(DP_{current}, o_i, o_j)$;
         $c_{work} = \text{COST}(DP_{work})$;
         $CurrentGain = c_{current} - c_{work}$;

         if $CurrentGain > BestGain$ **then**
          $BestGain = CurrentGain$;
          $BestOp1 = o_i$; $BestOp2 = o_j$;
        endif
       endfor
    endfor

    if $BestGain > 0$ **then**
       $DP_{current} = \text{SWAP}(DP_{current}, BestOp1, BestOp2)$;
    endif
until $BestGain \leq 0$

variables assigned to r_k. Clearly, the iterative refinement process has to approach the problem at a coarser level by considering multiple objects simultaneously. We must take into account the relationship between entities of different types. For example, the gain obtained in operation reallocation may be much higher if its input variables are also reallocated simultaneously. The strategy of reallocating a group of different types of entities can be as simple as a greedy constructive algorithm (Section 8.4) or as sophisticated as a branch-and-bound search (e.g., STAR [TsHs90]).

8.7　Summary and Future Directions

In this chapter, we described the datapath allocation problem, which consists of four basic subtasks: unit selection, functional-unit binding, storage binding and interconnection binding. We outlined the features of some realistic architectures and their impact on the interconnection complexity for allocation. We described the basic techniques for allocation using a simple model that assumes only a straight-line code behavioral description and a simple point-to-point interconnection topology. We discussed the interdependencies among the subtasks that can be performed in an interleaved manner using a greedy constructive approach, or sequentially, using a decomposition approach. We applied three graph-theoretical algorithms to the datapath allocation problem: clique partitioning, left-edge algorithm and weighted bipartite-matching. We also showed how to iteratively refine a datapath by a selective, controlled reallocation process.

The greedy constructive approach (Algorithm 8.1) is the simplest amongst all the approaches. It is both easy to implement and computationally inexpensive. Unfortunately, it is liable to produce inferior designs. The three graph theoretical approaches solve the allocation tasks separately. The clique-partitioning approach (Algorithm 8.2) is applicable to storage, functional and interconnection unit allocation. The left-edge algorithm (Algorithm 8.3) is well suited for storage allocation. The bipartite matching approach (Algorithm 8.4) is applicable to storage and functional unit allocation. Although they all run in polynomial time, only the left-edge and the bipartite matching algorithms guarantee optimal usage of registers, while only the clique-partitioning and the bipartite-matching algorithms are able to take into account the impact on interconnection during storage allocation. The iterative refinement approach (Algorithm 8.5) achieves a high quality design at the expense of excessive computation time. The selection and reallocation of a set of behavioral entities have a significant impact on the algorithm's convergence as well as the datapath cost.

Future work in datapath allocation will need to focus on improving the allocation algorithms in several directions. First, the allocation algorithms can be integrated with the scheduler in order to take advantage of the coupling between scheduling and allocation. As we have pointed out, the number of control steps and the required number of functional units cannot accurately reflect the design quality. A fast allocator can quickly provide the scheduler with more information (e.g., storage sharing between operands of different operations, interconnection cost and data-transfer delays) than just these two numbers. Consequently, the scheduler will be able to make more accurate decisions. Second, the algorithms must use cost functions based on physical design characteristics. This will give us more realistic cost functions that closely match the actual design. Finally, allocation of more complex datapath structures must be incorporated. For example, the variables and arrays in the behavioral description could be partitioned into memories, a task that is complicated by the fact that memory accesses may take several clock cycles.

8.8   Exercises

1. Using components from a standard datapath library, compare the delay times for the following datapath components:
   (a) a 32-bit 4-to-1 multiplexer,
   (b) a 32-bit 2-to-1 multiplexer,
   (c) a 32-bit ALU performing addition,
   (d) a 32-bit ALU performing a logic OR,
   (e) a 32-bit floating-point multiplier,
   (f) a 32-bit fixed-point multiplier, and
   (g) a 32-bit 16-word register file performing a read.

2. Suppose the result of operation o_i is needed by operation o_j and the two operations have been scheduled into two consecutive control steps. In a shared-bus architecture such as that of Figure 8.6, o_j will be reading its input operands before o_i has written its output operand. What changes are required in the target architecture to prevent o_j from getting the wrong values?

3. Extend the design in Figure 8.7 to support chaining of functional units. Discuss the implications of this change on allocation algorithms.

4. *Extend the left-edge algorithm [KuPa87] to handle inputs with conditional branches. Does it still allocate a minimal number of registers?

5. Prove or disprove that the bipartite-matching method [HCLH90] uses the same number of registers as the left-edge algorithm does in a straight-line code description.

6. Extend the weighted bipartite-matching algorithm to handle inputs with conditional branches. Does it still allocate a minimal number of registers?

7. Show that every multiplexer can be replaced by a bus and vice versa.

8. In a bus-based interconnection unit, assume that there is one level of tri-state buffers from the inputs to the buses and one level of multiplexers from the buses to the outputs. Redesign the interconnection unit of Figure 8.6 (which, by coincidence, uses no multiplexers at all), using the same number of buses so that the data transfer capability of our five-transfer example is preserved and the number of tri-state buffers is minimized (at the expense of more multiplexers, of course).

9. *Some functional units, such as adders, allow their two input operands to be swapped (e.g., $a + b = b + a$). Suppose both the functional-unit and storage allocation have been done. Design an interconnection-allocation algorithm that takes advantage of this property.

10. Using a component library and a timing simulator, measure the impact of bus loading on data-transfer delay. Draw a curve showing the delay as a function of the number of components attached to the bus.

11. Show an example similar to that of Figure 8.8, where swapping of a pair of variables reduces the worst case delay for the data transfer. Assume that latches exist at both the input and output ports of all functional units.

12. Given a partially allocated datapath in which both the functional-unit allocation and the register allocation have been done, design an interconnection-allocation algorithm that minimizes the maximal bus load for a non-shared bus-based architecture.

13. *Design an interconnection-first algorithm for datapath allocation targeted towards a bus-based architecture. That is, assign data transfers to buses before register and functional-unit allocation. Compare this algorithm with the interconnection-last algorithms. Does this approach simplify or complicate the other two tasks?

14. Given a scheduled DFG, derive a lower bound on the number of multiplexer inputs needed for a datapath using a point-to-point architecture.

15. Compare the storage requirements for scheduling a loop in the following two cases:
    (a) when the loop is unrolled,
    (b) when the loop is folded.

16. *Develop a synthesis procedure for the allocation of arrays in the behavioral description to memories.

17. **Design a synthesis procedure that combines scheduling and allocation. That is, the scheduling of operations and allocation of storage, functional and interconnection units are to be interleaved. Hint: A possible strategy is to iteratively schedule a set of operations and then allocate hardware units for that set of operations [BaMa89, BePa90].

Chapter 9

Design Methodology for High-Level Synthesis

9.1   Basic Concepts in Design Methodology

In the previous chapters we have defined synthesis, presented target architectures and design-quality measures, discussed description languages and design representations and presented algorithms for partitioning, scheduling and allocation for synthesis on higher-than-logic abstraction levels. In this chapter we will combine those ideas into a design methodology for synthesis on system and chip levels. System-level synthesis takes a behavioral description of the complete system and generates another behavioral description of the same system in which each custom, semicustom or standard chip is described separately. Chip synthesis converts the behavioral description of each chip into a structural description with register-transfer (RT) components. We will discuss several alternatives for synthesis systems and the technical justification behind approaches taken in those alternatives. The work on design methodology is scarce and no high-level synthesis (HLS) systems are widely used in practice. In this chapter, we mainly discuss requirements for HLS systems and propose possible solutions, sometimes based on our own work but most often based on speculations on the nature of other work in industry and academia.

A design methodology must clearly specify:

(a) the syntax and semantics of the input and output descriptions,

(b) the set of algorithms for translating input into output descriptions,

(c) the set of components to be used in the design implementation,

(d) the definition and ranges of design constraints,

(e) the mechanism for selection of design styles, architectures, topologies and components, and

(f) control strategies (usually called scenarios or scripts) that define synthesis tasks and the order in which they are executed.

Usually, requirements (a) and (b) are defined by the choice of a description language and a set of synthesis tools, whereas requirements (c) and (d) are the consequence of chosen languages and algorithms in (a) and (b). Requirements (e) and (f) are usually not defined at all and are assumed to be the designer's responsibility. We will intuitively explain the meaning of requirements (a) through (f) with respect to the high-level synthesis design methodology on one simple example and derive three simple synthesis systems of increasing complexity to satisfy those requirements.

Let us consider a trivial description of an infinite loop that computes

$$y = (a + b) * (c - d)$$

in every iteration. Many digital signal processing (DSP) algorithms can be described with infinite loops since continuous signals are sampled at a constant rate with the same computation performed on each sample. In our example, we assume that the input description consists of only variable assignments using arithmetic operators. This description is compiled into a data-flow graph (DFG) that explicitly identifies data dependencies (Figure 9.1(a)). We will assume that new data arrives in each clock cycle and that each loop iteration executes in two clock cycles. We will also assume that each operation is executed in one clock cycle by a single-function functional unit, all functional units and registers have the same bit width, and the number of functional units is not limited. With the hardware resources and design constraints specified, we can apply the scheduling and allocation algorithms described in Chapters 7 and 8. These algorithms annotate the DFG with additional attributes to indicate bindings to states and components. Each operator is assigned to a control state and to a functional unit to execute the operation in the given clock cycle. Each edge in the DFG is assigned a latch, a register or a wire. The annotated DFG, shown in Figure 9.1(b), is used for generation of the datapath shown in Figure 9.1(c). Note that the annotated DFG and derived datapath have the same topology, which indicates the simplicity of the target architectures and the mapping algorithms.

The synthesis system for this simple case (Figure 9.1(d)) would consist of a compiler to convert the input behavioral description into the DFG representation, a scheduler to assign control states, an allocator to select and assign functional units to operations and storage units to data values, and a netlist generator to generate the datapath netlist for use by other tools. The datapath netlist contains interconnected RT components, namely, an adder, a subtracter, a multiplier and registers. Since we assumed that all units perform only a single function and have the same bit width, we can also assume that a single designer would predesign all functional and storage units for semicustom or custom technologies and store them in a set of files accessible by all other designers. Each file would contain the width and the height of each unit for floorplanning and the position of I/O ports for routing. Thus, the task of physical design would consist of performing placement and routing of the datapath netlist using predefined unit sizes and port locations.

The synthesis system described in Figure 9.1(d) is too limited to be useful in practical situations for several reasons. First, the assumption that all units be of the same bit width and same propagation delay is too restrictive. A designer usually has a choice between units of different bit widths, functionality and propagation delays. Second, the data-flow style of design, in which the number of functional units equals the number of operators in the behavioral description, is prohibitively expensive for all but very small descriptions. On the other hand, a target architecture with very few functional units requires a control unit to assure proper sequencing of operator execution and proper sharing of the functional units. For such an architecture, a different design methodology must be used. Third, the assumption that input and output data streams match input and output datapath rates is also too restrictive. Data may come from different sources, in different orders and at different times not

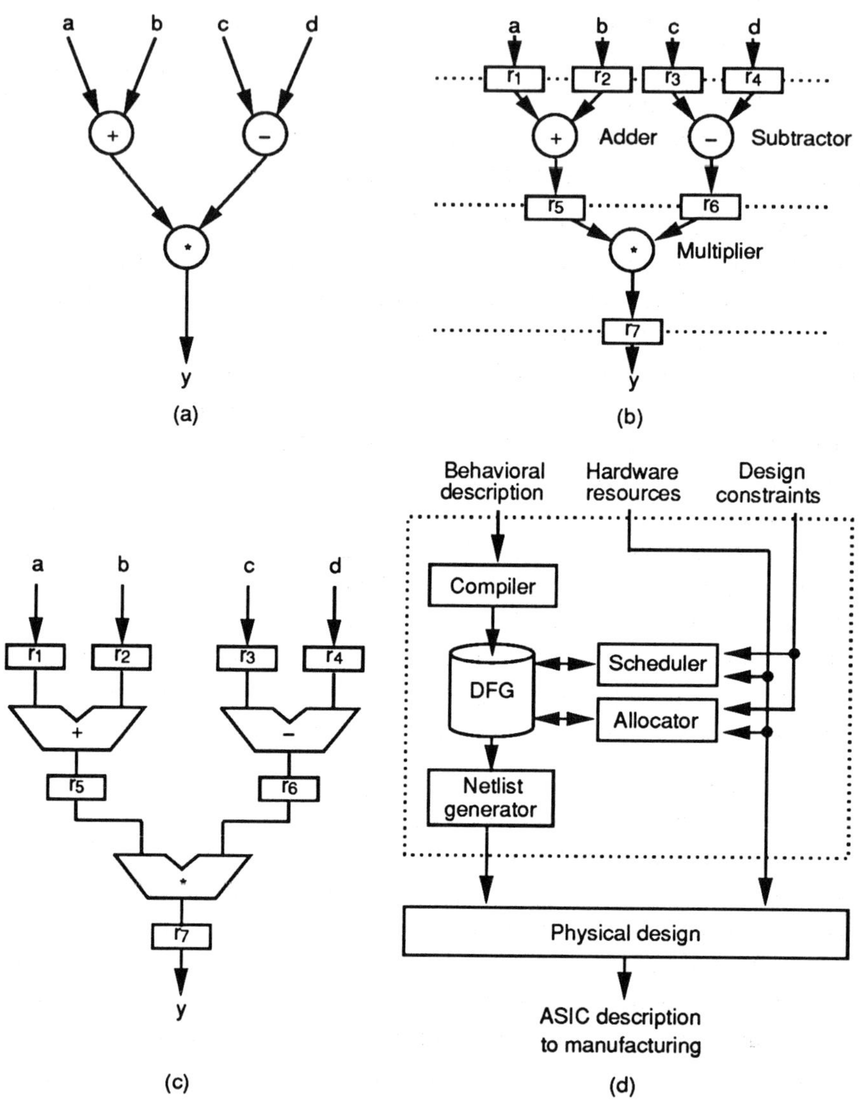

Figure 9.1: Trivial synthesis system: (a) DFG, (b) annotated DFG, (c) datapath structure, (d) synthesis system.

necessarily synchronized with the datapath clock. For that reason, we need input and output queues or memories to synchronize data streams with different rates and protocols.

To discuss a synthesis system for a broader class of architectures, we will consider the same example but with different design constraints. We will require that the computation $y = (a+b)*(c-d)$ be performed in 300 ns, using only two functional units, a multiplier and an adder/subtractor with propagation delays less than 100 ns each.

The DFG for the computation of y is shown in Figure 9.2(a). We can schedule the DFG into three control states using the given multiplier and adder/subtracter. Since we can reuse registers r_2 and r_4, the final design needs only five registers in comparison with seven registers in Figures 9.1(b) and (c). The final design now requires four 2-input multiplexers. The annotated DFG is shown in Figure 9.2(b) and the design structure is shown in Figure 9.2(c). In addition to the datapath, the final design requires a control unit to select and load values to registers and to select operations in the functional units. The *Control unit* also controls the loading of data values from *Memory1* and storing of data values into *Memory2*, and requires another datapath for the generation of the load and store addresses. The resource utilization is shown in Figure 9.2(d). Each operation takes one clock cycle to execute, including loading and storing of data, whereas each iteration takes five clock cycles. A new iteration is started every two cycles. Thus, it takes $2(n-1)+5$ clock-cycles to complete n iterations.

We see that the ALU is a critical resource, used in every clock-cycle, while the multiplier and each of the memories are used in only one out of every two cycles. We can trade off this low utilization of some components for silicon area. For example, we can replace the multiplier with a slower one that takes two clock cycles but less area. Similarly, we may use slower memories, or replace them with memories with fewer ports or less bits per word. For example, *Memory1* with four 16-bit ports can be replaced with a memory component with two 16-bit ports. Thus, it will take two clock cycles to deliver four 16-bit operands to the datapath. We can also use only one 16-bit port in *Memory2* and take two clock cycles to store the 32-bit result. Therefore, if we substitute the multiplier and both memories with components that take two clock cycles instead of one, all parts of the design would be fully utilized (Figure 9.2(e)). Note

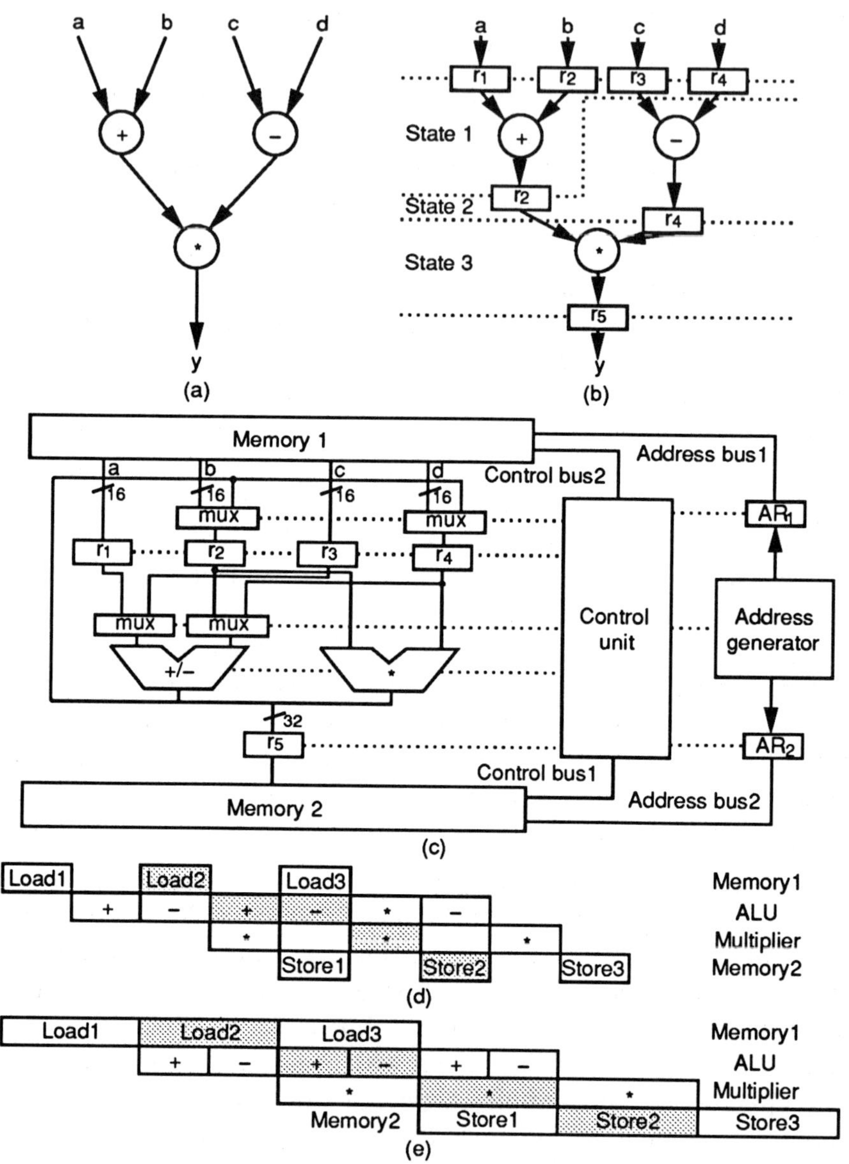

Figure 9.2: A 3-operations example: (a) DFG, (b) annotated DFG, (c) FSMD implementation, (d) resource utilization, (e) improved resource utilization.

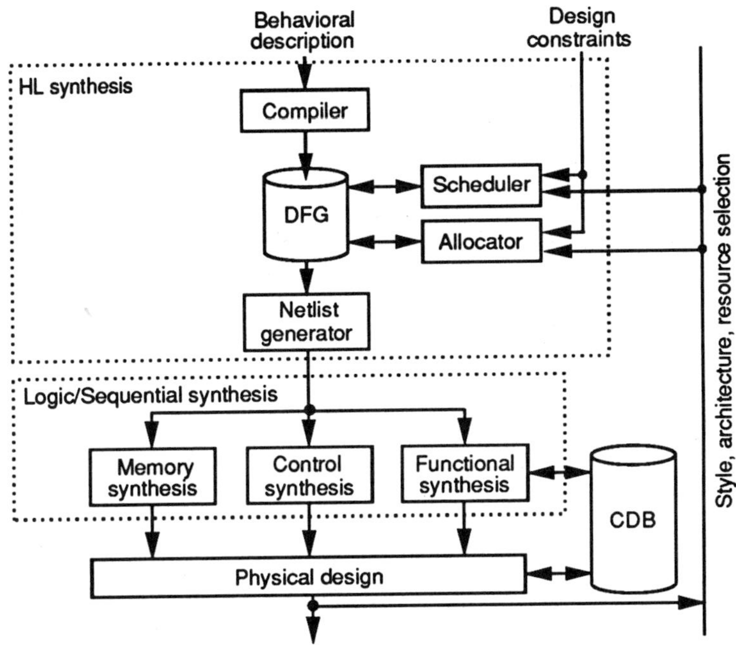

Figure 9.3: Synthesis system with a component database.

that the total loop execution time has increased by only three clock cycles to $2(n-1)+8$ clock cycles for n iterations.

The previous example demonstrates that high-level synthesis generates a structural description that contains components such as ALUs, multipliers, selectors, registers, memories and controllers. Each of these components must be synthesized using logic or sequential synthesis techniques from the description of each component stored in a component database (CDB). The CDB may also provide the size of each component for floorplanning and the position of each component's pins for routing during physical design. Figure 9.3 shows a block diagram of the synthesis system with a CDB.

A design can be improved iteratively by changing the target architecture, topology or component characteristics. We improved the design in our last example by increasing the utilization of under-utilized components. A human designer usually assesses the design quality and makes

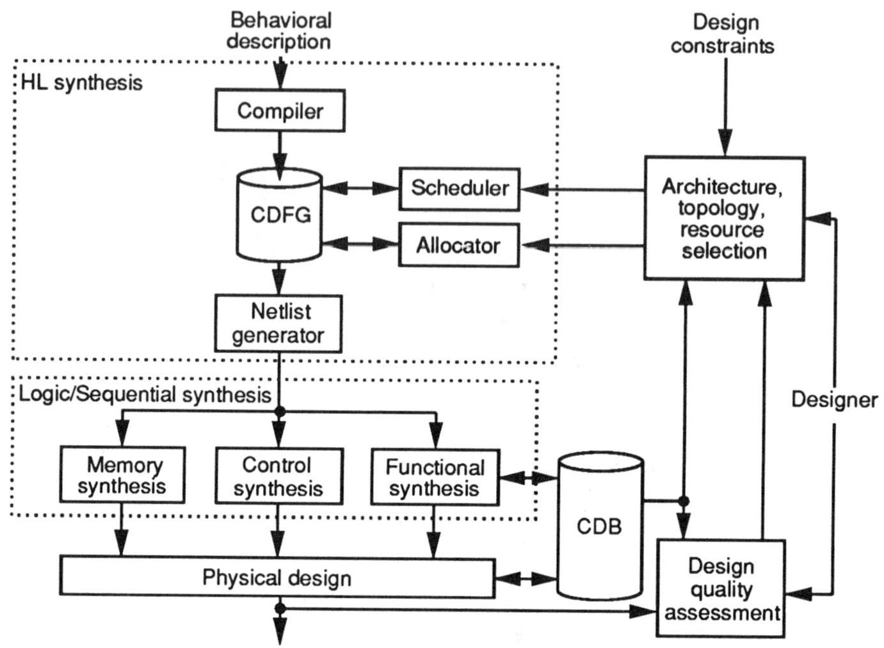

Figure 9.4: Synthesis system using iterative improvement.

changes in design style and resource constraints to iteratively improve the design. This task can be also automated by defining quality measures and developing a procedure for design-quality assessment that drives the selection of design characteristics (Figure 9.4). The CDB supports the assessment and selection tasks by providing alternate design implementations for each component.

We have shown some of the main issues of high-level synthesis in the 3-operation example. In the next section we will describe the design-process requirements and a hypothetical synthesis system, and discuss in detail each component of the system in subsequent sections.

9.2 Generic Synthesis System

In this section we describe a hypothetical generic synthesis system. It is generic because it brings together all the main concepts of HLS discussed in the previous chapters. It is hypothetical because no research or commercial synthesis system satisfies all of the following criteria [DuHG90, Gajs91, LNDP91]:

(a) Completeness. The system should provide synthesis tools for all levels of the design process from specification to manufacturing documentation. The synthesis tools in the system should be able to accommodate all the different target architectures.

(b) Extensibility. The system should be flexible enough to allow addition of new algorithms and tools. Similarly, addition of new architectural styles should be provided, if possible, through rewriting of the design-process scripts only. The system should also support the addition of new components and algorithms for mapping these components to specific technologies.

(c) Controllability. A designer should be able to control the types of tools applied to a specific description and the order in which they are executed. The same designer should be able to control design exploration by selection of different architectural styles and components. In order to assist the designer in the selection process, the system should provide different quality-assessment measures.

(d) Interactivity. A designer should be able to interact with synthesis tools by partially specifying the design structure or by modifying the design after synthesis. A designer should also be able to override each binding of behavioral objects to structural objects, as well as description and design transformations.

(e) Upgradability. The synthesis system should allow evolutionary upgrade from a capture-and-simulate to a describe-and-synthesize methodology and allow the possibility of mixing of both strategies on each level of abstraction.

The global diagram of a generic synthesis system is shown in Figure 9.5. It has a *Conceptualization environment* to support interactivity

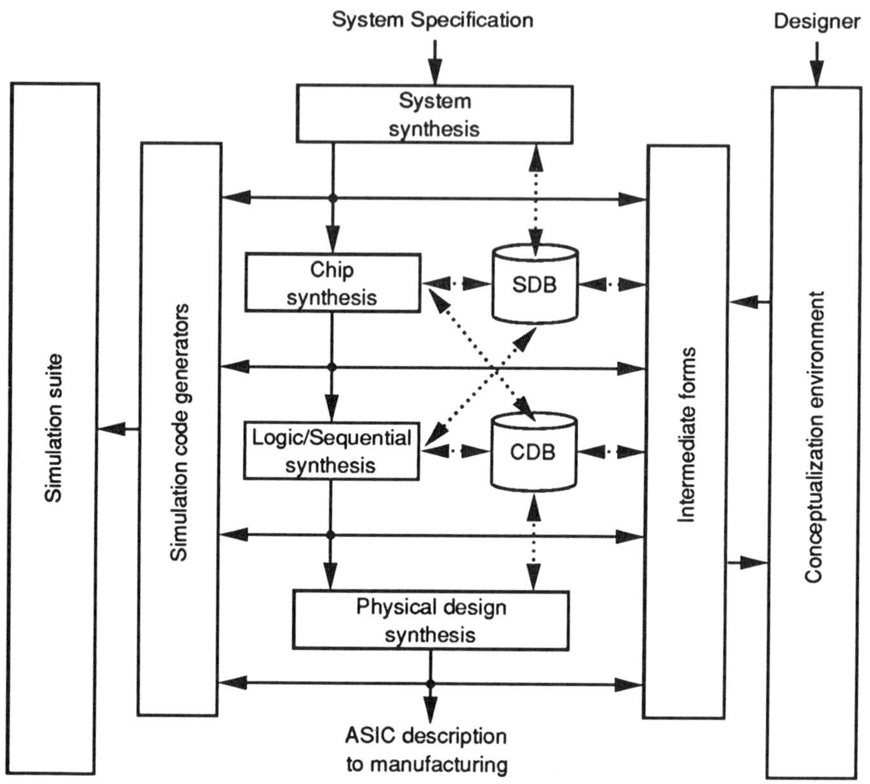

Figure 9.5: A hypothetical synthesis system.

and controllability. It is complete since it supports synthesis on the system, chip, logic and physical levels. The extensibility is supported through two databases. The *Component database (CDB)* supports addition of new components for synthesis, whereas the *System database (SDB)* supports new styles through its comprehensive design representation and facilitates addition of new algorithms by providing different types of information through different design views.

The system is also capable of supporting both the capture-and-simulate and describe-and-synthesize methodologies. The first methodology is supported by capture of the schematic through the *Conceptualization environment* and its simulation through one of the simulators in the *Simulation suite.* The second methodology is supported by capture of design descriptions through one or more intermediate form and synthesis with appropriate synthesis tools.

Different *Intermediate forms* are used for different design aspects on the same level or for the same design aspect on different levels of design. Instead of many intermediate forms, we could define one universal language for all levels and design styles, but such a language would be too cumbersome and inefficient. It would take many years to reach an agreement on a standard and probably many more years to teach designers how to use it. Instead, we could use a standard simulation language such as VHDL. However, such a language is burdened by constructs necessary for simulation and not for synthesis. Furthermore, its language constructs are not adequate to model many design styles, and designers are not really interested in writing design descriptions in VHDL. They are only interested in finding the output values for a set of test vectors and observing the timing relationships among selected signals. Thus, an output report generated by a simulator will be satisfactory no matter what the source of the input to the simulator, whether another language or any other form of captured design. Thus, we can use any intermediate form to capture design information, as long as we provide a simulation code generator to translate the intermediate form into a simulatable description. The *Simulation code generators* shown in Figure 9.5 are used for this purpose.

The concept of code generators fits also the capture-and-simulate methodology. For example, we can think of a schematic as an intermediate form that is translated by a netlist generator into a simulatable description. Similarly, intermediate forms on the logic, sequential, register-transfer and system levels can be translated into VHDL or some other simulation language. This approach allows the captured description to be close to a human designer's way of thinking and also makes it simulatable by any standard simulator for which a simulation code generator is available. Furthermore, the intermediate forms can be used for interactive synthesis to support manual partitioning, scheduling, binding, transformation and verification of designs and descriptions. They can also be used for design-quality assessment and for estimating of the effect of design changes on design quality.

9.3   System Synthesis

A system can be described with a set of communicating processes controlled by external or internal events [Hoar78]. The goal of system synthesis is to capture such a system description, verify its intent, evaluate some key characteristics such as performance, and partition it into a hierarchy of descriptions reflecting physical entities such as PC boards, multi-chip modules, chips or macrocells used to manufacture the system [HLNP88, NaVG91].

Partitioning of a system description into manufacturable or synthesizable entities is a difficult task. It is particularly difficult if some entities may be bound to standard components (such as processors and memories) while others may not be defined completely or defined on a different level of abstraction. The system-description language must be easy to understand, not only to simplify the initial description but also to make partitioning results and synthesis following partitioning easy to understand. Such a language should possess behavioral and structural hierarchy, abstract communication constructs and a description style that is preserved through several levels of partitioning. The language should be able to represent partial structural information, such as chips, ports and connections. It must also permit partial system specification so that synthesis can be performed without detailed knowledge of some portions of the system. The description must also be executable to allow verification of its intent before and after partitioning.

A typical tool for system synthesis is shown in Figure 9.6. The input description is compiled into a *System Representation* (*SR*) used by all other tools. The simulation code is also generated from *SR*. The system description is partitioned into subdescriptions that will execute sequentially on some hardware or exist concurrently in the same package. Two types of objects are subjected to partitioning: data structures and behavior. Data structures include global and local variables, arrays, lists, queues and records. They are eventually implemented using memory components. Therefore, we must group data structures that can be mapped into a single memory and then partition those memories among chips or other physical packages. Similarly, behaviors can be partitioned among different execution units or processors. We may assume that the input description is state based, with traversals between states triggered

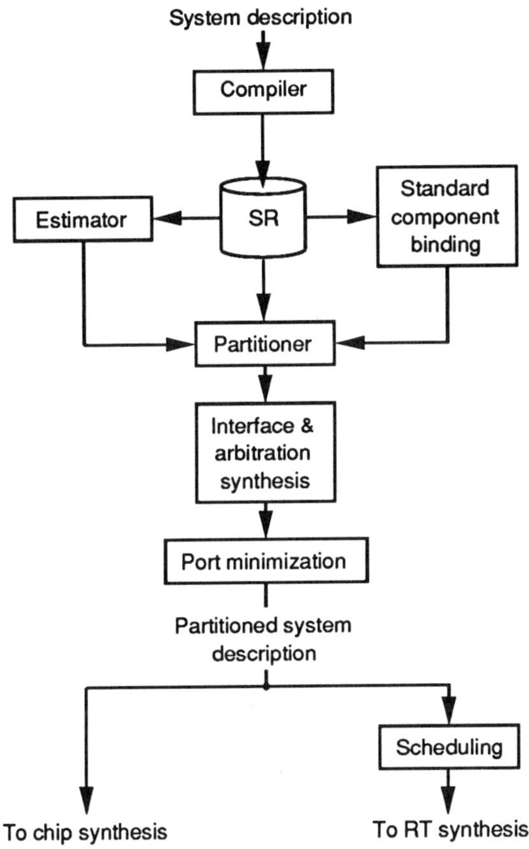

Figure 9.6: System synthesis methodology.

by events and with some computation executed in each state, including assignment statements, functions, procedures or processes. The behavior partitioning can be thought of as state partitioning. If the description is not based on states, we can divide the description into smaller chunks by partitioning it along block, process and procedure boundaries. The states whose behaviors are executed sequentially can be grouped for execution on a single FSMD architecture, which can then be further partitioned among physical packages.

In order to evaluate the quality of each partition we must estimate the clock period, performance, area, power, test and packaging costs. Since the total execution time depends on the input data, the estimation of the clock period and performance from a behavioral description is particularly difficult because the data is not available before execution. The estimation algorithms must use probabilistic techniques to define how many times each path in the description is traversed.

If standard components, such as processors or controllers, are to be used in the final design, then the partitioning task becomes very difficult. The partitioning algorithms must be able to recognize parts of descriptions that can be mapped onto standard components. Thus, a prepartitioning of design description into parts that are executable and non-executable on given standard components must precede partitioning.

Any partitioning may drastically change some of the system characteristics, particularly performance. Moving a data-structure location from a chip using that data structure to another chip may satisfy packaging goals but may also greatly increase the access time for that data structure. Furthermore, merging several data structures may require changes in their access protocols. Similarly, merging data structures accessed by several concurrent states requires the introduction of some arbitration mechanism. Thus, interfaces and arbitration logic must be synthesized after each iteration of partitioning.

After the system description is partitioned into entities satisfying physical constraints, communication channels and their protocols can be merged to minimize the number of ports in each partition. Merging of channels can be also thought of as multiplexing messages over the same bus. Obviously, this multiplexing is possible only when no two messages must be sent concurrently.

System synthesis results in a structure of subbehaviors in which each subbehavior defines a package or a module that must be further synthesized. This behavioral description of each package can be passed to a chip-synthesis tool, or it can be further decomposed into clock cycles by scheduling and passed directly to RT-synthesis tools.

9.4 Chip Synthesis

System synthesis produces a chip description in terms of subbehaviors that can be implemented with communicating FSMDs. FSMDs may be used to describe processors, caches, queues, memories, bus arbiters and interface logic. Chip synthesis deals with the transformation of a FSMD behavioral description into a structural description with RT components. A generic chip-synthesis system based on ideas from [Marw85, BCDO88, CaRo89, TLWN90, CaST91, Gajs91, KuDe91, LNDP91, NaON91] is shown in Figure 9.7.

The synthesis system consists of a compiler with a corresponding representation scheme, a set of HLS tools, a RT-component database, a technology mapper and an optimizer. The input behavioral description is compiled into a design representation (such as the CDFG described in Chapter 5) that exposes control and data dependencies needed for scheduling and allocation. Several HLS tools annotate this representation with information needed for logic and sequential synthesis. A *Scheduler* binds operations to control steps (as explained in Chapter 7), whereas *Storage, Functional,* and *Interconnect unit allocators* allocate scalar and array variables to registers and memories, operators to functional units, and connections to buses (as described in Chapter 8). The *Storage merger* groups registers into register files, and register files and memories into larger memories. The *Storage merger* also determines the location of each variable in the merged memory. The *Module selector* selects the RT components to be used in the design. It may partition the CDFG and perform some preliminary floorplanning to determine a proper mix of RT components that will satisfy the given constraints (as described in Chapter 6).

The *Component Database (CDB)* stores RT components to be used in the synthesis and answers queries about their characteristics (as described in Section 9.8). The *Technology mapper* maps generic components from the synthesized design description into component instances stored in the *CDB*. The estimated critical path through the design and delays along those paths may change after mapping. The *Microarchitecture optimizer* eliminates some critical paths and reduces delays by redistributing components and inserting faster components on critical paths and slower components on non-critical paths.

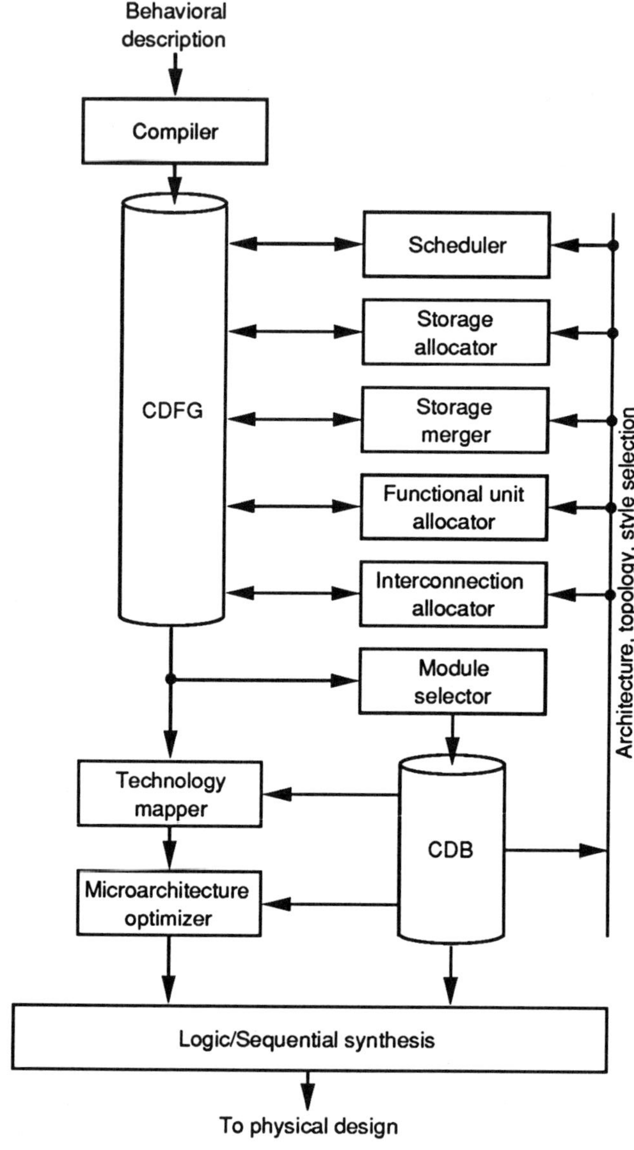

Figure 9.7: Chip synthesis.

The presence and function of the last four system components (*CDB, Module selector, Technology mapper,* and *Microarchitecture optimizer*) depend on the design methodology used for chip design. We will describe three different methodologies: top-down, meet-in-the-middle and bottom-up.

The top-down methodology starts with generic RT components assuming some nominal delays. Since scheduling and allocation use those generic components, they will appear in the resulting structural description. The optimization of the RT structure is performed on the logic level. Before logic optimization is performed, RT components are expanded into their corresponding gate-level netlists obtained from the *CDB.* Thus, technology mapping in this methodology consists of expanding component definitions, whereas logic optimization is used to finalize the design. This methodology is in tune with present-day logic synthesis tools and is very applicable to chips implemented with gate-arrays and standard-cells.

The meet-in-the-middle strategy starts from both the top and the bottom. It assumes that RT components have been predesigned manually or that module generators for layout generation of RT components exist. The predesigned components and module generators are stored in the *CDB.* The HLS tools use generic components for synthesis. Technology mapping for the resulting RT structure is achieved by substituting one or a group of RT generic components with technology-dependent components from the *CDB.* Because of differences between generic and real components, microarchitecture optimization is performed after technology mapping. It attempts to reduce propagational delays on critical paths and decrease area cost on non-critical paths. Technology mapping and microarchitecture optimization are similar to technology mapping and optimization used in logic-synthesis tools. The meet-in-the-middle strategy works well for application-specific designs with large custom components such as multipliers, execution units and memories.

The bottom-up strategy also assumes that predesigned RT components are available in the *CDB.* Furthermore, it assumes that the *CDB* can be queried about the availability of those components and their characteristics. HLS algorithms must be modified to work with a variety of real components and select the most appropriate ones. In other words, module selection has been shifted to the scheduler and allocators. By contrast, in the other two strategies module selection is performed before scheduling and allocation. Furthermore, in this strategy there is no technology mapping or microarchitecture optimization. The only task performed after HLS is the expansion of RT components into logic-level components for implementation in gate-arrays or standard-cells. In the case of full custom design, the layout of each RT component from the *CDB* is inserted during the physical-design phase.

The three strategies described are also applicable to chips consisting of many FSMDs. Some of them, such as memories and multipliers, are considered to be standard FSMDs or macrocells. These macrocells are synthesized during the physical-design phase. Thus, chip synthesis basically generates different types of macrocells for chip floorplanning, some being datapath-dominated FSMDs and others just fixed macrocells.

9.5   Logic and Sequential Synthesis

The output of HLS is a structural description in which each component is a FSMD described with a state table, a datapath component described with Boolean expressions, an interface component described with timing or waveform diagrams, or a memory specified with its size, number and type of ports, and access protocol. Each of these components requires different synthesis techniques, as shown in Figure 9.8 for a hypothetical logic-synthesis system.

A FSMD consists of a datapath and a control unit whose behavior can be modeled as a deterministic FSM. A FSM contains a state register for holding the present state and a combinational circuit for evaluating both the next state and output functions. A control-unit synthesizer should be capable of translating such a description into a hardware structure by performing state minimization, encoding, logic minimization and technology mapping.

State minimization reduces the number of states in a FSM by eliminating redundant states whose functions can be performed by other states. State minimization procedures partition states into classes of equivalent states. Two states s_i and s_j are equivalent if there is no input sequence that generates two different output sequences when applied to the FSM in states s_i and s_j. Since all equivalent states generate the same

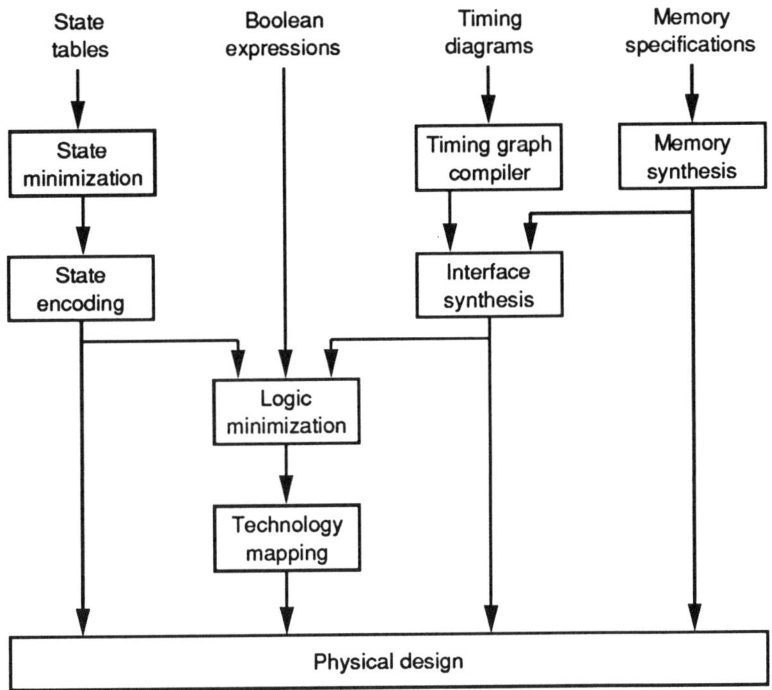

Figure 9.8: Logic-synthesis system.

output, each equivalent class can be replaced by only one state. State minimization is important since the number of states determines the size of the state register and control logic. State minimization also improves FSM testability since FSMs without redundant states are considerably more testable.

State encoding assigns binary codes to symbolic states. The goal of state encoding is to minimize the logic needed for the FSM implementation. Two approaches can be taken to satisfy this goal, one targeted toward two-level logic implementation and the other targeted toward multi-level logic implementation. First, in the two-level logic implementation approach [DeBS85], a multi-valued (i.e., symbolic) logic minimization is performed on the unencoded logic by grouping the set of inputs that correspond to each output symbol. This grouping also establishes a set of constraints for encoding each symbol. The symbols are encoded sequentially using the minimum-length code that satisfies all constraints. In the multi-level logic implementation approach [DMNS88], the encoding procedure tries to minimize the cost of the required multi-level implementation by minimizing the number of literals in the Boolean expressions of the next-state logic. The literal minimization is achieved through maximization of the number and size of common subexpressions which is achieved by assigning similar codes to states that generate similar outputs.

After all symbolic inputs, outputs and states are encoded, the logic minimizer reduces the cost of next-state and control-logic implementation. Different target implementations require the minimizer to focus on different objectives. For two-level logic, the objective is to minimize the number of product terms in the sum-of-product form of the Boolean equations. For multi-level logic, the objective is to minimize the number of literals in the Boolean equations by application of several techniques [BRSW87]. The extraction operation identifies common divisors in Boolean expressions that can be used to reduce the total number of literals in the network. The resubstitution operation checks whether an existing function itself is a divisor of other functions. The phase-assignment operation chooses whether to implement the function or its complement in order to minimize the total number of inverters needed to implement the network. The gate-factoring operation factors the logic equation for each complex gate in order to produce an optimal pull-down (and pull-up) network for the gate. The gate decomposer breaks down large functions into smaller pieces. The simplifier applies two-level minimization to each node in the network. The above mentioned logic operations are executed in the order specified in a script whose effectiveness greatly depends on the characteristics of the input logic network. User interaction in the short term and a self-adopting script in the long term can be used to achieve high-quality minimization.

While two-level logic is usually implemented with a PLA, multi-level logic can be implemented with standard-cell libraries. Technology mapping transforms a technology-independent logic network produced by the logic minimizer into a net-list of standard cells from a particular library. One among many approaches in technology mapping is DAGON [Keut87] which partitions the logic network into a forest of trees and finds a minimal-cost cover by library cells for each tree. The matching of the library cells with subtrees in each tree is performed by dynamic programming.

Since we assume that every design can be described with a set of communicating processes and implemented with a set of communicating FSMDs, then the design of communication logic is a necessary part of synthesis. The interface circuitry can be implemented as an integral part of each FSMD or as an independent FSMD. If a communication protocol is part of the FSMD description, then each response to a protocol signal takes an integral number of states. For example, we can implement the request-acknowledge protocol using a wait state and an acknowledge state. The assertion of the request signal forces the FSMD to leave the wait state and enter the acknowledge state in which the acknowledge signal is asserted. When request is removed, the FSMD will leave the acknowledge state. On the other hand, when two standard FSMDs with different protocols must be interfaced, then a third FSMD must be synthesized to convert one protocol into another [Borr91, NeTh86].

Interface synthesis is difficult because of the lack of interface description languages and synthesis methods for mixed synchronous and asynchronous sequential logic. The communication protocols are usually described by timing diagrams with imposed timing constraints as shown in the description of memory read-write cycles in commercial memory and microprocessor databooks. Three types of timing constraints exist: ordering, simultaneity and synchronicity [Borr91]. An ordering constraint simply defines the order in which events occur in some time range. A simultaneity constraint defines all the events that occur together, and a synchronicity constraint defines the setup and hold time for synchronous logic. Such a description must be converted into a timing graph, in which every node represents an event, that is, a logic transition on a signal wire. The arcs of the graph correspond to timing constraints that relate two events. These timing graphs can be constructed in a straightforward manner from the timing diagrams. Interface synthesis consists of mapping the timing graph into a minimum number of latches and flip-flops and reducing the logic for their setting and resetting. Depending on the protocol supported by the standard components, the interface logic may be synchronous or asynchronous.

Memory synthesis generates a memory description for given memory requirements such as the number of words, number of bits per word, number of ports, port types (e.g., read, write, or read/write), access rates, access protocol, and timing constraints. Synthesizing memories may be complex for certain requirements. For example, designing a 4-port memory using only single-port memory chips may be simple if the data can be partitioned into four groups and each group accessed only over one port (e.g., *Memory1* in Figure 9.2(c)). However, if the same data is accessed on two different ports or two ports access the data in the same single port memory chip, then conflict-resolution logic needs to be added; this increases the access time of the memory. Thus, the goal of the storage merger is to partition the data and assign each partition to a memory in such a way that the number of conflicts and the cost of the address-generation logic are minimized.

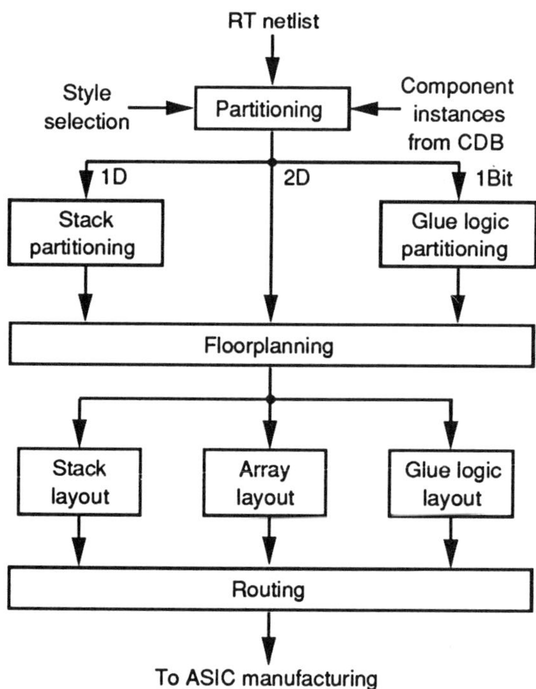

Figure 9.9: Physical-design methodology.

9.6   Physical-Design Methodology

The physical design methodology shown in Figure 9.9 starts with an RT netlist that contains three types of components, which are characterized by the number of dimensions in which they exhibit regularity. The largest components are 2-dimensional arrays such as memories, multipliers and register files. Medium size components are 1-dimensional arrays such as registers, counters, ALUs, bus drivers and logic units. Logic gates, latches and flip-flops are small components without regularity and are often called glue logic. Each group of components has a different layout style. 2-dimensional components are usually large and left as separate modules. 1-dimensional components can be grouped together into bit-sliced layout stacks (Figure 9.10(a)), in which data is routed in one direction and control in another direction on different metal layers as described in Chapter 3. Components in the stack may not be as strongly connected as they would be in a datapath. As an efficient placement architecture style, stack architecture significantly minimizes the layout area and wire length [LuDe89, TrDi89, CNSD90, WuGa90] However, stack placement may result in a significant waste of area because of unequal component dimensions. For example, we may have a 16-bit ALU and a 4-bit counter placed next to each other in the stack. In such a case, the empty area must be filled with glue logic using the standard cell layout style, in which cells are placed in rows and connected together in routing channels between the rows. For example, the empty area in the stack in Figure 9.10(a) is divided into three rectangular areas. Two of them are filled with control logic for the components in the stack, but the third one is filled with unrelated glue logic.

Thus, after partitioning the components into three groups according to their dimensionality, they are further partitioned to satisfy the spatial constraints imposed by the selected floorplan (Figure 9.10(b)). Glue logic is partitioned to fill the areas between large modules. The available area may be in the shape of an arbitrary polygon and thus must itself be partitioned into rectangular areas with estimated transistor capacities. Glue logic is then partitioned into groups that do not exceed the transistor capacities of their assigned area rectangles. Similarly, 1-dimensional components may be partitioned into several stacks for a better fit in the floorplan.

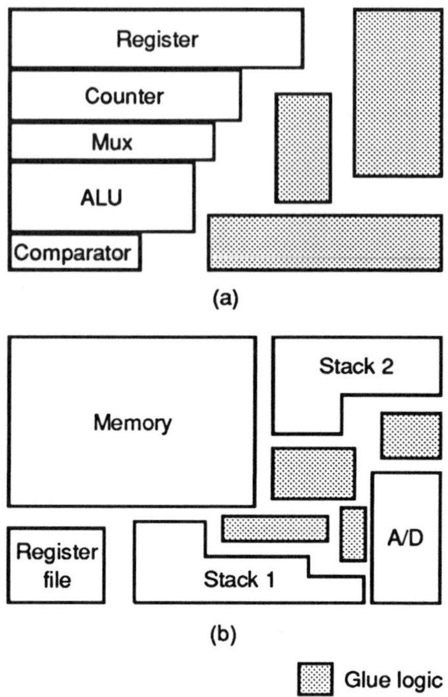

Figure 9.10: Generic floorplans: (a) stack floorplan, (b) chip floorplan.

Partitioning and floorplanning should be performed iteratively since each partition influences floorplanning, and conversely, each floorplan influences partitioning. After floorplanning is completed, the layout description for each of the components must be synthesized using one of three layout architectures: array, stack, or standard cells. The final task in physical design is the routing of the floorplan with instantiated component layouts. The details of partitioning, placement and routing algorithms can be found in [PrLo87, Leng90].

9.7   System Database

Traditionally, synthesis systems start as a loose collection of stand-alone design tools. Such systems suffer from mismatch in data representations, and input and output formats. The second generation of such a synthesis system can be developed as a tightly or loosely integrated system. A tightly integrated system uses a common data representation, with all tools using the same procedures to access it. Such a system is very efficient but extremely rigid since a change in format influences all the tools. A loosely integrated system separates design data from the tools. The data is stored in a database and each tool accesses only the information that it needs and uses its own representation for processing the retrieved information.

Another motivation for databases is the existence of many different versions of the same design during design exploration. In many systems, designers are left to manage the design hierarchy, its revisions and configurations, although these management tasks could be assigned to the database. A database could store and manage design data for the designer using a design data model (DDM) that consists of two parts (Figure 9.11).

The first level of DDM captures the overall administrative organization of the design data through the *Design Entity Graph*, which supports design hierarchy, version control and configuration management. A design entity is the smallest unit of data manipulated by the database. Most of the work on CAD databases focuses on this administrative level [Katz82, Siep89, CaNS90, RuGa90, WSBD90, BBEF91].

The second level of DDM maintains the actual design data. This design data is modeled by the behavioral and the physical graphs. The behavioral graph model captures the hierarchical description of the design behavior on the operation level, which is augmented with control and data dependencies, timing and physical constraints, and state and component bindings. The physical graph model captures the hierarchical structure of interconnected components augmented by timing and physical design information [CaTa89, LGPV91, LNDP91, RuGa91].

As shown in Figure 9.11, each object in the design-entity graph is a collection of related design objects. The graph contains domain-specific attributes for assessment of compactness and quality of design data. For example, some of the behavioral-domain attributes represent the number of states, storage and operator needs, and the frequency of storage accesses, whereas the structural-domain attributes represent the estimated area, the maximum delay and the number of components. Associated

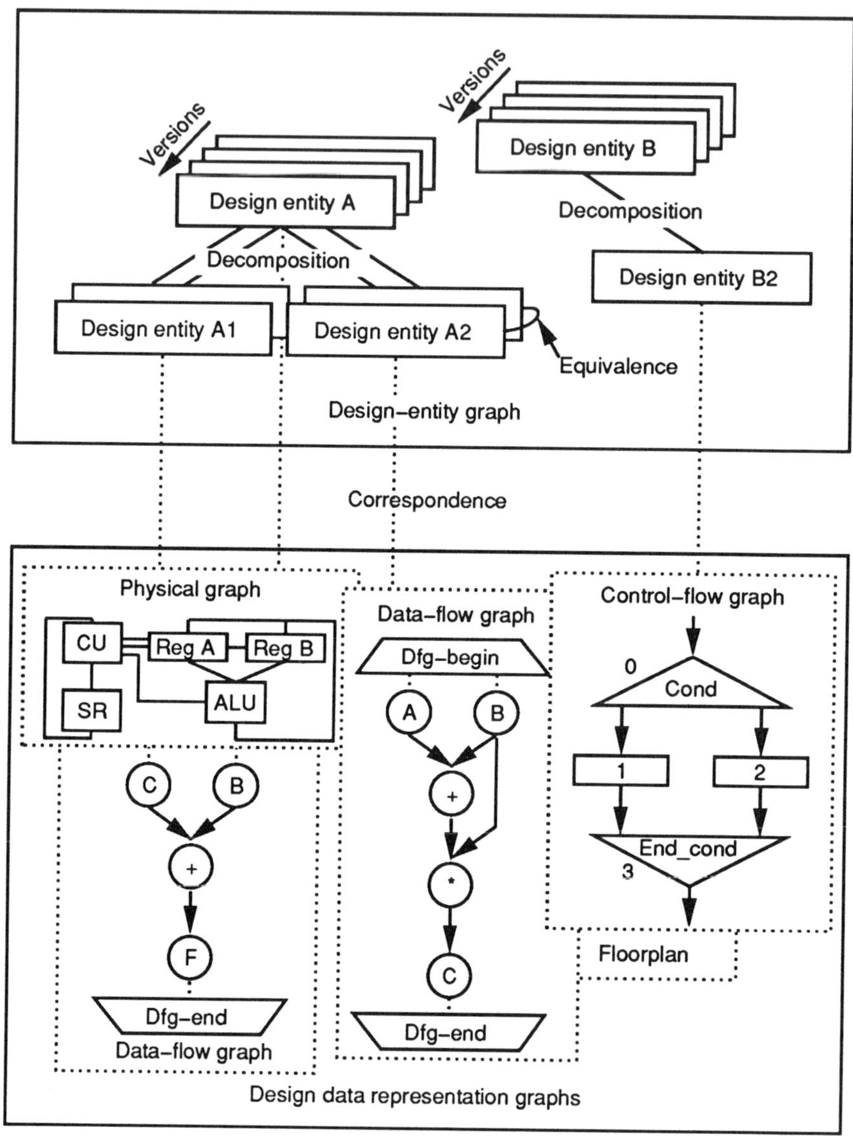

Figure 9.11: Design-data model.

with each attribute is an application-specific procedure that automatically calculates the value of the attribute from the underlying design data. This ensures the consistency between the attributes kept in the design-entity graph and the design data in the behavioral and physical graphs.

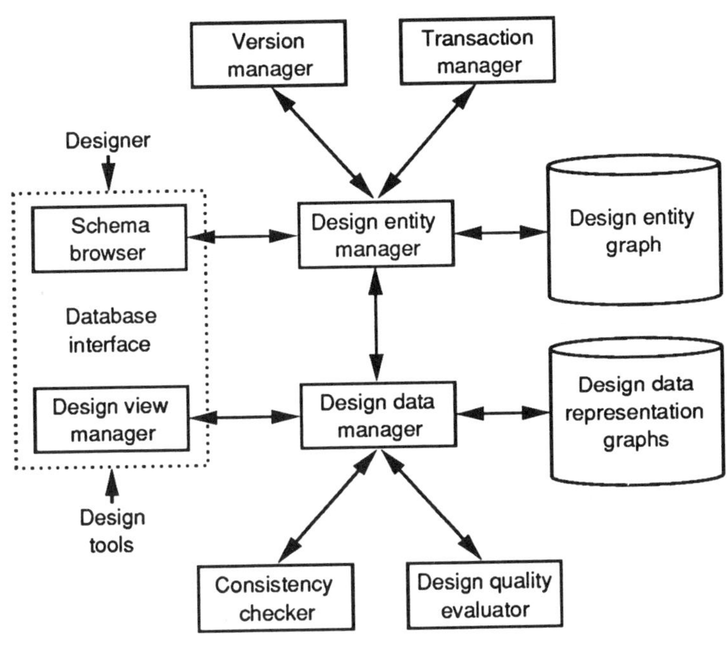

Figure 9.12: System database architecture.

A typical architecture of a system design database (SDB) is shown in Figure 9.12. A *Schema browser* allows a designer to view and possibly manipulate design entities and their relationships. The *Version manager* implements the version derivation policy chosen by the designer. In particular, it interprets the database requests for check-out and check-in of design data and then issues an appropriate command to the *Design-entity manager* for creating and manipulating relationships between design entities. The *Transaction manager* controls the access to design entities, thereby preventing simultaneous access to the same design data by several design tools. It locks data when it is checked out by a design tool, and only unlocks it when the design tool terminates a transaction via a check-in request.

To support the interaction of foreign tools with the database, SDB must provide customized interfaces for these tools. These customized interfaces, also called design views, contain a subset of the information in DDM supplied in the format specified by the tools. SDB provides a view description language for the specification of these design views. The *Design-view manager* compiles the design-view specification into a set of commands for the *Design-data manager* to access design data and formats it to comply with the design-view specification. This organization provides flexibility since new customized views can be created rapidly, and guarantees extensibility since new information can be added to the design data without disturbing existing views.

Consistency checking is performed on two different levels. First, the *Consistency checker* verifies that tools are applied in the proper order and that the data required by each tool is available in the database. Second, it verifies that new added data is consistent with the original data. For example, binding operators to units must not change the number of control steps and a change in the number of control steps must not change the given initial behavior. The *Design-quality evaluator* provides quality measures as summary attributes in the design-entity graph after each update so that different design versions can be retrieved by one or a combination of quality measures.

9.8   Component Database

Behavioral synthesis tools generate a microarchitectural design from behavioral or register-transfer descriptions. The generated microarchitecture consists of register-transfer components such as ALUs, multipliers, counters, decoders, and register files. Unlike basic logic components, register-transfer components have many options that can be parameterized. One of the parameters is the component size, given in the number of bits. Other parameters are related to the functionality or electrical and geometrical properties of the component. For example, counters may have increment and decrement options and load, set and reset functions. Also, each component may have different delays and may drive different loads on each of its output pins. In addition, each component may have several different options of aspect ratios for layout, as well as the positioning of I/O ports on the boundary of the module.

The component database (CDB) generates components that fit specific design requirements and provides information about a component's electrical and layout characteristics for possible design tradeoffs.

Thus, the CDB must provide two types of information: estimates of component characteristics for each available component type and a component-instance description for inclusion into RT, logic, and layout chip descriptions. Component estimates are used by HLS tools for scheduling and allocation, and by physical-design tools for floorplanning. Component descriptions are used on three different levels. First, component and component pin names are used in RT descriptions generated by HLS tools. Second, logic level descriptions are used during logic synthesis and optimization. And third, a layout descriptions are used during layout assembly. In addition to component descriptions, the CDB must supply three different models for simulation on the RT, logic and layout levels.

Not much attention has been given to component generation and component databases for HLS. Some preliminary work has been reported in [Wolf86, RDVG88, ChGa90, TLWN90]. The generic CDB shown in Figure 9.13 consists of two parts: *Component server* and *Knowledge server*. *Component server* provides estimates and RT, logic and layout descriptions when answering component queries for component instances. *Knowledge server* allows addition of new component instances, component generators or component optimization tools. Database engineers use it to expand the libraries of available components. The component may itself be stored as a component instance when handcrafted for some particular implementation, or as a parameterized description that describes a class of component instances. *Component generators* are tool sets that generate component instances from parameterized descriptions. Logic or layout-optimization tools may be used as a part of the component generation process. All handcrafted or generated instances are stored in the *Component store* for use by a particular designer or a particular synthesis tool.

The CDB is a crucial component of HLS methodology since it provides a clean separation of system and chip synthesis from logic and layout synthesis. The existence of a CDB allows designers to concen-

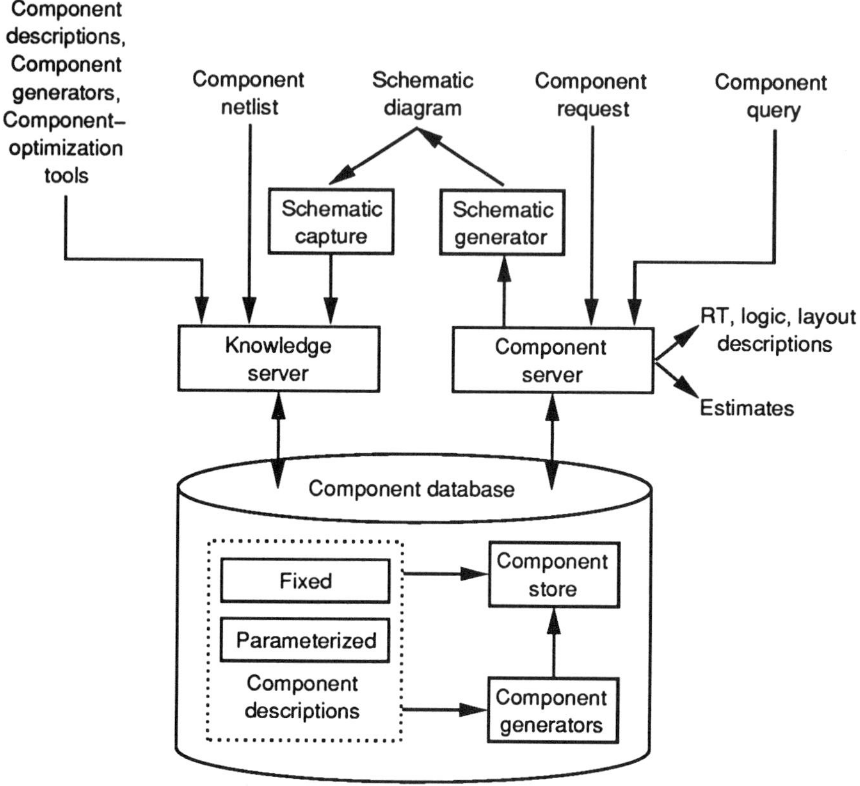

Figure 9.13: Component database.

trate on system design without being involved in logic or layout design. It also allows rapid growth of HLS tools by providing standard interfaces to component synthesis and isolation from changes in fabrication and tool technologies.

9.9 Conceptualization environment

The complete automation of the design process from higher levels of description is an important goal, although not practical immediately. Even if all HLS tools were available today, designers would have to gain familiarity and confidence in using them. It will also take time for synthesis algorithms to achieve uniform quality over different design styles. Therefore, it is necessary to provide an environment in which designers can verify the design decisions and manually override synthesis algorithms. Even when algorithms are perfected, designers should be able to make high-level decisions such as the selection of style, architecture, topology and component requirements and constraints. Thus, a conceptualization environment must allow control of design decisions and design strategy through every stage of the design process, which requires the conceptualization environment to provide rapid feedback of usable quality measures to every level of abstraction [BuEl89, CPTR89, YoHs90, HaGa91]. The conceptualization environment works in tandem with system and component databases. These databases store all versions of the design data over long periods of time, whereas the conceptualization environment accesses only small parts of the data for data modification or refinement over short periods of time. The conceptualization environment can be thought as short-time storage (i.e., a scratch pad) used by designers during design refinement.

The conceptualization environment consists of five parts: data and design manager, displays and editors, design-quality estimators, synthesis algorithms, and design-consistency checkers. The design manager manipulates local design data, and displays allow viewing and editing of design data. Design quality is estimated from partial or complete design data. Quality measures, as introduced in Chapter 3, can be computed relatively easily from the final design, but must use different estimation techniques on different levels of abstraction. Synthesis algorithms can be applied to the total design or to just some of its parts. A designer should be able to divide synthesis into small steps and apply algorithms to each step separately. We may think of synthesis tools as providing hints to the designer during manual refinement. For example, in the floorplanning process, a designer may assign components to silicon one at a time; the placement algorithm may suggest the best position for each component, which a designer may accept or reject. The final piece of the conceptualization environment is a consistency checker which ensures that manual design changes did not produce any change in the behavioral description. For example, when a designer takes a register-transfer, $y \leftarrow a+b+c$, that requires two adders and breaks it into two register-transfers, $y \leftarrow a + b$, and $y \leftarrow y + c$, that require only one adder but two states, the consistency checker must ensure that the two register-transfers generate the same result as the original register-transfer.

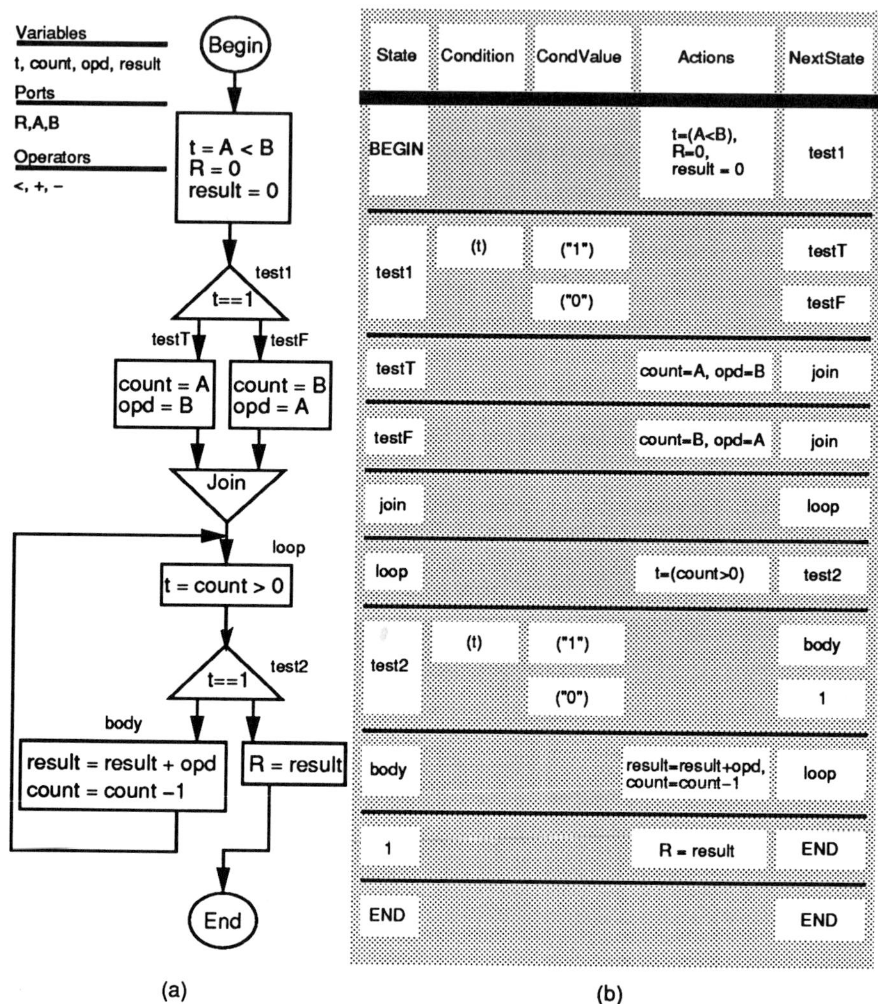

State	Condition	CondValue	Actions	NextState
BEGIN			t=(A<B), R=0, result = 0	test1
test1	(t)	("1")		testT
		("0")		testF
testT			count=A, opd=B	join
testF			count=B, opd=A	join
join				loop
loop			t=(count>0)	test2
test2	(t)	("1")		body
		("0")		1
body			result=result+opd, count=count−1	loop
1			R = result	END
END				END

(b)

Figure 9.14: A behavioral description: (a) flowchart, (b) behavioral state table.

The complete chip design can be captured in two forms: behavioral state table and design floorplan. Every behavioral description expressed in terms of "if", "case" and "loop" statements can be naturally partitioned along "if", "case" and "loop" boundaries. For example, the conditional test in an "if" statement can be assigned to one state, with the straight-line code in each branch assigned to one state each. If there is more than one variable assignment statement in each branch, they may be further divided manually or automatically into several states. For example, a behavioral description drawn in the form of a flowchart is shown in Figure 9.14(a). Its corresponding partitioned behavior is shown in the form of a behavioral state table in Figure 9.14(b) from [HaGa91]. Note that the join node in Figure 9.14(a) unnecessarily occupies an entire state but simplifies the readability of the description and allows verification of changes. In the final design, such a description is compacted into the allocated state table. However, it is not possible to understand the meaning of the design from a compacted and unstructured state table.

The second display is the design floorplan that shows the connectivity and positioning of each component (Figure 9.15). The floorplan display is hierarchical since the floorplan of each component, such as the datapath, can be also displayed. Each component has ports on its boundary. The connections between ports can be displayed selectively as port-to-port straight-line connections or as routed connections when placed in channels between components. Other displays facilitate module selection, floorplanning, allocation and scheduling. Module selection is achieved by browsing through the available components in the CDB by increasingly specific categories, such as class (e.g., combinatorial, clocked), functionality (e.g., ALU, comparator, counter), the available bit width, and style (e.g., ripple or look-ahead carry). Physical information about each component is displayed after the previous categories have been specified. Floorplanning is achieved through the floorplan display and editor. Designers may add, delete, rotate and reposition modules. Also, they can position module ports on the module boundaries. Similarly, designers may add and delete connections or position them into channels around modules. The quality of a floorplan, in terms of height, width, routing area, transistor density, wire length of any subset of nets, and clock rate, is displayed for every floorplan to guide the designer.

Allocation is performed by selecting units and assigning behavioral objects (e.g., variables and operations) to RT components, and connections to multiplexers and buses. This assignment can be accomplished using a matrix display with behavioral objects on one side and comp-

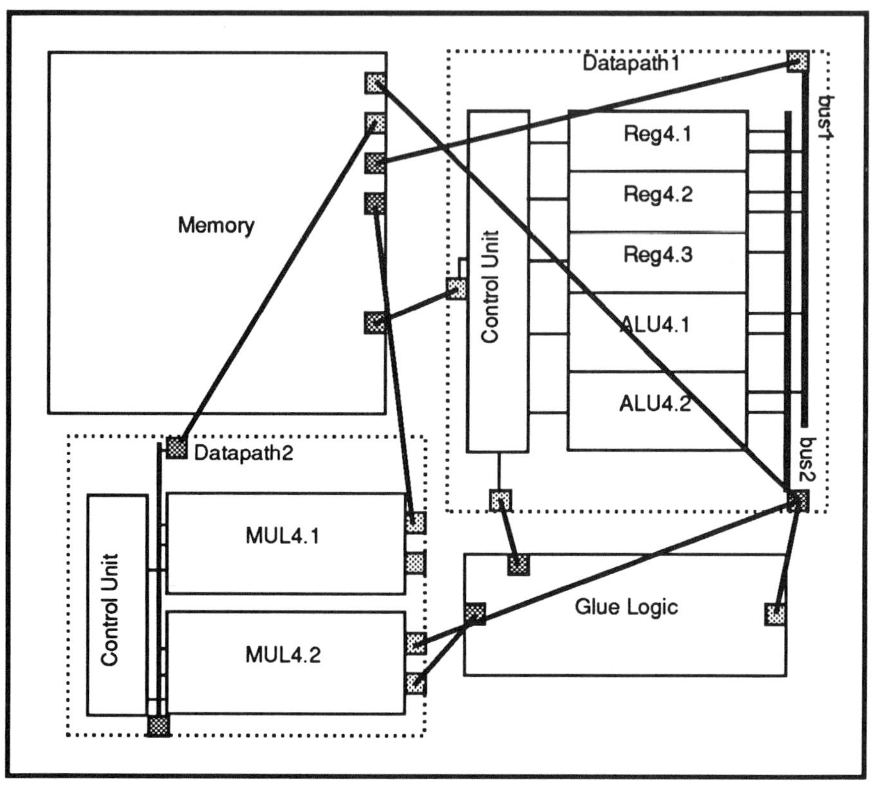

Figure 9.15: Floorplan display.

onents on the other side. The quality of an assignment is estimated by
metrics, such as component utilization, register-to-register delays and
chip-routing areas.

Scheduling is performed by repartitioning the initial behavioral description in the behavioral state table. A designer may add and delete
states; split and merge expressions, conditions and loops; and reallocate
assignments from one state to another. The quality of scheduled behavior is estimated by metrics such as the number of components needed,
the sum of all component areas, register-to-register delays, the number
of states, and total execution time.

In order to allow rapid estimation, the behavioral description and floorplans are linked through a common data structure so that changes in one are propagated to the other. Sometimes the propagation of a change is not completely possible and both structures must be updated. For example, if two additions are scheduled into one control step of a one-adder design, then this change will result in a new design with an additional adder not connected to any other component. This adder must be connected manually to make the behavioral description correct.

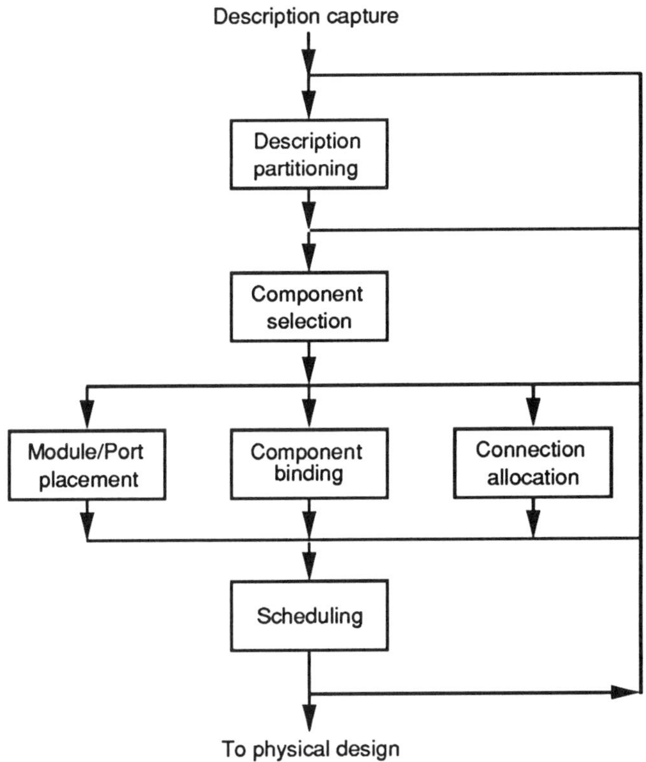

Figure 9.16: Possible scenarios for interactive HLS.

The conceptualization environment allows many different methodologies to be used. A designer should be able to iterate through the same task as many times as necessary to satisfy the requirements. Figure 9.16 shows some possible scenarios. A designer may start with a captured description and partition it into smaller and smaller pieces. Since the behavior is entered through a state table, the description always has an initial schedule. After component selection and binding, the designer may discover that the schedule violates certain design constraints and so may reschedule the design. After the initial schedule becomes satisfactory, the designer may browse through the CDB via the component-selection display to identify the key components for the final design. The remaining components can be bound later either manually or automatically. The designer can prebind some important behavioral objects to selected components before a complete allocation is performed. After component selection, a designer may proceed with floorplanning. Since the key components usually occupy most of the area, having a floorplan early speeds up design exploration. A designer will may specify key interconnections, such as data and address buses. After floorplanning, the designer may return to scheduling since resource and timing constraints are now better known. A designer will usually try to improve physical or performance quality measures through several iterations until a satisfactory point is reached. It is reasonable to expect that a designer will make most of the high-level decisions but manually synthesize only a small portion of the design. The rest of the design will be generated by automatic synthesis tools that are not expected to drastically affect high-level decisions made by the designer.

In summary, a conceptualization environment allows designers to manually perform the same tasks that all designers do when working on higher abstraction levels. However, it also allows designers to partially complete those tasks with automatic synthesis tools. As HLS algorithms improve, more of the design work will be shifted to HLS tools.

9.10   Summary and Further Research

In this chapter, we introduced the concepts of HLS systems and presented a design methodology for synthesis from specification to layout. We briefly discussed system requirements and CAD tools to be used in those systems. In particular, we discussed design methodologies on the system, chip, logic and sequential, and physical levels. We also described a support environment for synthesis consisting of system and component databases and a conceptualization environment for interactive synthesis.

Research in HLS has been in existence for many years, but it has not concentrated on complete synthesis systems. More work is needed before synthesis for systems and chips comes to a designer's desk. Of particular importance is the synthesis of complex hardware/software systems. System descriptions do not necessarily distinguish between software and hardware implementations. Some portions of the description may be implemented in software running on a standard processor component, some portions may be mapped directly into system components such as device controllers, while others may be implemented directly in custom hardware. To enable mapping of descriptions into standard components, we must develop a mapping technology. For software implementations, we need generic compilers that will compile system descriptions into a given standard component specified by its structure rather than by an instruction set. Such compilers would use a set of register-transfers or a set of microoperations in lieu of instruction sets. More progress in formal specification of target architectures is needed so that compilation can be optimized for a given architecture.

Since chip and system tools will not be available soon, the emphasis on hybrid systems for interactive synthesis is justified. Such systems must allow manual control of the synthesis process with the help of good quality measures. The automation of design exploration can only be achieved after the definition of design-quality measures and an understanding of the influence of styles, architectures topologies and design methodology on the design quality. Thus, work on comparative analysis of design quality on all levels of design is needed.

Databases to support system- and chip-level designs are not readily available since very little attention is given to system databases. Similarly, more work is needed on component databases and component generators for generating component instances before HLS is widely accepted by design community.

In summary, serious work is needed not just in the development of algorithms but also in the development of the infrastructure for system and chip synthesis.

9.11 Exercises

1. Design the 4-port memory in Figure 9.2(c) that delivers four 16-bit operands

   (a) each clock cycle,

   (b) every two clock cycles.

2. Design *Memory 1* and *Memory 2* in Figure 9.2(c) as independent FSMDs communicating with the control unit through a simple handshaking protocol.

3. *Develop a synthesis procedure for designing memories with an arbitrary number of read and write ports and using a request-acknowledge protocol.

4. Define five different strategies to improve (a) the performance and (b) the cost of an arbitrary design.

5. *Develop an algorithm for clock-period estimation from behavioral descriptions.

6. Design an interface between two FSMDs for the message passing primitives "send" and "receive". Is this interface a separate FSMD or just part of the descriptions of two communicating FSMDs ?

7. *Develop an algorithm for interface synthesis from timing diagrams describing a protocol on a given communication channel.

8. *Given several channels between two chips, develop an algorithm for channel merging and protocol selection on the merged channel.

9. **Define a technique for standard component binding from a behavioral description.

10. *Develop a procedure for technology mapping for a library of RT components. Hint: see [DuKi91].

11. *Develop an algorithm to minimize the register-to-register propagation delay in a register-transfer structure where the gate-level structure of each component is not known.

12. Design an algorithm for module selection from a behavioral description for low cost implementations of the given description.

13. *Develop a methodology for chip synthesis for the following applications: (a) DSP processors, (b) RISC processors and (c) Interface controllers.

14. *Develop an algorithm for merging buses with different protocols.

15. *Design an algorithm for partitioning 1-dimensional components into several stacks of a given size in order to minimize (a) area, (b) wire length or (c) sum of stack boundaries.

16. Design an algorithm for merging arrays with (a) linear and (b) non-linear indices into single-port memories minimizing memory conflict.

17. Develop an algorithm for merging register into register files to minimize the interconnection cost. Hint: see [GrDe90].

18. Define two design views for a CDFG that will provide (a) control information for design of the control unit and (b) register-transfer information for design of the datapath.

19. **Develop procedures for checking data consistency before and after (a) a change in scheduled behavior caused by state splitting or merging, (d) a change in allocation after swapping allocation of two variables or two operators, (c) scheduling, and (d) allocation.

20. *Develop a language to describe combinational and sequential RT components on the logic level.

21. Develop a methodology to generate RT components from a parameterized component descriptions.

22. *Develop a procedure for manual scheduling from flowcharts such as the one shown in Figure 9.14.

23. Develop an algorithm to find an optimal rotation for a component in a floorplan.

Bibliography

[AbBi91]   J.M. Abramson and W.P. Birmingham, "Binding and Scheduling under External Timing Constraints: The Ariel Synthesis System," *Fifth ACM/IEEE International Workshop on High-Level Synthesis*, Germany, March 1991.

[Agne91]   D. Agnew, "VHDL Extensions Needed for Synthesis and Design," *Proceedings of the 10th International Symposium on Computer Hardware Description Languages and their Applications*, 1991.

[AhSU86]   A.V. Aho, R. Sethi and J.D. Ullman, *Compilers: Principles, Techniques and Tools*, Addison-Wesley, 1986.

[AiNi88]   A. Aiken and A. Nicolau, "Optimal Loop Parallelization," ACM SIGPLAN 88, *Conference on Programming Languages Design and Their Implementation*, June 1988.

[AmBS91]   T. Amon, G. Borriello and C. Sequin, "Operation/Event Graphs: A Design Representation for Timing Behavior," *Proceedings of the 10th International Symposium on Computer Hardware Description Languages and their Applications*, 1991.

[Arms89]   J.R. Armstrong, *Chip Level Modeling with VHDL*, Prentice-Hall, 1989.

[ArWC90]   A. Arsenault, J.J. Wong and M. Cohen, "VHDL Transition from System to Detailed Design," VHDL User's Group Meeting, Boston, April 1990.

[BaMa89]   M. Balakrishnan and P. Marwedel, "Integrated Scheduling and Binding: A Synthesis Approach for Design Space Exploration," *Proceedings of the 26th Design Automation Conference*, pp. 68–74, 1989.

[BBEF91]   S. Banks, C. Bunting, R. Edwards, L. Fleming and P. Hacket, "A Configuration Management System in a Data Management Framework", *Proceeding of the 28th Design Automation Conference* , pp. 699-703, 1991.

[Bann79]   U. Bannerjee, "Speedup of Ordinary Programs," Tech. Rpt. UIUCDCS-R-79-989, University of Illinois at Urbana-Champaign, Computer Science Department, October 1979.

[Barb81]   M. Barbacci, "Instruction Set Processor Specifications (ISPS): The Notation and its Applications," *IEEE Transactions on Computers*, vol C-30, no. 1, January 1981.

[BePa90]   N. Berry, and B.M. Pangrle, "SHALLOC: An Algorithm for Simultaneous Scheduling and Connectivity Binding in a Datapath Synthesis System," *Proceedings of the European Design Automation Conference*, pp. 78–82, 1990.

[BCDO88]   R.K. Brayton, R. Camposano, G. De Micheli, R.H.J.M. Otten, and J.T.J. van Eijndhoven, "The Yorktown Silicon Compiler System," in D.D. Gajski, Editor, *Silicon Compilation*, Addison-Wesley, 1988.

[Berg83]   N. Bergmann, "A Case Study of the F.I.R.S.T. Silicon Compiler," *Third Caltech Conference on VLSI*, 1983.

[BhLe90]   J. Bhasker and H-C Lee, "An Optimizer for Hardware Synthesis," *IEEE Design and Test of Computers*, pp. 20–36, October 1990.

[Borr88]   G. Borriello, "Combining Event and Data Flow Graphs in Behavioral Synthesis," *Proceedings of the International Conference on Computer-Aided Design*, pp. 56–59, 1988.

[Borr91]   G. Borriello, "Specification and Synthesis of Interface Logic," in Camposano and Wolf, Editors, *High-Level VLSI Synthesis*, Kluwer, 1991.

[Bout91]   R. Boute, "Declarative Languages — Still a Long Way to Go," Invited Paper, *Proceedings of the 10th International Symposium on Computer Hardware Description Languages and their Applications*, 1991.

[BRSW87]   R. K. Brayton, R. Rudell, A. Sangiovanni-Vincentelli and A. R. Wang, "MIS: A Multiple-Level Logic Optimization System," *IEEE Transactions on Computer-Aided Design of Integrated Circuits and Systems*, vol. 6, no. 6, pp. 1062–1080, November 1987.

[BSPJ90]   R. Beckman, W. Schenk, D. Pusch and R. Johnk, "The TREEMOLA Language Reference Manual Version 4.0," Report No. 364, Lehrstuk Informatik, University of Dortmund, December 1990.

[BuEl89]   O. A. Buset and M. I. Elmasry, "ACE: A Hierarchical Graphical Interface for Architectural Synthesis," *Proceeding of the 26th Design Automation Conference*, pp. 537–542, 1989.

[CaNS90]   A. Casotto, A.R. Newton and A. Sangiovanni-Vincentelli, "Design Management Based on Design Traces," *Proceeding of the 27th Design Automation Conference*, pp. 136–141, 1990.

[Camp91]   R. Camposano, "Path-Based Scheduling for Synthesis," *IEEE Transactions on Computer-Aided Design of Integrated Circuits and Systems*, vol. 10, no. 1, pp. 85–93, January, 1991.

[CaKu86]   R. Camposano and A. Kunzmann, "Considering Timing Constraints in Synthesis from a Behavioral Description," *Proceedings of the International Conference on Computer-Aided Design*, pp. 6–9, 1986.

[CaRo85]   R. Camposano and W. Rosenstiel, "A Design Environment for the Synthesis of Integrated Circuits," *11th EUROMICRO Symposium on Microprocessing and Microprogramming*, 1985.

[CaRo89] R. Camposano and W. Rosenstiel, "Synthesizing Circuits from Behavioral Descriptions," *IEEE Transactions on Computer-Aided Design of Integrated Circuits and Systems*, vol. 8, no. 2, pp. 171–180, February 1989.

[CaST91] R. Camposano, L.F. Saunders and R.M. Tabet, "High-Level Synthesis from VHDL," *IEEE Design and Test of Computers*, March, 1991.

[CaTa89] R. Camposano and R.M. Tabet, "Design Representation for the Synthesis of Behavioral VHDL Models," in J.A. Darringer, F.J. Rammig, Editors, *Proceedings of the 9th International Symposium on Computer Hardware Description Languages and their Applications*, 1989.

[CaWo91] R. Camposano and W. Wolf, Editors, *High-Level VLSI Synthesis*, Kluwer Academic Publishers, Boston, 1991.

[ChGa90] G.D. Chen and D.D. Gajski, "An Intelligent Component Database for Behavioral Synthesis," *Proceedings of the 27th Design Automation Conference*, pp. 150–155, 1990.

[Clar73] C.R. Clare, *Designing Logic Systems using State Machines*, McGraw-Hill Inc., 1973.

[CNSD90] H. Cai, S. Note, P. Six, H. De Man, "A Datapath Layout Assembler for High Performance DSP Circuits," *Proceeding of the 27th Design Automation Conference*, pp. 306–311, 1990.

[Coel89] D.R. Coelho, *The VHDL Handbook*, Kluwer Academic Publishers, 1989.

[Comp74] IEEE Computer, Special Issue on Hardware Description Languages, December 1974.

[Comp77] IEEE Computer, Special Issue on Hardware Description Languages, June 1977.

[CHDL79] *Proceedings of the 3rd International Symposium on Computer Hardware Description Languages and their Applications*, 1979.

[CHDL81]   *Proceedings of the 5th International Symposium on Computer Hardware Description Languages and their Applications*, 1981.

[CHDL83]   *Proceedings of the 6th International Symposium on Computer Hardware Description Languages and their Applications*, 1983.

[CPTR89]   C. M. Chu, M. Potkonjak, M. Thaler and J. Rabaey, "HYPER: An Interactive Synthesis Environment for High Performance Real Time Applications", *Proceeding of the International Conference on Computer Design*, pp. 432–435, 1989.

[Cytr84]   R. Cytron, "Compiler-Time Scheduling and Optimization for Asynchronous Machines," Ph.D. thesis, Department of Computer Science, University of Illinois, Urbana-Champaign, 1984.

[DaJo80]   J.A. Darringer and W.H. Joyner, "A New Look at Logic Synthesis," *Proceedings of the 17th Design Automation Conference*, 1980.

[DBRI88]   P.J. Drongowski, J. Bammi, R. Ramaswamy, S. Iyengar and T. Wang, "A Graphical Hardware Design Language," *Proceedings of the 25th Design Automation Conference*, 1988.

[DeJo91]   G. DeJong, "Data Flow Graphs: System Specification with the Most Unrestricted Semantics," *Proceedings of the European Design Automation Conference*, 1991.

[DeNe89]   S. Devadas and A.R. Newton, "Algorithms for Allocation in Data Path Synthesis," *IEEE Transactions on Computer-Aided Design of Integrated Circuits and Systems*, vol. 8, pp. 768–781, July 1989.

[DeBS85]   G. De Micheli, R. Brayton and A. Sangiovanni-Vincentelli, "Optimal State Assignment for Finite State Machines," *Transactions on Computer-Aided Design of Integrated Circuits and Systems*, vol. 4, no. 3, pp. 269–284, 1985.

[DuDi75]   J.R. Duley and D.L. Dietmeyer, "Register Transfer Languages and Their Applications," in M.S. Breuer, Editor, *Digital System Design Automation: Languages, Simulation and Data Base*, Computer Science Press, 1975.

[DKMT90]   G. De Micheli, D. Ku, F. Mailhot and T. Truong, "The Olympus Synthesis System," *IEEE Design and Test of Computers*, October 1990.

[DMNS88]   S. Devadas, H.K.T. Ma, A.R. Newton and A. Sangiovanni-Vincentelli, "MUSTANG: State Assignment of Finite State Machines Targeting Multilevel Logic Implementations," *IEEE Transactions on Computer-Aided Design of Integrated Circuits and Systems*, Vol. 7, No. 12, pp. 1290–1299, 1988.

[DuCH91]   N.D. Dutt, J.H. Cho and T. Hadley, "A User Interface for VHDL Behavioral Modeling," *Proceedings of the 10th International Symposium on Computer Hardware Description Languages and their Applications*, 1991.

[DuGa89]   N.D. Dutt and D.D. Gajski, "Designer Controlled Behavioral Synthesis," *Proceedings of the 26th Design Automation Conference*, pp. 754–757, 1989.

[DuHG90]   N.D. Dutt, T. Hadley and D.D. Gajski, "An Intermediate Representation for Behavioral Synthesis," *Proceedings of the 27th Design Automation Conference*, pp. 14–19, 1990.

[DuKe85]   A.E. Dunlop and B.W. Kernighan, "A Procedure for Placement of Standard-Cell VLSI Circuits," *IEEE Transactions on Computer-Aided Design of Integrated Circuits and Systems*, vol. CAD–4, no. 1, pp. 92–98, January, 1985.

[DuKi91]   N.D. Dutt and J.R. Kipps, "Bridging High-Level Synthesis to RTL Technology Libraries," *Proceedings of the 28th Design Automation Conference*, pp. 526–529, 1991.

[EsWu91]   B. Eschermann and H-J. Wunderlich, "A Unified Approach for the Synthesis of Self-Testable Finite State Machines," *Proceedings of the 28th Design Automation Conference*, pp. 372–377, 1991.

[Ewer90]   C. Ewering, "Automatic High Level Synthesis of Partitioned Busses," *Proceedings of the International Conference on Computer-Aided Design*, pp. 304–307, 1990.

[FiMa82]   C.M. Fiduccia and R.M. Mattheyses, "A Linear-Time Heuristic for Improving Network Partitions," *Proceedings of the 19th Design Automation Conference*, pp. 175–181, 1982.

[Gajs91]   D.D. Gajski, "Essential Issues and Possible Solutions in High-Level Synthesis," in Camposano and Wolf, Editors, *High-Level VLSI Synthesis*, Kluwer Academic Publishers, Boston, 1991.

[GaKu83]   D.D. Gajski and R. Kuhn, "Guest Editors' Introduction: New VLSI Tools," *IEEE Computer*, vol. 16, no. 12, pp. 11–14, December 1983.

[GeEl90]   C.H. Gebotys and M.I. Elmasry, "A Global Optimization Approach for Architectural Synthesis," *Proceedings of the International Conference on Computer-Aided Design*, pp. 258–261, 1990.

[GiBK85]   E.F. Girczyc, R.J.A. Buhr and J.P. Knight, "Applicability of a Subset of Ada as an Algorithmic Hardware Description Language for Graph-Based Hardware Compilation," *IEEE Transactions on Computer-Aided Design of Integrated Circuits and Systems*, vol. CAD–4, no. 2, April 1985.

[Gord86]   M. Gordon, "Why Higher Order Logic is a Good Formalism for Specifying and Verifying Hardware," in G. Milne and P.A. Subrahmanyam, Editors, *Formal Aspects of VLSI Design*, North Holland, 1986.

[GoVD89]   G. Goosens, J. Vandewalle and H. De Man, "Loop Optimization in Register-Transfer Scheduling for DSP-Systems," *Proceedings of the 26th Design Automation Conference*, 1986.

[GrDe90]   D.M. Grant and P.B. Denyer, "Memory, Control and Communication Synthesis for Scheduled Algorithms," *Proceed-

ings of the 27th Design Automation Conference, pp. 162–167, 1990.

[GuDe90]   R. Gupta and G. De Micheli, "Partitioning of Functional Models of Synchronous Digital Systems," *Proceedings of the International Conference on Computer-Aided Design*, pp. 216–219, 1990.

[HaCa89]   R. Hartley and A. Casavant, "Tree-Height Minimization in Pipelined Architectures," *Proceedings of the International Conference on Computer-Aided Design*, pp. 112–115, 1989.

[HaGa91]   T. Hadley and D.D. Gajski, "A Decision Support Environment for Behavioral Synthesis," Technical Report 91-17, Department of Information and Computer Science, University of California, Irvine, CA, 1991.

[HaKl89]   P. Harper, S. Krolikoski and O. Levia, "Using VHDL as a Synthesis Language in the Honeywell VSYNTH System," *Proceedings of the 9th International Symposium on Computer Hardware Description Languages and their Applications*, 1989.

[Hare87]   D. Harel, "Statecharts: A Visual Formalism for Complex Systems," *Science of Computer Programming*, no. 8, 1987.

[Hart87]   "Hardware Description Languages," in R. W. Hartenstein, Editor, *Advances in CAD for VLSI*, Volume 7, North Holland, 1987.

[HaSt71]   A. Hashimoto and J. Stevens, "Wire Routing by Optimizing Channel Assignment within Large Apertures," *The 8th Design Automation Conference Workshop*, pp. 155–169, 1971.

[HaSt91]   R.E. Harr and A.G. Stanculescu, Editors, *Applications of VHDL to Circuit Design*, Kluwer Academic Publishers, 1991.

[HCLH90]   C-Y. Huang, Y-S. Chen, Y-L. Lin and Y-C. Hsu, "Data Path Allocation Based on Bipartite Weighted Matching," *Proceedings of the 27th Design Automation Conference*, pp. 499–504, 1990.

[Hilf85]   P.N. Hilfinger, "A High Level Language and Silicon Compiler for Digital Signal Processing," *Proceedings of the IEEE Custom Integrated Circuits Conference*, pp. 213–216, 1985.

[HLNP88]   D. Harel, H. Lachover, A. Namaad, A. Pnueli, M. Politi, R. Sherman and A. Shtul-Trauring, "Statemate: A Working Environment for the Development of Complex Reactive Systems," *Proceeding of International Conference on Software Engineering*, 1988.

[Hoar78]   C. Hoare, "Communicating Sequential Processes," *Communication of ACM*, August 1978.

[HuRS86]   M.D. Hung, F. Romeo and A. Sangiovanni-Vincentelli, "An Efficient General Cooling Schedule for Simulated Annealing," *Proceedings of the International Conference on Computer-Aided Design*, pp. 381–384, 1986.

[IEEE88]   *Standard VHDL Language Reference Manual*, New York: The Institute of Electrical and Electronics Engineers, Inc., 1988.

[JMSW91]   R. Jain, A. Mujumdar, A. Sharma and H. Wang, "Empirical Evaluation of Some High-Level Synthesis Scheduling Heuristics," *Proceedings of the 28th Design Automation Conference* pp. 210–215, 1991.

[John67]   S.C. Johnson, "Hierarchical Clustering Schemes," *Psychometrika*, pp. 241–254, September, 1967.

[Katz82]   Katz, R. H., "A Data Base Approach for Managing VLSI Design Data," *Proceeding of the 19th Design Automation Conference*, pp. 274–282, 1982.

[KeLi70]   K.H. Kernighan and S. Lin, "An Efficient Heuristic Procedure for Partitioning Graph," *Bell System Technical Journal*, vol. 49, no. 2, pp. 291–307, February, 1970.

[Keut87]   K. Keutzer, "DAGON: Technology Binding and Logic Optimization by DAG Matching," *Proceeding of the 24th Design Automation Conference*, pp. 341–347, 1987.

[KiGV83]   S. Kirkpatrick, C.D. Gelatt and M. P. Vecchi, "Optimization by Simulated Annealing," *Science*, vol. 220, no. 4598, pp. 671–680, 1983.

[KnPa85]   D.W. Knapp and A.C. Parker, "A Unified Representation for Design Information," *Proceedings of the 7th International Symposium on Computer Hardware Description Languages and their Applications*, pp. 337–353, 1985.

[Kris84]   B. Krishnamurthy, "An Improved Min-Cut Algorithm for Partitioning VLSI Networks," *IEEE Transactions on Computer-Aided Design of Integrated Circuits and Systems*, vol. C-33, pp. 438–446, May, 1984.

[KuDe91]   D. Ku and G. De Micheli, "Synthesis of ASICs with Hercules and Hebe," in Camposano and Wolf, Editors, *High-Level VLSI Synthesis*, Kluwer Academic Publishers, Boston, 1991.

[KuPa87]   F.J. Kurdahi and A.C. Parker, "REAL: A Program for Register Allocation," *Proceedings of the 24th Design Automation Conference*, pp. 210–215, 1987.

[KuPa90]   K. Kucukcakar and A. C. Parker, "Data Path Tradeoffs using MABAL", *Proceedings of the 27th Design Automation Conference*, pp. 511–516, 1990.

[KuPa91]   K. Kucukcakar and A.C. Parker, "CHOP: A Constraint-Driven System-Level Partitioner," *Proceedings of the 28th Design Automation Conference*, pp. 514–519, 1991.

[LaTh91]   E. Lagnese and D. Thomas, "Architectural Partitioning for System Level Synthesis of Integrated Circuits," *IEEE Transactions on Computer-Aided Design of Integrated Circuits and Systems*, vol. 10, no. 7, pp. 847–860, July, 1991.

[LeHL89]   J. Lee, Y. Hsu, and Y. Lin, "A New Integer Linear Programming Formulation for the Scheduling Problem in Data-Path Synthesis," *Proceedings of the International Conference on Computer-Aided Design*, pp. 20–23, 1989.

[Leng90]   T. Lengauer, *Combinatorial Algorithms for Integrated Circuit Layout,* Wiley, 1990.

[LGPV91]   D. Lanneer, G. Goossens, M. Pauwels, J. Van Meerbergen and H. De Man, "An Object-Oriented Framework Supporting the Full High-Level Synthesis Trajectory," *Proceedings of the 10th International Symposium on Computer Hardware Description Languages and their Applications 1991,* pp. 281–300, 1991.

[LiGa88]   J.S. Lis and D.D. Gajski, "Synthesis from VHDL," *Proceedings of the International Conference on Computer Design,* pp. 378–381, 1988.

[LiGa91]   J.S. Lis and D.D. Gajski, "Behavioral Synthesis from VHDL using Structured Modeling," ICS Technical Report 91-05, University of California at Irvine, January 1991.

[LiSU89]   R. Lispett, C. Shaefer and C. Ussery, *VHDL: Hardware Description and Design,* Kluwer Academic Publishers, 1989.

[Lipo77]   G.J. Lipovski, "Hardware Description Languages: Voices from the Tower of Babel," *IEEE Computer,* vol. 10, no. 6, June 1977.

[LKVW88]   M. Lightner, K. Keutzer, M. Vancura and W. Wolf, "A Survey of Synthesis Techniques: Design Demographics and Studies in Operator Synthesis," *The ACM/IEEE Workship on High Level Synthesis,* Orcas Island, January 1988.

[LNDP91]   D. Lanneer, S. Note, F. Depuydt, M. Pauwels, F. Catthoor, Gert Goossens and H. De Man, "Architectural Synthesis for Medium and High Throughput Signal Processing with the new CATHEDRAL environment," in Camposano and Wolf, Editors, *High-Level VLSI Synthesis,* Kluwer Academic Publishers, Boston, 1990.

[LoPa91]   D.A. Lobo and B.M. Pangrle, "Redundant Operator Creation: A Scheduling Optimization Technique," *Proceedings of the 28th Design Automation Conference,* pp. 775–778, 1991.

[LuDe89]   W. K. Luke and A. A. Dean, "Multi-Stack Optimization for Datapath Chip Layout," *Proceeding of the 26th Design Automation Conference*, pp. 110-115, 1989.

[Marw85]   P. Marwedel, "The MIMOLA Design System: A Design System which Spans Several Levels," *Methodologies of Computer System Design*, B.D. Shriver, Editor, North Holland, pp. 223–237, 1985.

[McFa78]   M.C. McFarland, *The Value Trace: A Data Base for Automated Digital Design*, Report DRC-01-4-80, Design Research Center, Carnegie-Mellon University, December 1978.

[McFa86]   M.C. McFarland, "Using Bottom-Up Design Techniques in the Synthesis of Digital Hardware from Abstract Behavioral Descriptions," *Proceedings of the 23rd Design Automation Conference*, pp. 474–480, 1986.

[McKo90]   M.C. McFarland and T.J. Kowalski, "Incorporating Bottom-Up Design into Hardware Synthesis," *IEEE Transactions on Computer-Aided Design of Integrated Circuits and Systems*, vol. 9, no. 9, pp. 938–950, September 1990.

[NaON91]   Y. Nakamura, K. Oguri and A. Nagoya, "Synthesis from Pure Behavioral Descriptions," in Camposano and Wolf, Editors, *High-Level VLSI Synthesis*, Kluwer Academic Publishers, Boston, 1990.

[NaVG91]   S. Narayan, F. Vahid and D.D. Gajski, "System Specification and Synthesis with the SpecChart Language", *Proceeding of the International Conference on Computer-Aided Design*, 1991.

[NeTh86]   J.A. Nestor and D.E. Thomas, "Behavioral Synthesis with Interfaces," *Proceedings of the International Conference on Computer-Aided Design*, pp. 112–115, 1986.

[NiPo91]   A. Nicolau and R. Potasman, "Incremental Tree Height Reduction for High Level Synthesis," *Proceedings of the 28th Design Automation Conference*, pp. 770–774, 1991.

[OrGa86]   A. Orailogulu and D.D. Gajski, "Flow Graph Representation," *Proceedings of the 23rd Design Automation Conference*, pp. 503–509, 1986.

[OtGi84]   R. Otten and L. van Ginncken, "Floorplan Design using Simulated Annealing," *Proceedings of the International Conference on Computer-Aided Design*, pp. 96–98, 1984.

[PaKy91]   I-C. Park and C-M. Kyung, "Fast and Near Optimal Scheduling in Automatic Data Path Synthesis," *Proceedings of the 28th Design Automation Conference*, pp. 680–685, 1991.

[PaKn89]   P.G. Paulin and J.P. Knight, "Force-Directed Scheduling for the Behavioral Synthesis of ASIC's," *IEEE Transactions on Computer-Aided Design of Integrated Circuits and Systems*, vol. 8, no. 6, pp. 661–679, June 1989.

[PaSt82]   C.H. Papadimitriou and K. Steiglitz, *Combinatorial Optimization*, Prentice-Hall, 1982.

[Perr90]   D. Perry, *VHDL*, McGraw Hill, Inc., 1991.

[PiBo85]   R. Piloty and D. Borrione, "The Conlan project: Concepts, Implementations, and Applications," *Computer*, no. 18(2). pp. 6–8. February 1985.

[PrLo87]   B. Preas and M. Lorenzetti, Editors *Physical Design Automation of VLSI Systems*, Benjamin/Cummings, 1988.

[Pilo91]   R. Piloty, "Some Basic Issues in Behavioral Modeling of Hardware," Keynote Address, *Proceedings of the 10th International Symposium on Computer Hardware Description Languages and their Applications*, 1991.

[PoNi91]   R. Potasman and A. Nicolau, "Incremental Tree Height Reduction for High-Level Synthesis," *Proceedings of the 28th Design Automation Conference*, June 1991.

[RaGa91]   L. Ramachandran and D.D. Gajski, "An Algorithm for Component Selection in Performance Optimized Scheduling," *Proceedings of the International Conference on Computer-Aided Design*, 1991.

[RDVG88]   J. Rabaey, H. DeMan, J. Vanhoff, G. Goossens and F. Catthoor, "Cathedral II: A Synthesis System for Multiprocessor DSP Systems," in D.D. Gajski, Editor, *Silicon Compilation*, Addison-Wesley, pp. 311–360, 1988.

[RoSa85]   F. Romeo and A. Sangiovanni-Vincentelli, "Probabilistic Hill Climbing Algorithms: Properties and Applications," *Proceedings of the 1985 Chapel Hill Conference on VLSI*, pp. 393–417, 1985.

[RuGa90]   E.A Rundensteiner and D.D. Gajski, "A Design Representation for High-Level Synthesis", Info. & Computer Science Dept., UCI, Tech. Rep. 90-27, Sep. 1990.

[RuGa91]   E.A Rundensteiner and D.D. Gajski, "A Design Data Base for Behavioral Synthesis", *High Level Synthesis Workshop*, 1991.

[RuGB90]   E.A. Rundensteiner, D.D. Gajski and L. Bic, "The Component Synthesis Algorithm: Technology Mapping for Register Transfer Descriptions," *Proceedings of the International Conference on Computer-Aided Design*, pp. 208–211, 1990.

[Rose86]   W. Rosenstiel, "Optimizations in High Level Synthesis," *Microprocessing and Microprogramming (18)*, pp. 543–549, 1986.

[ScGH84]   J. Schuck, M. Glesner and H. Joepen, "ALGIC — A Flexible Silicon Compiler System for Digital Signal Processing," in *VLSI Signal Processing*, IEEE Press, 1984.

[Shiv79]   S.G. Shiva, "Computer Hardware Description Languages — A Tutorial," *Proceedings of the IEEE*, vol. 67, no. 2, December, 1979.

[Siep89]   E. Siepmann and G. Zimmermann, "An Object-Oriented Data Model for the VLSI Design System PLAYOUT", *Proceeding of the 26th Design Automation Conference*, pp. 814–817, 1989.

[Snea57]   P.H.A. Sneath, "The Application of Computers to Taxonomy," *J. gen, Microbiology 17*, pp. 201–226, 1957.

[Stok91]   L. Stok, "Architectural Synthesis and Optimization of Digital Systems," Ph.D. Thesis, Eindhoven University, July 1991.

[TaKK89]   T. Tanaka, T. Kobayashi and O. Karatus, "HARP: Fortran to Silicon," *IEEE Transactions on Computer-Aided Design of Integrated Circuits and Systems*, vol. 8, no. 6, pp. 649–660, June 1989.

[ThBR87]   D.E. Thomas, R.L. Blackburn and J.V. Rajan, "Linking the Behavioral and Structural Domains of Representation for Digital System Design," *IEEE Transactions on Computer-Aided Design of Integrated Circuits and Systems*, vol. CAD-6, no. 1, January 1987.

[TLWN90]   D.E. Thomas, E.D. Lagnese, R.A. Walker, J.A. Nestor, J.V. Rajan and R.L. Blackburn, *Algorithmic and Register-Transfer Level Synthesis: The System Architect's Workbench*, Kluwer Academic Publishers, Boston, 1990.

[TrDi89]   M. T. Trick and S. Director, "LASSIE: Structure to Layout for Behavioral Synthesis Tools," *Proceeding of the 26th Design Automation Conference*, pp. 104-109, 1989.

[Tric87]   H. Trickey, "Flamel: A High-Level Hardware Compiler," *IEEE Transactions on Computer-Aided Design of Integrated Circuits and Systems*, vol. CAD-6, no. 2, pp. 259–269, March 1987.

[TsHs90]   F-S. Tsai and Y-C. Hsu, "Data Path Construction and Refinement," *Proceedings of the International Conference on Computer-Aided Design*, pp. 308–311, 1990.

[TsSi86]   C.J. Tseng and D.P. Siewiorek, "Automated Synthesis of Data Paths on Digital Systems," *IEEE Transactions on Computer-Aided Design of Integrated Circuits and Systems*, vol. CAD-5, no. 3, pp. 379–395, July 1986.

[VAKL91]   W.F.J. Verhaegh, E.H.L. Aarts, J.H.M. Korst and P.E.R. Lippens, "Improved Force-Directed Scheduling," *Proceedings of the European Design Automation Conference*, pp. 430–435, 1991.

[VaNG90]   F. Vahid, S. Narayan and D.D. Gajski, "Synthesis from Specifications: Basic Concept," Technical Report 90-03, Department of Information and Computer Science, University of California, Irvine, CA, January 1990.

[WaCa91]   R.A. Walker and R. Camposano, Editors, *A Survey of High-Level Synthesis*, Kluwer Academic Publishers, 1991.

[WaTh89]   R.A. Walker and D.E. Thomas, "Behavioral Transformations for Algorithmic Level IC Design," *IEEE Transactions on Computer-Aided Design of Integrated Circuits and Systems*, vol. 8, No. 10, pp. 1115–1128, October 1989.

[Waxm86]   R. Waxman, "Hardware Design Languages for Computer Design and Test," *IEEE Design and Test of Computers*, vol. 19, no. 4, April 1986.

[Wolf86]   W. Wolf, "An Object Oriented Procedural Database for VLSI Chip Planning," *Proceedings of the 23rd Design Automation Conference*, 1986.

[WSBD90]   P. Van der Wolf, G.W. Sloof, P. Bingley and P. Dewilde, "Meta Data Management in the NELSIS CAD Framework", *Proceeding of the 27th Design Automation Conference*, pp. 142-145, 1990.

[WuGa90]   C-H.A. Wu and D.D. Gajski, "Partitioning Algorithms for Layout Synthesis from Register-Transfer Netlists," *Proceedings of the International Conference on Computer-Aided Design*, pp. 144–147, 1990.

[YaIs89]   H. Yasuura and N. Ishiura, "Semantics of a Hardware Design Language for Japanese Standardization," *Proceedings of the 26th Design Automation Conference*, 1989.

[YoHs90]   H. S. Yoo and A. Hsu, "Debbie: A Configurable User Interface for CAD Frameworks," *Proceeding of the International Conference on Computer Design*, pp. 135–140, 1990.